TEA TREE

Medicinal and Aromatic Plants – Industrial Profiles
Individual volumes in this series provide both industry and academia with in-depth coverage of one major medicinal or aromatic plant of industrial importance.

Edited by Dr Roland Hardman

Volume 1
Valerian
edited by Peter J. Houghton

Volume 2
Perilla
edited by He-Ci Yu, Kenichi Kosuna and Megumi Haga

Volume 3
Poppy
edited by Jenő Bernáth

Volume 4
Cannabis
edited by David T. Brown

Volume 5
Neem
H.S. Puri

Volume 6
Ergot
edited by Vladimír Křen and Ladislav Cvak

Volume 7
Caraway
edited by Éva Németh

Volume 8
Saffron
edited by Moshe Negbi

Volume 9
Tea Tree
edited by Ian Southwell and Robert Lowe

Other volumes in preparation

Allium, edited by K. Chan
Artemisia, edited by C. Wright
Basil, edited by R. Hiltunen and Y. Holm
Cardamom, edited by P.N. Ravindran and K.J. Madusoodanan
Chamomile, edited by R. Franke and H. Schilcher
Cinnamon and Cassia, edited by P.N. Ravindran and S. Ravindran
Colchicum, edited by V. Šimánek
Curcuma, edited by B.A. Nagasampagi and A.P. Purohit
Eucalyptus, edited by J. Coppen

Please see the back of this book for other volumes in preparation in Medicinal and Aromatic Plants – Industrial Profiles

TEA TREE

The Genus *Melaleuca*

Edited by

Ian Southwell *and* Robert Lowe
Wollongbar Agricultural Institute, Wollongbar, Australia

harwood academic publishers
Australia • Canada • China • France • Germany • India
Japan • Luxembourg • Malaysia • The Netherlands • Russia
Singapore • Switzerland

Printed in Singapore.

Amsteldijk 166
1st Floor
1079 LH Amsterdam
The Netherlands

British Library Cataloguing in Publication Data

Tea tree : the genus melaleuca. – (Medicinal and aromatic plants : industrial profiles ; v. 9)
1. Melaleuca 2. Melaluca alternifolia oil 3. Melaleuca – Industrial applications 4. Melaleuca – Therapeutic use
I. Southwell, Ian II. Lowe, Robert
583.7'65

ISBN 90-5702-417-9
ISSN 1027-4502

The cover illustration depicts a *Melaleuca alternifolia* mature tree drawn by Lexanne Leach.

CONTENTS

PREFACE TO THE SERIES

There is increasing interest in industry, academia and the health sciences in medicinal and aromatic plants. In passing from plant production to the eventual product used by the public, many sciences are involved. This series brings together information which is currently scattered through an ever increasing number of journals. Each volume gives an in-depth look at one plant genus, about which an area specialist has assembled information ranging from the production of the plant to market trends and quality control.

Many industries are involved such as forestry, agriculture, chemical, food, flavour, beverage, pharmaceutical, cosmetic and fragrance. The plant raw materials are roots, rhizomes, bulbs, leaves, stems, barks, wood, flowers, fruits and seeds. These yield gums, resins, essential (volatile) oils, fixed oils, waxes, juices, extracts and spices for medicinal and aromatic purposes. All these commodities are traded world-wide. A dealer's market report for an item may say "Drought in the country of origin has forced up prices".

Natural products do not mean safe products and account of this has to be taken by the above industries, which are subject to regulation. For example, a number of plants which are approved for use in medicine must not be used in cosmetic products.

The assessment of safe to use starts with the harvested plant material which has to comply with an official monograph. This may require absence of, or prescribed limits of, radioactive materials, heavy metals, aflatoxins, pesticide residue, as well as the required level of active principle. This analytical control is costly and tends to exclude small batches of plant material. Large scale contracted mechanised cultivation with designated seed or plantlets is now preferable.

Today, plant selection is not only for the yield of active principle, but for the plant's ability to overcome disease, climatic stress and the hazards caused by mankind. Such methods as *in vitro* fertilisation, meristem cultures and somatic embryogenesis are used. The transfer of sections of DNA is giving rise to controversy in the case of some end-uses of the plant material.

Some suppliers of plant raw material are now able to certify that they are supplying organically-farmed medicinal plants, herbs and spices. The Economic Union directive (CVO/EU No 2092/91) details the specifications for the **obligatory** quality controls to be carried out at all stages of production and processing of organic products.

Fascinating plant folklore and ethnopharmacology leads to medicinal potential. Examples are the muscle relaxants based on the arrow poison, curare, from species of *Chondrodendron*, and the antimalarials derived from species of *Cinchona* and *Artemisia*. The methods of detection of pharmacological activity have become increasingly reliable and specific, frequently involving enzymes in bioassays and avoiding the use of laboratory animals. By using bioassay linked fractionation of crude plant juices or extracts, compounds can be specifically targeted which, for example, inhibit blood platelet aggregation, or have antitumour, or antiviral, or any other required activity. With the assistance of robotic devices, all the members of a genus may be readily screened. However, the plant material must be **fully** authenticated by a specialist.

The medicinal traditions of ancient civilisations such as those of China and India have a large armamentarium of plants in their pharmacopoeias which are used throughout South East Asia. A similar situation exists in Africa and South America. Thus, a very high percentage of the world's population relies on medicinal and aromatic plants for their medicine. Western medicine is also responding. Already in Germany all medical practitioners have to pass an examination in phytotherapy before being allowed to practise. It is noticeable that throughout Europe and the USA, medical, pharmacy and health related schools are increasingly offering training in phytotherapy.

Multinational pharmaceutical companies have become less enamoured of the single compound magic bullet cure. The high costs of such ventures and the endless competition from "copy" compounds from rival companies often discourage the attempt. Independent phytomedicine companies have been very strong in Germany. However, by the end of 1995, eleven (almost all) had been acquired by the multinational pharmaceutical firms, acknowledging the lay public's growing demand for phytomedicines in the Western World.

The business of dietary supplements in the Western World has expanded from the Health Store to the pharmacy. Alternative medicine includes plant based products. Appropriate measures to ensure the quality, safety and efficacy of these either already exist or are being answered by greater legislative control by such bodies as the Food and Drug Administration of the USA and the recently created European Agency for the Evaluation of Medicinal Products, based in London.

In the USA, the Dietary Supplement and Health Education Act of 1994 recognised the class of phytotherapeutic agents derived from medicinal and aromatic plants. Furthermore, under public pressure, the US Congress set up an Office of Alternative Medicine and this office in 1994 assisted the filing of several Investigational New Drug (IND) applications, required for clinical trials of some Chinese herbal preparations. The significance of these applications was that each Chinese preparation involved several plants and yet was handled as a **single** IND. A demonstration of the contribution to efficacy, of **each** ingredient of **each** plant, was not required. This was a major step forward towards more sensible regulations in regard to phytomedicines.

My thanks are due to the staff of Harwood Academic Publishers who have made this series possible and especially to the volume editors and their chapter contributors for the authoritative information.

Roland Hardman

CONTRIBUTORS

Gary Baker
Wollongbar Agricultural Institute
Wollongbar NSW 2477
Australia

S. Bouraïma-Madjebi
Laboratoire de Biologie et Physiologie Végétales Appliquées
Université Française du Pacifique
Centre Universitaire de Nouvelle Calédonie
BP 4477 Noumea
New Caledonia

Joseph J. Brophy
School of Chemistry
University of NSW
Sydney NSW 2052
Australia

A.J. Campbell
Tropical Fruit Research Station
NSW Agriculture
PO Box 72
Alstonville NSW 2477
Australia

Robert T. Colton
NSW Agriculture
Locked Bag 21
Orange NSW 2800
Australia

Lyn A. Craven
CSIRO Division of Plant Industry
Australian National Herbarium
GPO Box 1600
Canberra ACT 2601
Australia

Geoffrey R. Davis
21 Rosemead Road
Hornsby NSW 2077
Australia

Richard L. Davis
GR Davis Pty Ltd
3/9 Apollo Street
Warriewood NSW 2102
Australia

John C. Doran
Australian Tree Seed Centre
CSIRO Division of Plant Industry
PO Box 4008 Queen Victoria Terrace
Canberra ACT 2600
Australia

C.D.A. Maddox
Tropical Fruit Research Station
NSW Agriculture
PO Box 72
Alstonville NSW 2477
Australia

Julie L. Markham
School of Applied & Environmental Sciences
University of Western Sydney Hawkesbury
Richmond NSW 2753
Australia

G. John Murtagh
Agricultural Water Management
2 Sunnybank Avenue
Goonellabah NSW 2480
Australia

Don Priest
Australian Tea Tree Oil Research Institute
Southern Cross University
Lismore NSW 2480
Australia

James S. Rowe
Technical Consultancy Services Pty Ltd
20 King Street
Rockdale NSW 2216
Australia

Michael Russell
Wollongbar Agricultural Institute
Wollongbar NSW 2477
Australia

Ian Southwell
Wollongbar Agricultural Institute
Wollongbar NSW 2477
Australia

B. Trilles
Laboratoire de Biologie et Physiologie Végétales Appliquées
Université Française du Pacifique
Centre Universitaire de Nouvelle Calédonie
BP 4477 Noumea
New Caledonia

G. Valet
Laboratoire de Biologie et Physiologie Végétales Appliquées
Université Française du Pacifique
Centre Universitaire de Nouvelle Calédonie
BP 4477 Noumea
New Caledonia

John G. Virtue
Sustainable Resources Group
Primary Industries SA
GPO Box 1671
Adelaide SA 5001
Australia

INTRODUCTION

IAN SOUTHWELL

Wollongbar Agricultural Institute, Wollongbar, NSW, Australia

The Australian flora is rich in trees and shrubs from the family Myrtaceae. *Eucalyptus*, is well known for profuse oil glands which contain a diverse range of constituents, of which 1,8-cineole is the most abundant and most commercially utilized (Doran 1991). The genus *Melaleuca* also contains hundreds of individual species with a myriad of oil constituents present in the leaf (Brophy and Doran 1996). Both genera extend beyond Australia to neighbouring regions of SE Asia and the Pacific. *Eucalyptus* is now grown extensively in many parts of the world. Consequently Australia produces only about five percent of the world's eucalyptus oil. Although *Melaleuca* has not yet been as extensively dispersed, plantings of *M. alternifolia* have been established in the United States, Zimbabwe, New Zealand, China, India and other countries. Provenances of *M. quinquenervia*, a species native to Australia, New Caledonia and Papua New Guinea, have been grown in Madagascar (Ramanoelina *et al.* 1992, 1994) and, along with the New Caledonian provenance, have been used as a source of niaouli oil. *M. cajuputi*, native to Australia, Indonesia, Papua New Guinea, Malaysia, Thailand, Cambodia and Vietnam, along with some provenances of *M. quinquenervia* have sometimes been described as *M. leucadendron* (Todorova and Ognyanov 1988; Motl *et al.* 1990). In addition, *M. quinquenervia* is grown elsewhere as an ornamental species and for swamp reclamation and erosion control. In Florida this species has colonized vast areas to the detriment of the environmentally important Everglades (Weiss 1997) but in Hawaii, plantings of an estimated two million trees have not produced a weed problem (Geary 1988).

Although some other uses are found for *Melaleuca,* this aromatic and medicinal plant genus is best known for the production of medicinal essential oils. Non-medicinal uses (Boland *et al.* 1984; Wrigley and Fagg 1993) include broom fence manufacture from the branches, bark paintings, sealing and insulation from their many coloured barks, fuel and construction materials from the wood and honey from the nectar.

Oil production figures indicate increasing *M. alternifolia* (tea tree) and *M. cajuputi* (cajuput) volumes with steady but smaller outputs of *M. quinquenervia* (niaouli) oil. Production estimates are shown in the Table 1. The space allocated to each of these species in this volume reflects their industrial significance.

The name tea tree arose when Captain James Cook, on his exploratory voyage of Australia in 1770, encountered a myrtaceous shrub (possibly a *Leptospermum*) with leaves that were used by his sailors as a substitute for tea (*Camellia sinensis*). The naturalist, Sir Joseph Banks, collected specimens of tea tree during this voyage. Subsequently these myrtaceous shrubs, now known as the genera *Leptospermum*, *Melaleuca*, *Kunzea* and *Baeckea*, were collectively known as "tea trees", not to be confused with the Maori or Samoan derived "ti-tree" or "ti-palm" names given to plants of the *Cordyline* genus (Weiss 1997).

Table 1 Estimated production (tonnes per annum) for *Melaleuca oils*

Species	*M. alternifolia*	*M. cajuputi*	*M. quinquenervia*
Common name	tea tree	cajuput	niaouli
Chemotype	terpinen-4-ol	1,8-cineole	1,8-cineole
Countries	Australia	Indonesia, Vietnam	New Caledonia, Madagascar
1930–1950 (max)	5[a]	145[b]	20[b]
1984	8[c]	50[c]	4[c]
1997	300[a]	600[a]	8–12[a]

[a] Industry estimate; [b] Guenther (1950); [c] Lawrence (1985).

Banks, on his 1770 voyage with Cook, also collected specimens from taller broadleaved tea trees. The long standing tendency to treat these large leaved species collectively as *M. leucadendron* was in many ways overcome when Blake (1968) distinguished *M. leucadendron, M. cajuputi, M. quinquenervia* and *M. viridiflora*. The first of these is now known as *M. leucadendra*. Both *M. cajuputi* and *M. quinquenervia* have been distilled since last century for the production of cajuput and niaouli oils respectively (Penfold and Morrison 1950). Both oils were used in medicinal preparations, especially for the treatment of internal disorders, stomach and intestinal problems including worms, and for the relief of headache, toothache, laryngitis and bronchitis. The oils also have reputed insecticidal properties. The leaves of these plants have been reported to have been used by the Australian aborigines for colds, influenza, fever and congestion, by inhaling the vapour from crushed leaves in boiling water or by sipping the infused water (Aboriginal Communities 1988).

The aboriginal use of the smaller leaved tea trees like *M. alternifolia* is not as easy to confirm although it is possible that the leaves were tied as a poultice to a wound. The resurgence of the tea tree oil industry in the 1980s has prompted statements such as "the aborigines are sure to have used the species" which have then been repeated as "the aborigines used the plant" and then as "when Captain Cook came to Australia he found that tea tree oil was already in use". With the last statement obviously erroneous because of the absence of distillation facilities, one tends to even question the validity of the earlier quotes. Confirmation of the aboriginal use of medicinal plants some 200 years ago is difficult in a culture where such remedies are preserved by oral and not written tradition.

Throughout this volume, the term "tea tree oil" will be reserved for "Oil of *Melaleuca*—Terpinen-4-ol Type" derived from *M. alternifolia, M. linariifolia, M. dissitiflora* and other species of *Melaleuca*, giving comparable oils conforming to the ISO Standard (International Standards Organisation 1996). The broad leaved tea tree oils, normally known by their specific names of cajuput or niaouli when derived from *M. cajuputi* and *M. quinquenervia* respectively are named as such in this volume. Non-cineole chemical varieties of these two species will be clearly identified as being atypical of the usual commercial varieties.

The "tea tree" oil industry based around the distillation of *M. alternifolia* and associated species is at an exciting stage of development as this volume goes to print. The last ten years have seen a ten-fold increase in production to around 300 tonnes per annum, coming from in excess of three thousand hectares of plantations (Plate 1) and natural stands. During this time, plantation production has increased from a seemingly insignificant contribution with respect to bush oil production, to being the predominant source of oil.

With this development have come the problems associated with a new monoculture crop. Right from the plantation planning stage, matters like financial costing, soil suitability, site requirements, land preparation, choice of genetic material, propagation method, nursery establishment, transplanting, plant density and plant configuration are issues that need addressing (Colton and Murtagh 1990). Weed and insect control, fertiliser application, irrigation, harvest machinery, season and height of harvest all require prior planning. Appropriate processing equipment and operating procedures must all be in place before the first harvest and quality control methods established before market outlets can be located and maintained. For the continuing bush industry, the operation begins with hand harvesting of private or licensed forest lands and continues with the processing, quality assessment, value adding and marketing aspects of the industry.

Production of tea tree oil is but one of the many areas covered by this volume. For example, confusing aspects of the taxonomy of the commercial *Melaleucas* are clarified by the botanist undertaking a complete taxonomic revision of the genus. The increasing commercial value of the industry has initiated similar in-depth studies in all other areas of *Melaleuca* oil production.

The chemistry of the oil has been thoroughly investigated by several researchers, some of whom have discovered *in situ* leaf precursors which transform during distillation to the commercial oil constituents (Southwell and Stiff 1989). Other investigators have used ^{13}C nuclear magnetic resonance (NMR), chiral gas chromatography (GC) and ^{18}O incorporation to add to our knowledge of the chemistry of the oil and its *in vivo* and *in vitro* formation (Leach *et al.* 1993; Cornwell *et al.* 1995).

With leaf and oil yields of utmost importance for producer profit, much research has gone into factors that effect biomass and oil yields. Parameters such as oil gland anatomy and density, nutrients, temperature, irrigation, time of harvest, post-harvest drying, wind breaks and others have been studied (Murtagh and Etherington 1990; Murtagh 1991; List *et al.* 1995; Murtagh 1996; Murtagh and Smith 1996; Whish and Williams 1996).

Weeds and insects flourish in the warm regions where tea tree thrives. Both chemical and non-chemical means of weed control have been assessed and recommendations made available to producers (Virtue 1997; Storrie *et al.* 1997). Similarly, insect and pathogen attacks on tea tree have been documented, pest species listed and control procedures recommended (Campbell and Maddox 1997). Pesticides have been recommended and residue carry over into the distilled oil has been monitored. The fate of the volatile oil in some insect species has been studied by examining frass volatiles and reasons for the metabolism of cineole advanced (Southwell *et al.* 1995).

Harvesting tea tree from natural stands leaves one exposed to the genetic and chemical variation that occurs in nature. Plantation establishment provides an opportunity to eliminate this variation by the selection of genetically improved planting stock. Although propagation by tissue culture and cuttings have been investigated, the quantities of seedlings required have, for economic reasons, meant that most plantations have been established from seed. The establishment of seed orchards to facilitate cross pollination of selected genotypes is important for producing high quality and high oil yielding strains. Investigations selecting for these superior trees, although long term projects, are beginning to show yield gains (Doran *et al.* 1997).

Steam distillation is the preferred method for processing tea tree leaf for oil production. The cost of alternate processes like supercritical fluid extraction or the application of microwave extraction methods is prohibitive for such a high volume—low value product. Based on eucalyptus distillation technology (Davis and House 1991), tea tree is distilled either by hydro or steam distillation in plants ranging from state-of-the-art gas or oil fired boilers, stainless steel bins and tubular horizontal condensers, to the more primitive hydrodistillation bins with "coiled-pipe-in-the-water-tank" type condensers.

Tea tree oil gained early popularity because of strong antimicrobial activity measured in the 1920s. It has been established that this activity is caused chiefly by terpinen-4-ol, the major component in tea tree oil (Southwell *et al.* 1993). Early uses however included flavoring applications where it was added to citrus oil to enhance the terpinen-4-ol content. Antimicrobial zones of inhibition (ZOI) and minimum inhibitory concentration (MIC) values have been recorded for the oil and numerous oil constituents against many bacteria, fungi and plant pathogens (Carson and Riley 1995; Southwell *et al.* 1997b; Bishop 1995; Bishop and Thornton 1997). Clinical trials have determined the activity *in vivo* for conditions such as acne, tinea and vaginitis (Bassett *et al.* 1990; Tong *et al.* 1992; Bélaiche 1985) and confirmed the non-irritant nature of the oil when applied topically at less than 25% concentration (Southwell *et al.* 1997a). Investigations such as these have led to improved *in vivo* and *in vitro* testing methods for tea tree oil (Carson and Riley 1995; Mann and Markham 1998) including cell-line testing of cytotoxicity as an alternative to animal testing (Hayes *et al.* 1997).

Entrepreneurs have capitalised on this bioactivity by either selling the oil neat or formulating it into a myriad of value added products where it acts as a simple preservative, antiseptic or antibacterial soothing agent or as an active ingredient in medicinal products. Such products include antiseptic creams, soaps, mouthwashes, toothpastes, bath oils, body lotions, lip balms, acne creams and washes, tinea creams and vaginitis creams and douches. Legislation governs how, where and in what concentration tea tree oil can be used, what claims can be made about its activity and how bottles must be sealed, stored and labelled. Use in medicinal products is legislated in Australia, for example by the Therapeutic Goods Administration (TGA) which either "lists" or "registers" such therapeutic goods depending on the testing that the product and ingredients have undergone. The TGA then controls the claims that can then be made about the product. Basic toxicological and efficacy investigations have established the safety and effectiveness of use by accumulating toxicological data and measuring both *in vitro* and *in vivo* activity (Altman 1991; Tisserand and Balacs 1995). Data on LD_{50} values, dermal toxicity, dermal irritation, mutagenicity etc. have been acquired from animal testing and on skin irritation and allergy response from human panellists. Although not a skin irritant, especially in formulations containing less than 25% of the oil, a very small number of people react with allergy to the application of tea tree oil and tea tree oil products (Southwell *et al.* 1997a). Investigations are determining which tea tree oil constituents are the most allergenic and suggestions for their possible removal are being made.

Over a period of ten years, the volume of production at the farm gate has increased ten-fold. This oil has, in most cases, been sold with only minor delays and a minimum of stock-piling. The price has varied somewhat to reflect the supply and demand equilibrium. Prices peaked in the late eighties when demand exceeded supply. More recently however the price has fallen and stabilised gradually to reflect increasing production. The establishment of more large-scale plantations, if successful, will increase production beyond demand and consequently decrease the international price of the oil. An increasing quantity of the oil is being sold in value-added products by the smaller producers rather than being sold in bulk to commercial formulators. Some believe that markets will have to expand in existing buyer countries, spread into new territories, and be formulated into a wider range of pharmaceutical type products, for the world market to absorb the projected increase in oil volumes.

Although traditionally longer established on world markets, cajuput and niaouli oils from the large leaved *M. cajuputi* and *M. quinquenervia* respectively (Brophy and Doran 1996) have not enjoyed the surge in market popularity associated with the *M. alternifolia* group. These species are still popular in their local producing regions of SE Asia and New Caledonia respectively, with the latter also growing in Madagascar. The potential exists for many other *Melaleuca* species to be distilled and their oils traded commercially. *M. stipitata*, for example, combines the antimicrobially active ingredients of the *M. alternifolia* group with the pleasant aromatic notes of lemon grass. Higher yielding varieties of these chemotypes would be needed for their commercial production to be viable. Other large leaved varieties have potential as sources of linalool, nerolidol and viridiflorol (*M. quinquenervia*), methyl cinnamate and ocimene (*M. viridiflora*) and lemon oil (*M. citrolens*). Other small leaved species could also be used as sources of eudesmol (*M. uncinata*), methyl eugenol and methyl isoeugenol (*M. bracteata*).

In this volume we attempt to review the wealth of information on the commercial and potentially commercial *Melaleuca* species that exists in the scientific literature, in agricultural bulletins, in government legislation and in the manuals of *Melaleuca* oil entrepreneurs.

REFERENCES

Aboriginal Communities of the Northern Territory. (1988) *Traditional Bush Medicines. An Aboriginal Pharmacopoeia.* Greenhouse, Richmond, Victoria, pp. 160–165.

Altman, P. (1991) Assessment of the skin sensitivity and irritation potential of tea tree oil. Report for Rural Industries Research and Development Corporation. Pharmaco Pty Ltd., Sydney, Australia.

Bassett, I.B., Pannowitz, D.L. and Barnetson, R.St-C. (1990) A comparative study of tea-tree oil versus benzoylperoxide in the treatment of acne. *Med. J. Aust.*, **153**, 455–458.

Bélaiche, P. (1985) Traitement des infections vaginales a *Candida albicans* par l'huile essentielle de *Melaleuca alternifolia* (Cheel). *Phytotherapy*, **15**, 13–14.

Bishop, C.D. (1995) Antiviral activity of the essential oil of *Melaleuca alternifolia* (Maiden & Betche) Cheel (Tea Tree) against tobacco mosaic virus. *J. Essent. Oil Res.*, **7**, 641–644.

Bishop, C.D. and Thornton, I.B. (1997) Evaluation of the antifungal activity of the essential oils of *Monarda citriodora* var. *citriodora* and *Melaleuca alternifolia* on post-harvest pathogens. *J. Essent. Oil Res.*, **9**, 77–82.

Blake, S.T. (1968) A revision of *Melaleuca leucadendron* and its allies (Myrtaceae) *Contr. Queensland Herb.*, **1**, 1–114.

Boland, D.J., Brooker, M.I.H., Chippendale, G.M., Hall, N., Hyland, B.P.M., Kleinig, D.A. and Turner, J. (1984) *Forest Trees of Australia*, 4th edition, Nelson, Melbourne.

Brophy, J.J. and Doran, J.C. (1996) Essential Oils of Tropical *Asteromyrtus*, *Callistemon* and *Melaleuca* Species. ACIAR Monograph No. 40, Canberra.

Campbell, A.J. and Maddox, C.D.A. (1997) Controlling insect pests in tea tree using pyrgo beetle as the basis. *RIRDC Research Paper Series*, **97/62**, Rural Industries Research and Development Corporation, Canberra.

Carson, C.F. and Riley, T.V. (1995) Antimicrobial activity of the major components of the essential oil of *Melaleuca alternifolia*. *J. Appl. Bact.*, **78**, 264–269.

Colton, R.T. and Murtagh, G.J. (1990) Tea-tree oil—plantation production. *Agfact P6.4.6*. NSW Agriculture and Fisheries, Orange, Australia.

Cornwell, C.P., Leach, D.N. and Wyllie, S.G. (1995) Incorporation of oxygen-18 into terpinen-4-ol from the $H_2{}^{18}O$ steam distillates of *Melaleuca alternifolia* (Tea Tree). *J. Essent. Oil Res.*, **7**, 613–620.

Davis, G.R. and House, A.P.N. (1991) Still design and distillation practice. In D.J. Boland, J.J. Brophy and A.P.N. House (eds.), *Eucalyptus Leaf Oils*, Inkata Press, Sydney, pp. 187–194.

Doran, J.C. (1991) Commercial sources, uses, formation, and biology. In D.J. Boland, J.J. Brophy and A.P.N. House (eds.), *Eucalyptus Leaf Oils*, Inkata Press, Sydney.

Doran, J.C., Baker, G.R., Murtagh, G.J. and Southwell, I.A. (1997) Improving tea tree yield and quality through breeding and selection. *RIRDC Research Paper Series*, **97/53**, Rural Industries Research and Development Corporation, Canberra.

Geary, T.F. (1988) *Melaleuca quinquenervia* (Cav.) S.T. Blake. In R.M. Burns and M. Mosquera (eds.), *Useful Trees of Tropical North America*. North American Forestry Commission Publ. No. 3, Washington DC.

Guenther, E. (1950) *The Essential Oils*, Van Nostrand, New York, Vol. 4, pp. 526–548.

Hayes, A.J., Leach, D.N. and Markham, J.L. (1997) *In vitro* cytotoxicity of Australian tea tree oil using human cell lines. *J. Essent. Oil Res.*, **9**, 575–582.

International Standards Organization (1996) Oil of *Melaleuca*, terpinen-4-ol type (Tea Tree Oil). *International Standard ISO 4730: 1996(E)*, International Standards Organization, Geneva.

Lawrence, B.M. (1985) A review of the world production of essential oils (1984). *Perfumer and Flavorist*, **10**(5), 1–16.

Leach, D.N., Wyllie, S.G., Hall, J.G. and Kyratzis, I. (1993) The enantiomeric composition of the principal components of the oil of *Melaleuca alternifolia*. *J. Agric. Food Chem.*, **41**, 1627–1632.

List, S., Brown, P.H. and Walsh, K.B. (1995) Functional anatomy of the oil glands of *Melaleuca alternifolia* (Myrtaceae). *Austral. J. Bot.*, **43**, 629–641.

Mann, C.M. and Markham, J.L. (1998) A new method of determining the MIC of essential oils. *J. Appl. Microbiology*, in press.

Motl, O., Hodacová, J. and Ubik, K. (1990) Composition of Vietnamese cajuput essential oil. *Flavour and Fragrance Journal*, **5**, 39–42.

Murtagh, G.J. (1991) Irrigation as a management tool for production of tea tree oil. *RIRDC Research Report DAN-19A*. Rural Industries Research and Development Corporation, Canberra.

Murtagh, G.J. (1996) Month of harvest and yield components of tea tree. I. Biomass. *Australian Journal of Agricultural Research*, **47**, 801–815.

Murtagh, G.J. and Etherington, R.J. (1990) Variation in oil concentration and economic return from tea tree (*Melaleuca alternifolia* Cheel) oil. *Australian Journal of Experimental Agriculture*, **30**, 675–679.

Murtagh, G.J. and Smith, G.R. (1996) Month of harvest and yield components of tea tree. II Oil concentration, composition and yield. *Australian Journal of Agricultural Research*, **47**, 817–827.

Penfold, A.R. and Morrison, F.R. (1950) Tea tree oils. In Guenther, E. (ed.), *The Essential Oils*, Van Nostrand Co. Inc., New York, Vol. 4, pp. 526–548.

Ramanoelina, P.A.R., Bianchini, J.P., Andriantsiferana, M., Viano, J. and Gaydou, E.M. (1992) Chemical composition of niaouli essential oils from Madagascar. *J. Essent. Oil Res.*, **4**, 657–658.

Ramanoelina, P.A.R., Viano, J., Bianchini, J.P. and Gaydou, E.M. (1994) Occurrence of variance chemotypes in niaouli (*Melaleuca quinquenervia*) essential oils from Madagascar using multivariate statistical analysis. *J. Agric. Food Chem.*, **42**, 1177–1182.

Southwell, I.A., Hayes, A.J., Markham, J. and Leach, D.N. (1993) The search for optimally bioactive Australian tea tree oil. *Acta Horticulturae*, **344**, 256–265.

Southwell, I.A., Freeman, S. and Rubel, D. (1997a) Skin irritancy of tea tree oil. *J. Essent. Oil Res.*, **9**, 47–52.

Southwell, I.A., Maddox, C.D.A. and Zalucki, M.P. (1995) Metabolism of 1,8-cineole in tea tree (*Melaleuca alternifolia* and *M. linariifolia*) by pyrgo beetle (*Paropsisterna tigrina*). *J. Chem. Ecol.*, **21**, 439–453.

Southwell, I.A., Markham, J. and Mann, C. (1997b) Why cineole is not detrimental to tea tree oil. *RIRDC Research Paper Series*, **97/54**, Rural Industries Research and Development Corporation, Canberra.

Southwell, I.A. and Stiff, I.A. (1989) Ontogenetical changes in monoterpenoids of *Melaleuca alternifolia* leaf. *Phytochemistry*, **28**, 1047–1051.

Storrie, A., Cook, T., Virtue, J., Clarke, B. and McMillan, M. (1997) *Weed Management in Tea Tree.* NSW Agriculture, Orange, Australia.

Tisserand, R. and Balacs, T. (1995) *Essential Oil Safety—A Guide for Health Care Professionals.* Churchill Livingstone, Edinburgh, pp. 45–55, 80, 82, 150, 187, 204, 219.

Todorova, M. and Ognyanov, I. (1988) Composition of Vietnamese essential oil from *Melaleuca leucadendron* L. *Perfumer and Flavorist*, **13**, 17–18.

Tong, M.M., Altman, P.M. and Barnetson, R.St-C. (1992) Tea tree oil in the treatment of *Tinea pedis. Australasian J. Dermatol.* **33**, 145–149.

Virtue, J.G. (1997) Weed interference in the annual regrowth cycle of plantation tea tree (*Melaleuca alternifolia*). PhD Thesis. The University of Sydney.

Weiss, E.A. (1997) *Essential Oil Crops*, CAB International, Oxford, pp. 302–319.

Whish, J.P.M. and Williams, R.R. (1996) Effects of post harvest drying on the yield of tea tree oil (*Melaleuca alternifolia*). *J. Essent. Oil Res.*, **8**, 47–51.

Wrigley, J.W. and Fagg, M. (1993) *Bottlebrushes, Paperbarks and Tea Trees and all other Plants in the Leptospermum Alliance*, Angus and Robertson, Sydney.

Plate 1 Tea tree plantation in New South Wales, Australia (R. Colton)

Plate 2 Mature *Melaleuca alternifolia* tree in a natural stand (I. Southwell)

Plate 3 *Melaleuca alternifolia* flower (I. Holliday)

Plate 4 *Melaleuca cajuputi* subsp. *cajuputi* flower (J. Brock)

Plate 5 *Melaleuca quinquenervia* flower (B. Trilles)

1. BEHIND THE NAMES: THE BOTANY OF TEA TREE, CAJUPUT AND NIAOULI

LYN A. CRAVEN

Australian National Herbarium, CSIRO Division of Plant Industry, Canberra, Australia

INTRODUCTION

"We know what plant *X* is. Why should we have to worry about its scientific name?" Well, sometimes calling a plant *X* is neither sufficiently precise to enable us to communicate unambiguously nor to find the information that we seek in the ever-burgeoning literature with which we must cope. Australians use the name *tea tree* to refer to many indigenous species of *Leptospermum*, *Melaleuca* and *Neofabricia* (Wrigley and Fagg 1993). These three genera belong to the family Myrtaceae so they do have something in common. Clearly though, the use of tea tree by itself will not suffice as the only name for the subject of this book and we must use its scientific name, *Melaleuca alternifolia*, to distinguish it unambiguously from all the other tea trees.

Apart from being a label, the scientific name of a plant can be considered a concise definition as to what the plant actually is, as the name can be related to a particular morphological circumscription. Similarly with the many ecological and chemical attributes that species possess. *Melaleuca leucadendra* sensu Bentham taxonomically, and ecologically, geographically and chemically, is very different to *M. leucadendra* sensu Blake. Unless the name is qualified such as in the previous sentence, it is not easy to determine which of the potentially several taxonomic concepts a particular author is using.

In itself a name is not especially useful as an indication to the position occupied by a species in taxonomic classifications, other than informing us of the genus to which the species belongs. The binomial nomenclatural system, i.e. the combination of a generic with a specific epithet (such as *Melaleuca alternifolia*), does not cater for taxa between the ranks of genus and species, such as section or series. This is unfortunate in large genera such as *Melaleuca* in which we need to know in which part of the genus our taxon or taxa of interest are classified. Notwithstanding this, the scientific name is our only sure password to the information that is available in the literature.

This contribution provides a brief outline of the genus *Melaleuca* and of the species groups to which tea tree, cajuput and niaouli belong.

THE GENUS *MELALEUCA*

Melaleuca was established as a genus by Linnaeus in 1767 with *M. leucadendra* as its only, and hence type, species. Linnaeus based his genus on the pre-Linnaean *Arbor alba* that was described by Rumphius from plants growing on Ambon in present day Indonesia

(Rumphius 1741). From 1767 until the mid-1800s, several further species were added to the genus but it was not until 1867 that the first comprehensive treatment of *Melaleuca* was published in Bentham's account of Myrtaceae in his classic flora of Australia, *Flora Australiensis* (Bentham 1867). Within *Melaleuca*, Bentham recognised 97 species in 7 series. His series include some groupings of undoubtedly closely related species but overall the classification is artificial. This is not a reflection upon the general quality of Bentham's treatment but is an indication of the difficulty inherent in classifying a large group of species that, while differing considerably in gestalt, is remarkably similar in essential structural details. To the present day, Bentham's work has remained the most monographic account of *Melaleuca* available.

Since 1867, many other species have been described in *Melaleuca* by botanists. These usually have been published as isolated descriptions of novelties or in accounts of flora collected by large expeditions; it has only been in recent decades that studies of *Melaleuca* within large regions have been undertaken and published. These more comprehensive studies include a revision of the broad-leaved paperbarks (Blake 1968), a revision of *Melaleuca* in South Australia (Carrick and Chorney 1979), a treatment of the northern and eastern Australian species (Byrnes 1984, 1985, 1986), treatments of several largely southwestern and eastern Australian species groups (Barlow 1987; Barlow and Cowley 1988; Cowley *et al.* 1990), and a revisionary level treatment of the New Caledonian species (Dawson 1992). The genus *Asteromyrtus*, included in *Melaleuca* by Bentham (1867) and Byrnes (1984, 1985) was resurrected by Craven (1989) to accommodate a constellation of species most of which had been treated under *Melaleuca* but of which one had been placed in the monotypic genus *Sinoga* by Blake (1958). *Asteromyrtus* is not closely related to *Melaleuca*; its relationships lie with *Agonis* in the *Leptospermum* group of genera. In passing it may be noted that at least one species of *Asteromyrtus*, *A. symphyocarpa*, has potential as a viable source of essential oil (Chapter 16, this volume). Preparatory work towards an account of *Melaleuca* for *Flora of Australia* is in train and a precursory paper enumerating all the Australian species and providing identification keys currently is being completed by L.A. Craven and B.J. Lepschi.

According to the classification of Briggs and Johnson (1979), the genera most closely related to *Melaleuca* are *Callistemon*, *Conothamnus* and *Lamarchea* while *Beaufortia*, *Calothamnus*, *Eremaea*, *Phymatocarpus* and *Regelia* are more distantly related. It seems that there is no especially close relationship with the genera clustered around *Leptospermum*, i.e. *Agonis*, *Angasomyrtus*, *Asteromyrtus*, *Homalospermum*, *Kunzea*, *Neofabricia* and *Pericalymma*. The conventional circumscription of *Melaleuca* given below may require amendment when research in progress by the author into the relationships of *Callistemon* and *Conothamnus* to *Melaleuca* is concluded. At least two species of Australian *Callistemon*, *C. glaucus* and *C. viminalis*, have their stamens grouped into 5 basally fused groups, one of the key generic characters of *Melaleuca*. Indeed, this was the basis for Byrnes' transfer of the eastern Australian species, *C. viminalis*, to *Melaleuca* (Byrnes 1984, 1986). Similarly, the endemic New Caledonian species of *Callistemon* are very close to the endemic New Caledonian *Melaleuca*, several of them having stamens fused into groups, and they may be transferred to *Melaleuca* as a result of research in progress.

As it is presently circumscribed, *Melaleuca* consists of about 230 species (Craven 1997). The great majority of the species, about 220, is endemic to Australia and Tasmania but several also occur in adjacent parts of Indonesia and Papua New Guinea and one species, *M. cajuputi*, extends from Australia northwards to the Asian mainland. There is one endemic species in Lord Howe Island, *M. howeana*, and three species in New Caledonia (eight including the *Callistemon* species) of which *M. quinquenervia* also occurs in eastern Australia and New Guinea. Within Myrtaceae, *Melaleuca* is characterised by possession of the following combination of features: Shrubs or trees; leaves spiral, decussate or ternate, small to medium-sized, the venation pinnate to parallel; flowers in spikes or clusters or sometimes solitary, the basic floral unit being a monad, dyad or triad; sepals 5 (rarely 0); petals 5; hypanthium fused to the ovary in the proximal region only; stamens few to numerous, the filaments fused for part of their length into 5 bundles, the anthers dorsifixed (or rarely basifixed) and versatile with two parallel cells that open via longitudinal slits; ovary 3-celled, the ovules few to numerous; fruit a capsule within an usually woody to subwoody fruiting hypanthium; seeds with a thin testa, generally obovoid-oblong to obovoid, unwinged, the cotyledons planoconvex to obvolute.

TEA TREE

Tea tree, *Melaleuca alternifolia* (Plates 2, 3), is very closely related to *M. linariifolia*, and it is not surprising that Maiden and Betche treated the plant as a variety of that species when they described it in 1905. It belongs in the *M. linariifolia* species group of which there are five currently accepted species: *M. alternifolia*, *M. dissitiflora*, *M. linariifolia*, *M. linophylla* and *M. trichostachya*. Maiden & Betche distinguished their variety *alternifolia* as having alternate leaves that are much narrower and usually shorter than those of *M. linariifolia* sensu stricto and having its flowers (Plate 3) less densely arranged in the inflorescences. Cheel considered that the differences between the two, taken with their apparent geographic isolation, were sufficient justification for the variety to be raised to specific rank (Cheel 1924).

The first-described, and best known, species of the group is *M. linariifolia*; this was described by Smith in 1797 from specimens collected in the Sydney region in 1795. *M. linariifolia* is widely cultivated in Australia as it is hardy and forms an attractive large shrub or small tree with masses of pure white flowers.

The next described species is *M. trichostachya*, described by Lindley in 1848 from specimens collected by the explorer Thomas Mitchell in 1846 in subtropical Queensland. Bentham, however, reduced *M. trichostachya* to a variety of *M. linariifolia* in 1867. Of the recent taxonomists who have worked on *Melaleuca*, Bentham's placement was followed by Byrnes (1985) but Quinn *et al.* (1989), who studied the complex in more detail than did Byrnes, recognised *M. trichostachya* as a distinct species. The remaining two species of the group, *M. linophylla* and *M. dissitiflora*, were described in 1862 and 1863, respectively, by the 19th Century Australian botanist, Ferdinand Mueller.

Geographically, the group is widespread and occurs in a correspondingly broad range of climates. The distributions of the species are shown in Figure 1. Given the occurrence of terpinen-4-ol chemotypes in *M. alternifolia*, *M. dissitiflora* and *M. linariifolia*, it would seem that there is scope for a more intensive survey of the *M. linariifolia* group to gain a greater understanding of the chemotypes and their distributions. Then it would

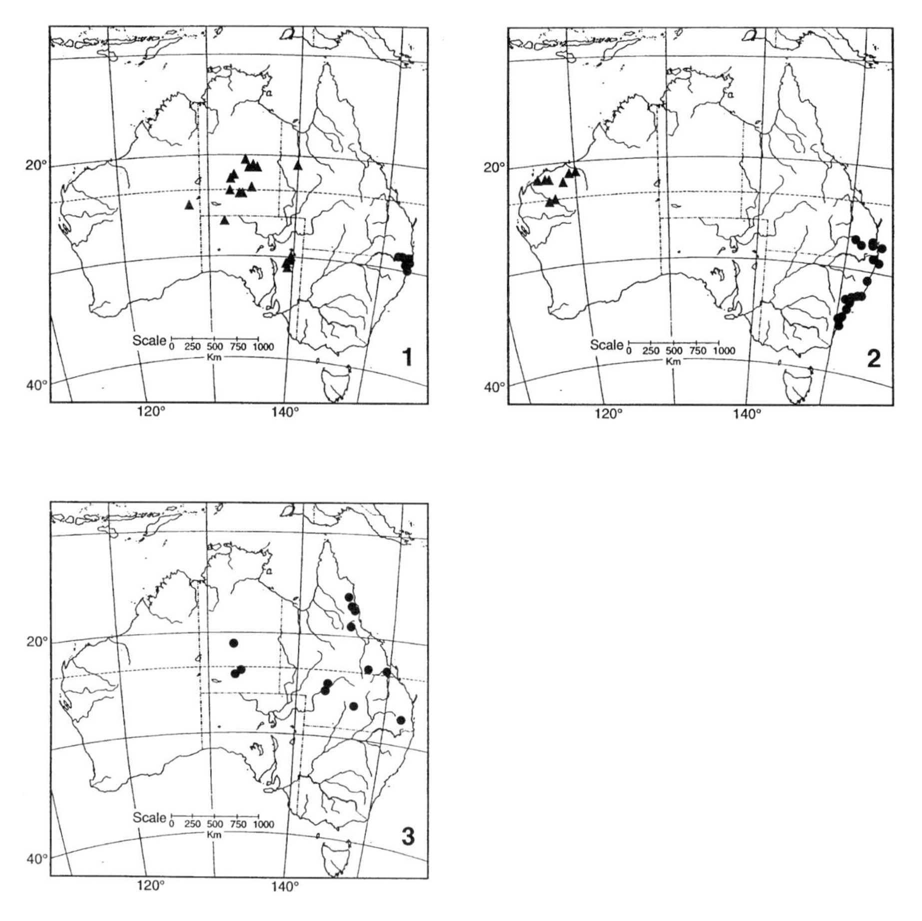

Figure 1 Distributions of *Melaleuca* species. 1, *M. alternifolia* ●; *M. dissitiflora* ▲. 2, *M. linariifolia* ●; *M. linophylla* ▲. 3, *M. trichostachya* ●

be possible to consider the prospects for expanding the tea tree oil industry, presently centred upon *M. alternifolia* in NE New South Wales, by growing other terpinen-4-ol rich genotypes in regions to which they are suited climatically and edaphically, by introgressing novel genes into *M. alternifolia* and *M. linariifolia* for crop improvement with the NE New South Wales region, or by both means.

Morphologically, the species need to be studied further, in the case of the central Australian representatives preferably with flowering and fruiting materials collected from the same trees, so that more comprehensive and possibly more definitive circumscriptions can be made. Until then, the species of the *M. linariifolia* species group can be distinguished by means of the key given below.

Future systematic studies should not be restricted to morphological aspects. Molecular systematics, based upon comparative studies of DNA, in particular will assist in establishing the evolutionary relationships of the species. Butcher *et al.* (1995) studied chloroplast

DNA (cpDNA) variation in *M. alternifolia*, *M. linariifolia* and *M. trichostachya* and in part their findings were that *M. linariifolia* and *M. trichostachya* were readily distinguished on cpDNA data. They also found that cpDNA data supported an earlier suggestion based upon leaf terpene data (Butcher *et al.* 1994) that hybridisation may have occurred between *M. alternifolia* and *M. linariifolia* in the southern part of the former species' range. Clearly, the incorporation of molecular data into the systematic synthesis will enable us to devise much more accurate and informative classifications than we have been able to do using morphological data alone.

List of species of the *M. linariifolia* group, including their synonyms and type specimens

1. *Melaleuca alternifolia* (Maiden & Betche) Cheel, J. Proc. Roy. Soc. New S. Wales 58: 195 (1924). *Melaleuca linariifolia* var. *alternifolia* Maiden & Betche, Proc. Linn. Soc. New S. Wales 29: 742 (1905). Typus: New South Wales: Coffs Harbour to Grafton, Nov. 1903, *Maiden and Boorman s.n.* (NSW, holo, *n.v.*).
2. *Melaleuca dissitiflora* F.Muell., Fragm. 3: 153 (1863). *Myrtoleucodendron dissitiflorum* (F.Muell.) Kuntze, Revis. gen. pl. 241 (1891). Typus: Northern Territory: between Bonney River and Mount Morphett, 1862, *Stuart s.n.* (MEL, holo, iso, *n.v.*).
3. *Melaleuca linariifolia* Sm., Trans. Linn. Soc. London, Bot. 3: 278 (1797). *Myrtoleucodendron linariifolium* (Sm.) Kuntze, Revis. gen. pl. 241 (1891). *Melaleuca linariifolia* var. *typica* Domin, Biblioth. Bot. 89: 456 (1928), *nom. inval.* Typus: New South Wales: 1795, *White s.n.* (LINN, holo, *n.v.*).

   Metrosideros hyssopifolia Cav., Icon. 4: 20, t.336. f.1. (1797). *Melaleuca hyssopifolia* (Cav.) Dum.Cours., Bot. cult. 2, 5: 375 (1811). Typus: The illustration accompanying the original description.
4. *Melaleuca linophylla* F.Muell., Fragm. 3: 115 (1862). *Myrtoleucodendron linophyllum* (F.Muell.) Kuntze, Revis. gen. pl. 241 (1891). Typus: Western Australia: northwest, 1861, *Gregory s.n.* (MEL, holo, *n.v.*; BRI, K, iso, *n.v.*).
5. *Melaleuca trichostachya* Lindl., in Mitchell, J. exped. trop. Australia 277 (1848). *Melaleuca linariifolia* var. *trichostachya* (Lindl.) Benth., Fl. Austral. 3: 141 (1867). *Myrtoleucodendron trichostachyum* (Lindl.) Kuntze, Revis. gen. pl. 242 (1891). Typus: Queensland: sandy bed of Belyando River, 16 Aug. 1846, *Mitchell 224* (CGE, holo, *n.v.*; K, MEL, NSW, iso, *n.v.*).

Key to the species of the *M. linariifolia* group

1. Leaves decussate
   2. Ovules 85–120 per locule; capsule persisting within the fruiting hypanthium; cotyledons planoconvex .... *M. linariifolia*
   2. Ovules 25–45 per locule; capsule at length deciduous and the empty fruiting hypanthium then persisting; cotyledons obvolute to almost planoconvex.......... *M. trichostachya*
1. Leaves spiral
   3. Hypanthium glabrous or effectively so (if distinctly hairy, then stamens more than 12 mm long)
      4. Stamens up to 12 mm long; capsule at length deciduous and the empty fruiting hypanthium then persisting.......... *M. trichostachya*

4. Stamens more than 12 mm long; capsule persisting within the fruiting hypanthium..*M. alternifolia*
3. Hypanthium distinctly hairy
5. Stamens 2.2–3.5 mm long, 7–15 per bundle..............................*M. linophylla*
5. Stamens 4 mm or more long, 15–35 per bundle.........................*M. dissitiflora*

CAJUPUT AND NIAOULI

It is ironic that, considering their prominent form and frequency in the landscape, and their utility as sources of timber, paperbark and essential oils, the tropical and subtropical tree species of *Melaleuca*, collectively known either as the *M. leucadendra* group or the broad-leaved paperbarks, have proved so difficult for botanists to classify. Numerous species and varieties have been proposed but presently it is considered that there are 15 species in the group, the majority being medium to large trees although a few are shrubs or small, poorly formed trees. The distributions of the species are shown in Figures 2–5.

An indication of the lack of appreciation of the taxonomic limits in the broad-leaved paperbarks can be found in the number of type specimens applicable to this group of plants. To date, there are 45 type specimens but only 15 accepted species, whereas in the *M. linariifolia* group there are 6 types and 5 species. Bentham (1867) dealt with the confusing array of names and specimens known to him by recognising only a single species in his *Flora Australiensis* account of the genus, i.e. *M. leucadendron*. [The majority of authors have used Linnaeus' later spelling "*leucadendron*"; it was only in 1966 that the original spelling "*leucadendra*" was re-established in the edition of the *International Code of Botanical Nomenclature* published in that year (Lanjouw *et al.* 1966).] Bentham was followed by most botanists, although some, notably Cheel (1917), endeavoured to partition the variation into varieties.

Blake (1968) published an extremely detailed revision of the *M. leucadendra* group, based upon his extensive field knowledge and an excellent appreciation of the discriminatory value of indumentum (hair type) and floral morphology in development of a workable taxonomy for this group of plants. The work by Blake has provided a sound basis for subsequent research. Byrnes (1984, 1985, 1986) generally followed Blake's treatment of the paperbarks although he combined *M. quinquenervia* and *M. viridiflora* on the basis that the two species overlapped in the quantitative key characters given by Blake (1968). Byrnes (1984) described several new taxa in the group, notably a number at varietal level; presumably in part this was an acknowledgment of the regional differentiation within some of the species that had been more broadly defined by Blake. In the 1980's, B.A. Barlow and co-workers reworked the taxonomy of the group but, although several new taxa were recognised by them (and a number of Byrnes' were not), their results unfortunately have not been published in a synthetic account.

The species from which cajuput oil is extracted, *M. cajuputi* (Plate 4, 18, 19) is widespread and occurs from southeast Asia to northern Australia. However, the source of commercial cajuput oil are populations apparently originating in a restricted part of Indonesia and now cultivated more widely in both Indonesia and other parts of southeast Asia. These populations belong to a distinct form of the species, *M. cajuputi* subsp. *cajuputi*.

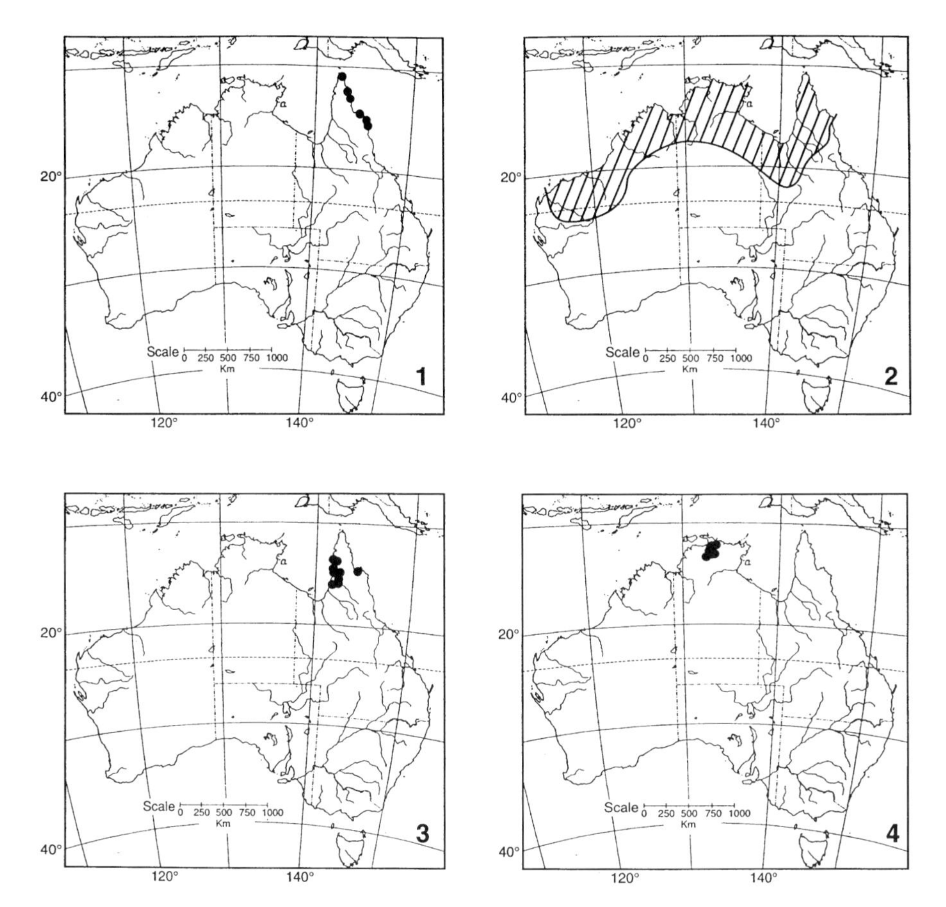

Figure 2 Distributions of *Melaleuca* species. 1, *M. arcana*. 2, *M. argentea*. 3, *M. clarksonii*. 4, *M. cornucopiae*

It is tempting to consider that the wide distribution of the species is a direct result of the fact that the species is of value for its therapeutic oil with man having carried the plant around to cultivate it. Once established in cultivation locally, it could then become naturalised and expand its range using natural means of dispersal. Barlow (1988) discussed patterns of differentiation in species of the *M. leucadendra* group, especially their geographic distributions. Barlow has recognised three subspecies within *M. cajuputi*: subsp. *cajuputi*, subsp. *cumingiana* and subsp. *platyphylla*. He considered that the western subspecies of *M. cajuputi*, subsp. *cumingiana*, represented a colonisation event across Wallace's Line from the east, an unusual event as most of the plant movements across this line have been from the west. This subject was taken up by Lum (1994). Lum concluded that genetic data supported the hypothesis that *M. cajuputi* had dispersed naturally westward. However, Lum's genetic data were derived from relatively few populations that do not sample the species across its range adequately; further work is required before an unequivocal assessment can be made.

3a

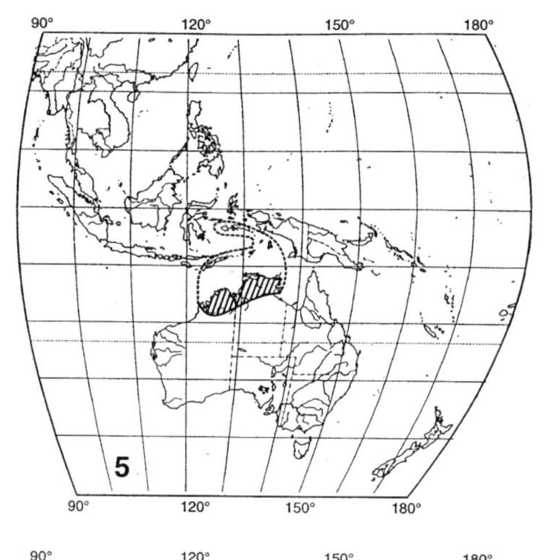

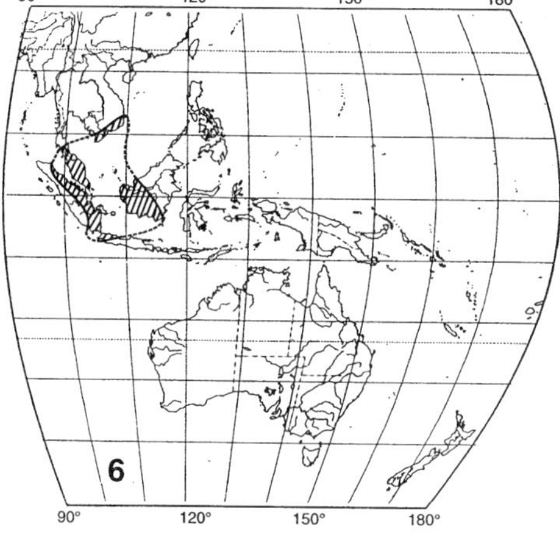

3b

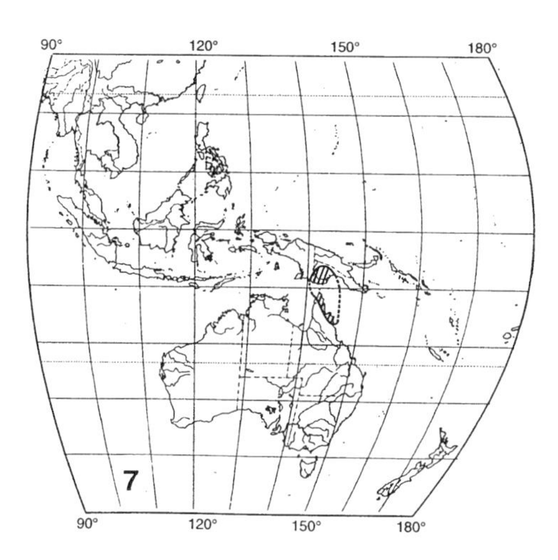

Figure 3 Distributions of *Melaleuca* species. 5, *M. cajuputi* subsp. *cajuputi*. 6, *M. cajuputi* subsp. *cumingiana*. 7, *M. cajuputi* subsp. *platyphylla*

The other species of the *M. leucadendra* group used as a source of essential oil is *M. quinquenervia* (Plate 5, 21–23), called niaouli in New Caledonia and the source of niaouli oil. This species occurs naturally in eastern Australia, southern New Guinea (including the southeast Papuan islands) and New Caledonia. It is cultivated for essential oil production in Madagascar (Ramanoelina *et al.* 1992, 1994) and is naturalised in the southeastern United States. Although *M. quinquenervia* could have arrived in New Caledonia from the southeast Papuan islands, it is more likely that it dispersed there from eastern Australia. As with *M. cajuputi*, very little is known about the degree of intraspecific variation. Molecular studies aimed at identifying markers that can be used to circumscribe provenance are required. Such studies may then enable light to be shed upon the origins of the New Caledonian population.

As far as future work on the *M. leucadendra* group is concerned, studies directed towards finding additional macrocharacters for distinguishing *M. cajuputi* from *M. quinquenervia* and *M. viridiflora* from *M. quinquenervia* need to be undertaken. Further investigations into essential oils are likely to be of interest; *M. viridiflora*, for example, has commercial potential as a source of methyl cinnamate from the oil (Chapter 16, this volume). Blake (1968) noted some examples of hybridisation between species in the *M. leucadendra* group and this needs further consideration. For example, is the difficulty in distinguishing *M. cajuputi* from *M. quinquenervia* due to hybridisation, and is the occasional occurrence of paper bark in the usually hard barked *M. clarksonii* due to hybridisation or is it part of the intrinsic variation of the species? Molecular techniques are likely to provide additional powerful tools in studying such questions.

List of species of the *M. leucadendra* group, including their synonyms and type specimens

1. *Melaleuca arcana* S.T.Blake, Contr. Queensland Herb. 1: 54, figs. 10, 15J. (1968). Typus: Queensland: NW of Cooktown and W of Cape Bedford, *Blake 20260* (BRI, holo, *n.v.*).

   Melaleuca leucadendron var. *albida* f. *ruscifolia* Cheel, in Ewart & Davies, Fl. N. Territory 302 (1917). *Melaleuca ruscifolia* Sol. ex Cheel, in Ewart & Davies, Fl. N. Territory 302 (1917), *nom. inval.* Typus: Queensland: Point Lookout, *Banks & Solander s.n.* (NSW, holo, *n.v.*; BM, E, K, MEL, P, W, *n.v.*, CANB, iso,).
2. *Melaleuca argentea* W.Fitzg., J. Proc. Roy. Soc. Western Australia 3: 187 (1918). Typus: Western Australia: base of Mt. Bartlett, Sep. 1905, *Fitzgerald 1258* (NSW, holo, *n.v.*; K *n.v.*, PERTH, iso).
3. *Melaleuca cajuputi* Powell.

3a. *Melaleuca cajuputi* Powell subsp. *cajuputi*, Pharm. Roy. Coll. Physic. London (Transl.) 1809, 22 (1809). *Myrtus saligna* J.F.Gmel, Syst. nat. 793 (1791), *nom. illeg.*, non Burm.f. *Melaleuca trinervis* Buch.-Ham., Mem. Wern. Nat. Hist. Soc. 6: 302 (1832), *nom. illeg.*, non Sm. in White. *Melaleuca saligna* (J.F.Gmel.) Blume, Mus. Bot. 1: 66 (1849), *nom. illeg.*, non Schauer. *Melaleuca leucadendron* var. *cajuputi* (Powell) Nied., Natürl. Pfl.Fam IV, 3. 7: 95 (1893). Typus: Rumphius, Herb. Amboin. 2: 76, t. 17, figs. 1, 2 (1741) (the figures and description).

   Melaleuca minor Sm., in Rees, Cycl. 23, no. 2 (1812). *Melaleuca leucadendron* var. *minor* (Sm.) Duthie, in Hooker, Fl. Brit. India 2: 465 (1878). Typus: Indonesia: Ceram, Amboina, Oct. 1796, *Chr. Smith 303* (LINN, holo, *n.v.*; BM, G, NSW, iso, *n.v.*).

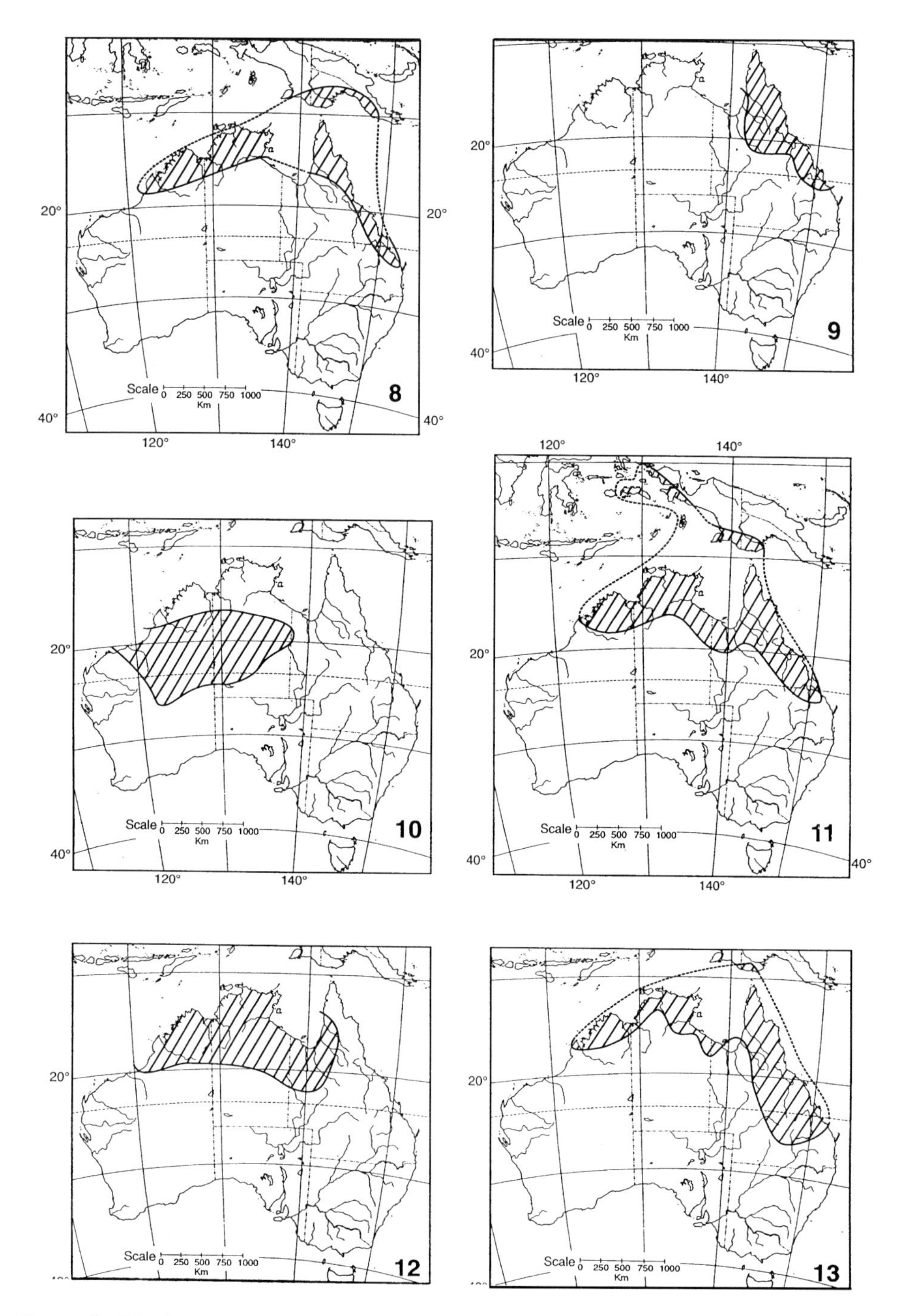

Figure 4 Distributions of *Melaleuca* species. 8, *M. dealbata*. 9, *M. fluviatilis*. 10, *M. lasiandra*. 11, *M. leucadendra*. 12, *M. nervosa* subsp. *crosslandiana*. 13, *M. nervosa* subsp. *nervosa*

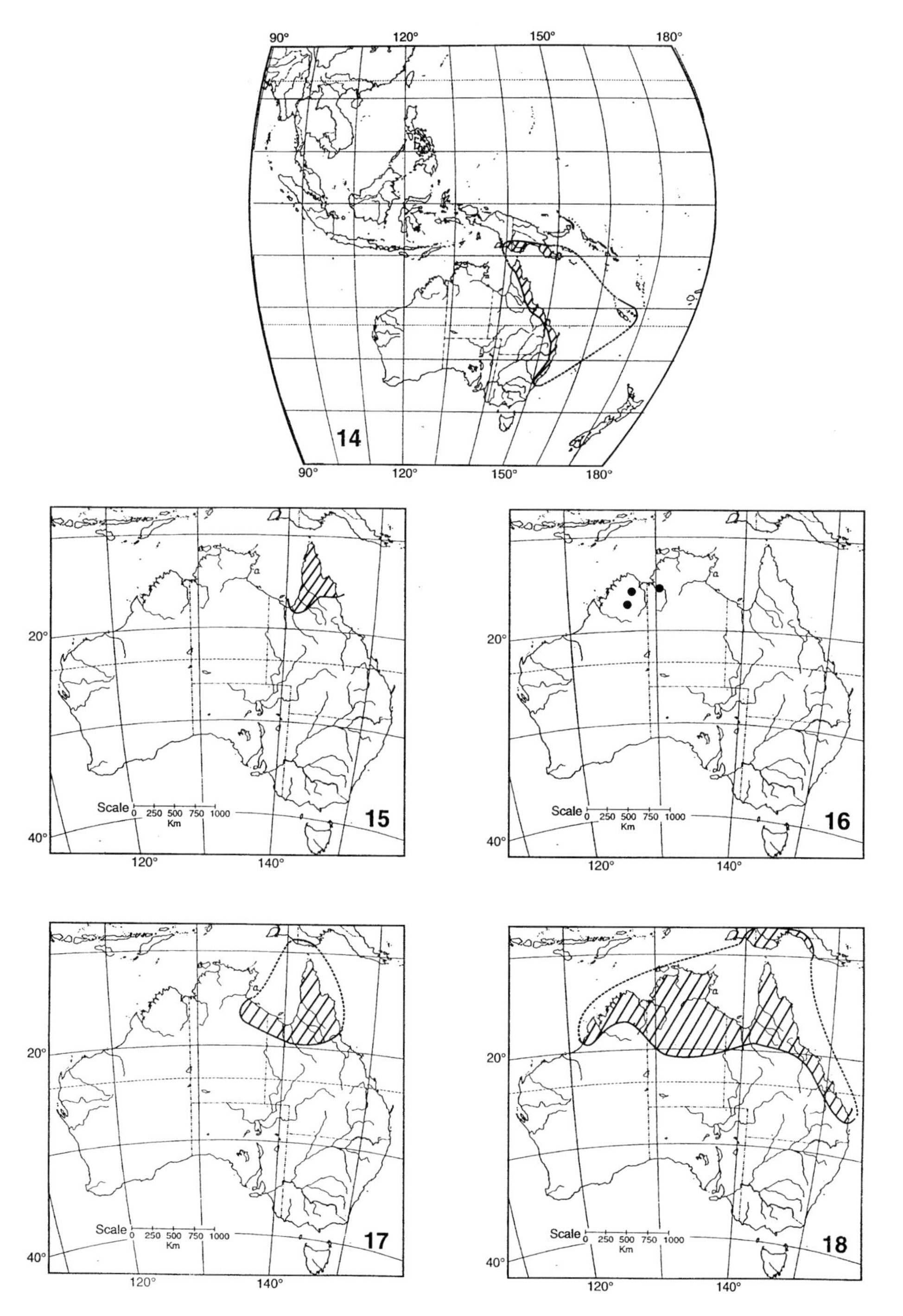

Figure 5 Distributions of *Melaleuca* species. 14, *M. quinquenervia*. 15, *M. saligna*. 16, *M. sericea*. 17, *M. stenostachya*. 18, *M. viridiflora*

Melaleuca viridiflora var. *angustifolia* Blume, Bijdr. fl. Ned. Ind. 1099 (1826). *Melaleuca angustifolia* (Blume) Blume, Mus. Bot. 1: 83 (1849), *nom. illeg.*, non Gaertn. Typus: Indonesia: Moluccas, Amboina, *herb. Blume s.n.* (L, holo, *n.v.*; BRI, fragm., *n.v.*).

Melaleuca lancifolia Turcz., Bull. Soc. Imp. Naturalistes Moscou 20: 164 (1847). *Melaleuca leucadendron* var. *lancifolia* (Turcz.) F.M.Bailey, Syn. Queensl. fl. 170 (1883). Typus: Indonesia: Sumatera, *Cuming 2427* (KW, holo, *n.v.*; CGE, FI, K, P, W, iso, *n.v.*).

3b. *Melaleuca cajuputi* subsp. *cumingiana* (Turcz.) Barlow, in Craven & Barlow, Novon 7: 113 (1997). *Melaleuca cumingiana* Turcz., Bull. Soc. Imp. Naturalistes Moscou 20: 164 (1847). Typus: Malaysia: Malaya, Malacca, *Cuming 2272* (KW, holo, *n.v.*; BRI, CGE, FI, K, L, LE, MEL, MO, P, W, iso, *n.v.*).

Melaleuca commutata Miq., Anal. bot. ind. 14 (1850). Typus: Borneo, *Korthals s.n.* (L, holo, *n.v.*; K, iso, *n.v.*).

3c. *Melaleuca cajuputi* subsp. *platyphylla* Barlow, in Craven & Barlow, Novon 7: 113 (1997). Typus: Papua New Guinea: Western Province: near Bula village, mouth of Morehead River, 4 Aug. 1967, *Pullen 6998* (CANB, holo; A, AD, BISH, BO, BRI, E, G, K, L, LAE, NSW, P, PNH, SING, TNS, US, iso, *n.v.*).

Notes

Myrtus saligna Burm.f. (Fl. Ind. 116 (1768)) was included in the synonymy of *M. cajuputi* by Blake (1968) but its type (Indonesia: Java, 1760, *leg. ign. s.n.* (G, holo, *n.v.*)) needs to be examined before the name can be assigned to one of the above subspecies.

Melaleuca eriorhachis Gand. (Bull. Soc. Bot. France 65: 26 (1918)) also was referred to *M. cajuputi* by Blake (1968). Its type (Singapore, *Ridley* (?) *s.n.* (LY, holo, *n.v.*)) similarly needs to be studied before confident placement to subspecies is possible.

4. *Melaleuca clarksonii* Barlow, in Craven & Barlow, Novon 7: 114 (1997). Typus: Queensland: Cape York Peninsula, 11.1 km SSE of Emu Lagoon, Alice River National Park, 7 May 1992, *Clarkson and Neldner 9582* (CANB, holo; BRI, DNA, K, L, MBA, MEL, NSW, PERTH, iso).

5. *Melaleuca cornucopiae* Byrnes, Austrobaileya 2: 74 (1984). Typus: Northern Territory: Koongarra, 16 Nov. 1975, *Dunlop 4030* (BRI, holo, *n.v.*; DNA, iso, *n.v.*).

6. *Melaleuca dealbata* S.T.Blake, Contr. Queensland Herb. 1: 41, figs. 5, 14E, 15E, 15N. (1968). Typus: Northern Territory: c. Lat. 12 40 S, Long. 131 25 E, *Blake 17000* (BRI, holo, *n.v.*; CANB, iso).

7. *Melaleuca fluviatilis* Barlow, in Craven & Barlow, Novon 7: 116 (1997). Typus: Queensland: sandy river bed, Bruce Highway, c. 50 km NW of Townsville, 13 Jul. 1985, *Barlow and Thiele 3940* (CANB, holo).

Melaleuca nervosa f. *pendulina* Byrnes, Austrobaileya 2: 74 (1984). Typus: Queensland: Coen River, *Brass 19778* (BRI, holo, *n.v.*; A, *n.v.*, CANB, iso).

8. *Melaleuca lasiandra* F.Muell., Fragm. 3: 115 (1862). *Myrtoleucodendron lasiandrum* (F.Muell.) Kuntze, Revis. gen. pl. 241 (1891). Syntypi: Northern Territory: Fitzmaurice River, Oct. 1855, *Mueller s.n.*; Victoria River, Jan.–May 1856, *Mueller s.n.* (both MEL, *n.v.*).

Melaleuca loguei W.Fitzg., J. Proc. Roy. Soc. Western Australia 3: 188 (1918). Typus: Western Australia: S of the Fitzroy River, *Logue in Fitzgerald s.n.* (NSW, holo, *n.v.*).

9. *Melaleuca leucadendra* (L.) L., Mant. pl. 1: 105 (1767). *Myrtus leucadendra* L., Sp. pl. ed. 2, 676 (1762). *Leptospermum leucadendron* (L.) J.R. & G.Forst., Char. gen. pl. 72 (1776). *Cajuputi leucadendron* (L.) Rusby ex A.Lyons, Pl. names Sci. Pop. 479 (1900). Lectotypus (*fide* Blake, Contr. Queensland Herb. 1: 17 (1968)): Rumphius, Herb. Amboin. 2: 72, t. 16 (1741) (the figure and description).

Melaleuca mimosoides Schauer, in Walpers, Repert. Bot. Syst. 2: 927 (1843). *Melaleuca leucadendron* var. *mimosoides* (Schauer) Cheel, in Ewart & Davies, Fl. N. Territory 295 (1917). Typus: Queensland: Rockingham Bay and Endeavour River, *Cunningham 253/1819* (BM, K, iso, *n.v.*).

Melaleuca amboinensis Gand., Bull. Soc. Bot. France 65: 26 (1918). Typus: Indonesia: Ceram, Amboina, *de Vriese s.n.* (LY, holo, *n.v.*).

10. *Melaleuca nervosa* (Lindl.) Cheel.

10a. *Melaleuca nervosa* (Lindl.) Cheel subsp. *nervosa*, J. Proc. Roy. Soc. New S. Wales 78: 65 (1944). *Callistemon nervosus* Lindl., in Mitchell, J. exped trop. Australia 235 (1848), as *nervosum*. *Melaleuca leucadendron* var.? *parvifolia* Benth., Fl. Austral. 3: 143 (1867), *pro parte* (as to *C. nervosus*). *Melaleuca leucadendron* var. *nervosa* (Lindl.) Domin, Biblioth. Bot. 89: 457 (1928), *nom. illeg*. Typus: Queensland: Balmy Creek, Jul. 1846, *Mitchell 241* (CGE, holo, *n.v.*; MEL, NSW, iso, *n.v.*).

Melaleuca nervosa f. *latifolia* Byrnes, Austrobaileya 2 (14 Jul. 1984) 74. Typus: Northern Territory: about SE of Brooks Creek, Jul. 1946, *Blake 16344* (BRI, holo, *n.v.*).

10b. *Melaleuca nervosa* subsp. *crosslandiana* (W. Fitzg.) Barlow ex Craven, comb. et stat. nov. *Melaleuca crosslandiana* W.Fitzg., Western Mail (Perth) (2 Jun. 1906), *basionym*. *Melaleuca leucadendron* var. *coriacea* f. *crosslandiana* (W.Fitzg.) Cheel, in Ewart & Davies, Fl. Northern Territory 298 (1917). Lectotypus (*fide* Blake, Contr. Queensland Herb. 1: 43 (1968)): Western Australia: base of Mt Harris, Jun. 1905, *Fitzgerald 1116* (BRI *n.v.*, NSW).

11. *Melaleuca quinquenervia* (Cav.) S.T.Blake, Proc. Roy. Soc. Queensland 69: 76 (1958). *Metrosideros quinquenervia* Cav., Icon. 4: 19, t. 333 (1797). Typus: New South Wales: Port Jackson, Apr. 1793, *Née s.n.* (MA, *n.v.*, specimen now lost but the illustration forming part of the protologue is available).

Metrosideros coriacea Poir., Encycl. Suppl. 3: 685 (1813), *nom. illeg.*, non Salisb. *Melaleuca leucadendron* var. *coriacea* (Poir.) Cheel, in Ewart & Davies, Fl. N. Territory 297 (1917). Typus: New Caledonia: *Labillardière s.n.* (FI, holo, *n.v.*; MEL, P, *n.v.*).

Melaleuca leucadendron β *angustifolia* L.f., Suppl. pl. 342 (1781). *Melaleuca viridiflora* var. *angustifolia* (L.f.) Byrnes, Austrobaileya 2: 74 (1984), *nom. illeg.*, non Bl. Typus: New Caledonia: *Forster s.n.* (LINN, holo, *n.v.*; K, iso, *n.v.*).

Melaleuca viridiflora var. β *rubriflora* Brongn. & Gris, Bull. Soc. Bot. France 11: 183 (1864). *Melaleuca rubriflora* Vieill. ex Brongn. & Gris, Bull. Soc. Bot. France 11: 183 (1864), *nom. inval. Melaleuca leucadendron* var. *rubriflora* (Brongn. & Gris) Guill., Ann. Inst. Bot.-Geol. Colon. Marseille 19: 73 (1911). Typus: New Caledonia: near Balade, *Vieillard 451* (P, holo, *n.v.*).

Melaleuca maideni R.T.Baker, Proc. Linn. Soc. New South Wales 38: 598 (1914). Lectotypus (*fide* Blake, Contr. Queensland Herb. 1: 28 (1968)): New South Wales: Port Macquarie, Jul. 1895, *Maiden s.n.* (NSW, lecto, *n.v.*).

Melaleuca smithii R.T.Baker, Proc. Linn. Soc. New South Wales 38: 599 (1914). Lectotypus (*fide* Blake, Contr. Queensland Herb. 1: 28 (1968)): New South Wales: Rose Bay, Jul. 1913, *Laseron s.n.* (NSW, lecto, *n.v.*).

Melaleuca leucadendron var. *albida* Cheel, in Ewart & Davies, Fl. Northern Territory 301 (1917), excl. forma *ruscifolia* Cheel. Lectotypus (*fide* Blake, Contr. Queensland Herb. 1: 28 (1968)): New South Wales: *Sieber Fl. Nov. Holl. 319* (MEL, M, K, isolecto, *n.v.*).

Melaleuca leucadendron var. vel forma *nana* Brongn. & Gris ex Guill., Bull. Soc. Bot. France 81: 6 (1934). Typus: New Caledonia: near Nouméa, *Balansa 99* (P, holo, *n.v.*).

Melaleuca leucadendron var. vel forma *latifolia* Guill., Bull. Soc. Bot. France 81: 6 (1934), *nom. illeg.*, non L.f. Typus: New Caledonia: near Nouméa, *Balansa 99* (P, holo, *n.v.*).

12. *Melaleuca saligna* Schauer, in Walpers, Repert. Bot. Syst. 2: 927 (1843). *Melaleuca leucadendron* var. *saligna* (Schauer) F.M.Bailey, Syn. Queensl. fl. 170 (1883). Typus: Queensland: swampy banks of the Endeavour River, *Cunningham 256/1819* (BM, iso, *n.v.*).

Melaleuca stenostachya var. *pendula* Byrnes, Austrobaileya 2: 74 (1984). Typus: Queensland: Jacky Jacky airstrip, Bamaga district, May 1962, *Webb & Tracey 5989* (BRI, holo, *n.v.*; CANB, QRS *n.v.*).

13. *Melaleuca sericea* Byrnes, Austrobaileya 2: 74 (1984). Typus: Western Australia: 15 km W of Tableland Station, Apr. 1955, *Lazarides 5133* (BRI, holo, *n.v.*; CANB, iso).

14. *Melaleuca stenostachya* S.T.Blake, Contr. Queensland Herb. 1: 50, figs. 8, 14 H, 15 H (1968). Typus: Queensland: Croydon, *Blake 19566* (BRI, holo, *n.v.*).

15. *Melaleuca viridiflora* Sol. ex Gaertn., Fruct. sem. plantarum 1 (Dec. 1788) 173, t. 35, fig. 1. *Myrtoleucodendron viridiflorum* (Sol. ex Gaertn.) Kuntze, Revis. gen. pl. 1: 241 (1891). *Cajuputi viridiflora* (Gaertn.) A. Lyons, Pl. names Sci. Pop. 74 (1900). *Melaleuca leucodendron* var. *viridiflora* (Gaertn.) Cheel, in Ewart & Davies, Fl. Northern Territory 299 (1917). Typus: Queensland: Endeavour River, Jul.–Aug. 1770, *Banks and Solander s.n.* (BM, MEL, NSW, iso, *n.v.*).

Melaleuca cunninghami Schauer, in Walpers, Repert. Bot. Syst. 2: 927 (1843). *Melaleuca leucadendron* var. *cunninghamii* (Schauer) F.M.Bailey, Syn. Queensl. fl. 171 (1883). Typus: Queensland: Endeavour River, *Cunningham s.n.* (K, iso, *n.v.*).

Melaleuca leucadendron var. *latifolia* Rivière, Bull. Soc. Nat. Acclim. France III, 9: 537 fig. 1 (1882). Syntypi: Northern Territory: Gulf of Carpentaria, *Mueller s.n.* (P, K, *n.v.*; MEL isosyn, *n.v.*), *nom. illeg.*, non L.f.

Melaleuca leucadendron var. *sanguinea* (Sol. ex Cheel) Cheel, in Ewart & Davies, Fl. N. Territory 296 (1917), *nom. illeg. Melaleuca sanguinea* Sol. ex Cheel, Fl. N. Territory 296 (1917), *nom. inval.* Typus: Queensland: Endeavour River, Jul.–Aug. 1770, *Banks & Solander s.n.* (NSW, holo, *n.v.*; BM, MEL, P, W, iso, *n.v.*).

Melaleuca cunninghamii var. *glabra* C.T.White, J. Arnold Arbor. 23: 87 (1942). *Melaleuca viridiflora* var. *glabra* (C.T.White) Byrnes, Austrobaileya 2: 74 (1986). Typus: Papua New Guinea: Tarara, *Brass 8485* (BRI, holo, *n.v.*; A, K, LAE, *n.v.*).

Melaleuca viridiflora var. *attenuata* Byrnes, Austrobaileya 2: 74 (1984). Typus: Queensland: outside Port Douglas, c. 11 km ESE of Mossman, Jul. 1967, *Moriarty 9* (BRI, holo, *n.v.*).

Melaleuca viridiflora var. *canescens* Byrnes, Austrobaileya 2: 74 (1984). Typus: Queensland: 48 km SSE of Strathleven homestead, Nov. 1965, *Pedley 1843* (BRI, holo, *n.v.*).

Key to the species of the *M. leucadendra* group

Note: Young growth may be needed to observe the leaf hairs as these are soon deciduous in some species.

1. Calyx lobes absent...*M. cornucopiae*
1. Calyx lobes present
    2. Staminal filaments hairy...*M. lasiandra*
    2. Staminal filaments glabrous
        3. Hypanthium distinctly hairy
            4. Leaf blade indumentum with at least some of the hairs lanuginulose or sericeous-lanuginulose (whether or not also with pubescent to sericeous or sericeous-pubescent hairs)
                5. Stamens 5–8 mm long
                    6. Calyx lobes 0.5–0.8 mm long; triads clustered (less than 1 hypanthium diameterapart)...*M. saligna*
                    6. Calyx lobes 0.9–1.5 mm long; triads scattered (more than 1 hypanthium diameter apart) or sometimes partly clustered.........*M. dealbata*
                5. Stamens 10–23 mm long
                    7. Leaves 1.6–10.2 times as long as wide, the blade 5–40 mm wide; hypanthium 1.8–3.5 mm long; stamens 3–7 per bundle
                        8. Leaf blade tardily glabrescent; leaves 9–40 mm wide, 1.6–8.8 times as long as wide.................................*M. nervosa* subsp. *nervosa*
                        8. Leaf blade soon glabrescent; leaves 5–30 mm wide, 2.8–10.2 times as long as wide............................*M. nervosa* subsp. *crosslandiana*
                    7. Leaves 5–20 times as long as wide, the blade 5–19 mm wide; hypanthium 1.3–2 mm long; stamens 3–9 per bundle.......*M. fluviatilis*
            4. Leaf blade indumentum without lanuginulose or sericeous-lanuginulose hairs (the hairs sericeous, sericeous-pubescent or pubescent)
                9. Inflorescence up to 30 mm wide
                    10. Inflorescence up to 20 mm wide
                        11. Triads distant (at least 1 hypanthium diameter apart, sometimes within an inflorescence some of the triads are closer)...... ....*M. stenostachya*
                        11. Triads clustered (less than 1 hypanthium diameter apart)
                            12. Leaf blade apex acuminate, narrowly acute or acute ....*M. sericea*
                            12. Leaf blade apex usually obtuse, sometimes acute, rounded, obtusely shortly acuminate or retuse....................*M. arcana*
                    10. Inflorescence more than 20 mm wide
                        13. Calyx lobes herbaceous in the proximal-central zone and scarious in a narrow marginal band; leaves 4.8–14 times as long as wide...*M. argentea*
                        13. Calyx lobes herbaceous in the proximal-central zone and scarious in a broad marginal band; leaves 1.3–9.7 times as long as wide

14. Older leaves with the secondary venation distinct and about as prominent as the major veins
    15. Leaves (17–)25–50(–60) mm wide (leaves 1.3–6.5 times as long as wide; stamens (8–)9–12(–15) per bundle) .... *M. cajuputi* subsp. *platyphylla*
    15. Leaves (6–)10–28(–39) mm wide
        16. Leaves (6–)10–16(–26) mm wide, 2.8–9.7 times as long as wide; stamens (6–)8–11(–14) per bundle .... *M. cajuputi* subsp. *cajuputi*
        16. Leaves (15–)19–28(–39) mm wide, 2.2–2.9 times as long as wide; stamens (4–)6–8(–10) per bundle .... *M. cajuputi* subsp. *cumingiana*
14. Older leaves with the secondary venation more or less obscure... *M. quinquenervia*

9. Inflorescence more than 30 mm wide
    17. Young shoots with the hairs completely appressed; hypanthium (1.8–)3–3.5 mm long; petals (2.7–)4–5.3 mm long; inflorescence axis sericeous or pubescent (occasionally glabrous).............. *M. viridiflora*
    17. Young shoots with at least some spreading-ascending to spreading hairs; hypanthium 1.5–2.5 mm long; petals 2.5–3.5 mm long; inflorescence axis pubescent... *M. quinquenervia*

3. Hypanthium glabrous or effectively so
    18. Calyx lobes puberulous on the abaxial surface........... *M. dealbata*
    18. Calyx lobes glabrous on the abaxial surface
        19. Calyx lobes herbaceous in the proximal-central zone and scarious in a narrow marginal band or the lobes herbaceous almost throughout
            20. Stamens 6–7 mm long; leaves 30–110 mm long, 3.3–9 times as long as wide; bark hard................ *M. clarksonii*
            20. Stamens 7–16 mm long; leaves 75–270 mm long, 3.5–16.1 times as long as wide; bark papery............ *M. leucadendra*
        19. Calyx lobes herbaceous in the proximal-central zone and scarious in a broad marginal band
            21. Leaf blade narrowly ovate, very narrowly ovate, rarely narrowly elliptic or very narrowly elliptic (often falcate to subfalcate); leaves 3.5–16 times as long as wide; petals with elliptic oil glands (occasionally long elliptic glands form an apparently linear gland).............. *M. leucadendra*
            21. Leaf blade elliptic to very narrowly elliptic, obovate to very narrowly obovate (rarely broadly elliptic or ovate or very narrowly ovate or approaching falcate); leaves usually 1.3–8.5 times as long as wide (in some forms of *M. viridiflora* 4.8–15.7 times as long as wide with the blade very narrowly elliptic or narrowly elliptic or narrowly

obovate or very narrowly obovate); petals with oil glands linear, elliptic, circular to subcircular, or oblong

22. Stamens 9.2–10 mm long, the bundle claw 0.2–0.4 times as long as the stamens................................ ....*M. cajuputi* subsp. *platyphylla*
22. Stamens (9.5–)10.5–23 mm long, the bundle claw 0.06–0.2(–0.5) times as long as the stamens
    23. Hypanthium 1.5–2.5 mm long; petals 2.5–3.5 mm long; leaves 10–30 mm wide; inflorescence axis pubescent............................*M. quinquenervia*
    23. Hypanthium (1.8–)3–3.5 mm long; petals (2.7–)4–5.3 mm long; leaves (8–)19–76 mm wide; inflorescence axis sericeous or pubescent (occasionally glabrous).........................*M. viridiflora*

ACKNOWLEDGMENTS

The assistance of Brendan Lepschi in collecting data and references, and of Julie Matarczyk in producing the distribution maps, is much appreciated. Bryan Barlow's specimen annotations and notes concerning type specimens were valuable in treating the *M. leucadendra* group.

REFERENCES

Barlow, B.A. (1987) Contributions to a revision of *Melaleuca* (Myrtaceae): 1–3. *Brunonia*, **1**, 163–177.

Barlow, B.A. (1988) Patterns of differentiation in tropical species of *Melaleuca* L. (Myrtaceae). *Proc. Ecol. Soc. Australia*, **15**, 239–247.

Barlow, B.A. and Cowley, K.J. (1988) Contributions to a revision of *Melaleuca* (Myrtaceae): 4–6. *Austral. Syst. Bot.*, **1**, 95–126.

Bentham, G. (1867, '1866') Myrtaceae. In, *Flora Australiensis*, Vol. 3, Lovell Reeve & Co., London, pp. 1–289.

Blake, S.T. (1958) New and critical genera and species of Myrtaceae subfamily Leptospermoideae from eastern Australia. *Proc. Roy. Soc. Queensland*, **69**, 75–88.

Blake, S.T. (1968) A revision of *Melaleuca leucadendron* and its allies (Myrtaceae) *Contr. Queensland Herb.*, **1**, 1–114.

Briggs, B.G. and Johnson, L.A.S. (1979) Evolution in the Myrtaceae—evidence from inflorescence structure. *Proc. Linn. Soc. New South Wales*, **102**, 157–256.

Butcher, P.A., Doran, J.C. and Slee, M.U. (1994) Intraspecific variation in leaf oils of *Melaleuca alternifolia* (Myrtaceae). *Biochem. Syst. Ecol.*, **22**, 419–430.

Butcher, P.A., Byrne, M. and Moran, G.F. (1995) Variation within and among the chloroplast genomes of *Melaleuca alternifolia* and *M. linariifolia* (Myrtaceae). *Pl. Syst. Evol.*, **194**, 69–81.

Byrnes, N.B. (1984) A revision of *Melaleuca* L. (Myrtaceae) in northern and eastern Australia, 1. *Austrobaileya*, **2**, 65–76.

Byrnes, N.B. (1985) A revision of *Melaleuca* L. (Myrtaceae) in northern and eastern Australia, 2. *Austrobaileya*, **2**, 131–146.

Byrnes, N.B. (1986) A revision of *Melaleuca* L. (Myrtaceae) in northern and eastern Australia, 3. *Austrobaileya*, **2**, 254–273.

Carrick, J. and Chorney, K. (1979) A review of *Melaleuca* L. (Myrtaceae) in South Australia. *J. Adelaide Bot. Gard.*, **1**, 281–319.

Cheel, E. (1917) Myrtaceae of Northern Territory (except *Eucalyptus*). In, A.J. Ewart and O.B. Davies (eds.), *Fl. N. Territory*, McCarron, Bird & Co., Melbourne, pp. 290–304.

Cheel, E. (1924) Notes on *Melaleuca*, with descriptions of two new species and a new variety. *J. Proc. Roy. Soc. New South Wales*, **58**, 189–197.

Cowley, K.J., Quinn, F.C., Barlow, B.A. and Craven, L.A. (1990) Contributions to a revision of *Melaleuca* (Myrtaceae): 7–10. *Austral. Syst. Bot.*, **3**, 165–202.

Craven, L.A. (1989) Reinstatement and revision of *Asteromyrtus* (Myrtaceae). *Austral. Syst. Bot.*, **1**, 373–385.

Craven, L.A. (1997) Australian National Herbarium, Canberra. Personal communication.

Dawson, J.W. (1992) *Melaleuca* Linné. In, Ph. Morat and H.S. Mackee (eds.), *Flore de la Nouvelle-Calédonie et dépendances*, Vol. 18, Muséum National d'Histoire Naturelle, Paris, pp. 216–229.

Lanjouw, J., Mamay, S.H., McVaugh, R., Robyns, W., Rollins. R.C., Ross, R., Rousseau, J., Schulze, G.M., Vilmorin, R. de and Stafleu, F.A. (1966) *International code of botanical nomenclature*, International Bureau for Plant Taxonomy and Nomenclature, Utrecht.

Lum, S.K.Y. (1994) *Dispersal of Australian plants across Wallace's Line: a case study of Melaleuca cajuputi (Myrtaceae)*, Ph.D. thesis, University of California, Berkeley.

Quinn, F.C., Cowley, K.J., Barlow, B.A. and Thiele, K.R. (1989) Contributions to a revision of *Melaleuca* (Myrtaceae): 11–15. Unpublished manuscript.

Ramanoelina, P.A.R., Bianchini, J.P., Andriantsiferana, M., Viano, J. and Gaydou, E.M. (1992) Chemical composition of niaouli essential oils from Madagascar. *J. Essent. Oil Res.*, **4**, 657–658.

Ramanoelina, P.A.R., Viano, J., Bianchini, J.P. and Gaydou, E.M. (1994) Occurrence of variance chemotypes in niaouli (*Melaleuca quinquenervia*) essential oils from Madagascar using multivariate statistical analysis. *J. Agric. Food Chem.*, **42**, 1177–1182.

Rumphius, G.E. (1741) Arbor alba, Arbor alba minor. In, *Herbarium Amboinese*, pp. 72, 76, t. 16, t. 17, figs. 1–2.

Wrigley, J.W. and Fagg, M. (1993) *Bottlebrushes, Paperbarks and Tea Trees*, Angus & Robertson, Pymble.

2. TEA TREE CONSTITUENTS

IAN SOUTHWELL

Wollongbar Agricultural Institute, Wollongbar, NSW, Australia

INTRODUCTION

As with most genera, *Melaleuca* contains a variety of primary and secondary metabolites. The volatile oil constituents of *M. alternifolia* and closely related species are the compounds responsible for the commercial development of *Melaleuca* as a medicinal and aromatic plant. Consequently the most frequently investigated aspects of the chemistry of tea tree concern the identification of these volatile oil constituents. As considerably less is known about the genus's primary metabolite tannins, polyphenols, waxes, amino acids and betaine constituents these will only receive brief mention. The constituents of the other *Melaleuca* species will be outlined in later chapters in this volume.

OIL CONSTITUENTS

Chronological Perspective

At the same time that *Melaleuca linariifolia* var. *alternifolia* (Maiden and Betche) was being raised to species status as *M. alternifolia* (Maiden and Betche) Cheel (Cheel 1924), the first chemical investigations of the taxon were being undertaken (Penfold 1925). Earlier, Baker and Smith (1906, 1907, 1910, 1911, 1913) had investigated the oils of *M. thymifolia, M. uncinata, M. nodosa, M. genistifolia* (*M. bracteata*), *M. gibbosa, M. pauciflora* and *M. leucadendron* in what must be considered, by todays standards, a most superficial way. Penfold (1925), investigated *M. linariifolia* and *M. alternifolia* concurrently and concluded that their essential oils were "practically identical". This investigation included measurement of oil yields (1.5–2.0%), specific gravity, optical rotation, refractive index, solubility in alcohol, fractional distillation and identification of chemical constituents by the preparation of derivatives and comparison of melting points and mixed melting points with authentic materials. In this manner pinene (2), α-terpinene (7), γ-terpinene (12), *p*-cymene (10), sabinene (?) (3), cineole (11), terpinen-4-ol (14) and sesquiterpenes including cadinene (18) were identified from *M. linariifolia* and all of these except sabinene identified from *M. alternifolia.* The methods involved were qualitative rather than quantitative for all constituents except 1,8-cineole (11) which was estimated to be present at 16–20% and 6–8% in *M. linariifolia* and *M. alternifolia* respectively.

In subsequent decades, Jones investigated *M. linariifolia* more thoroughly, discovered both high cineole (61%) (Jones 1936) and low cineole (Davenport *et al.* 1949) chemical varieties. Also α-thujene (1), β-pinene (4), myrcene (5), terpinolene (13), α-terpineol (15) and aromadendrene (16) were detected in the low cineole variety and dipentene (limonene) (8) in the high cineole form. These workers were able to provide an estimate of the percentage contributions made by each component identified that was remarkably close to the gas chromatographic measurements of recent workers (Southwell and Stiff 1990; Kawakami *et al.* 1990) (Table 1). The structures of these major components of tea tree oil are shown in Figure 1.

The next thorough investigation of tea tree oil chemistry involved reduced-pressure spinning-band column distillation and gas chromatographic (GC) analysis of resulting fractions (Laakso, 1966). This study did not report any new constituents. It does, however, seem to be the first published report on the GC examination of tea tree oil even though routine GC quality control of tea tree oil at the Museum of Applied Arts and Sciences in Sydney had commenced around 1960 (Museum of Applied Arts and Sciences, unpublished records). Laakso's study also seemed to be the first to describe a terpinolene(13)-rich chemical variety of *M. alternifolia.* Soon after this investigation Guenther (1968) reported a similar fractionation-GC study from the Fritzsche Brothers laboratories.

Table 1 Estimate of the percentage composition of the oil of the terpinen-4-ol chemical variety of *M. linariifolia* (Davenport *et al.* 1949) compared with the gas chromatographic results of Southwell and Stiff (1990) and Kawakami *et al.* (1990)

Constituent	Davenport *et al.* (1949)	Southwell *and* Stiff (1990)	Kawakami *et al.* (1990)†
α–thujene (1)	<1	4.1	4.2
α-pinene (2)	1–2	2.1	2.0
sabinene (3)	nr	2.7	10.9
β-pinene (4)	<1	0.6	0.5
myrcene (5)	<1	1.0	1.6
α-phellandrene (6)	nr	0.4	nr
α-terpinene (7)	40[a]	9.3	10.1
limonene (8)	nr	1.3	1.1
β-phellandrene (9)	nr	0.9	0.8
p-cymene (10)	<1	1.4	0.8
1,8-cineole (11)	4	6.0	6.0
γ-terpinene (12)	40[a]	18.9	17.2
terpinolene (13)	4	3.5	3.5
terpinen-4-ol (14)	37	38.2	32.1
α-terpineol (15)	4	1.5	2.7
aromadendrene (16)	6[b]	0.5	nr
ledene (viridiflorene) (17)	6[b]	0.4	tr
δ-cadinene (18)	6[b]	0.9	0.5
globulol (19)	2.5[c]	0.5	nr
viridiflorol (20)	2.5[c]	0.2	nr

†Clone 1, now known to be *M. linariifolia* (Southwell *et al.* 1992). nr, not recorded; tr, trace; [a,b,c]estimated collectively.

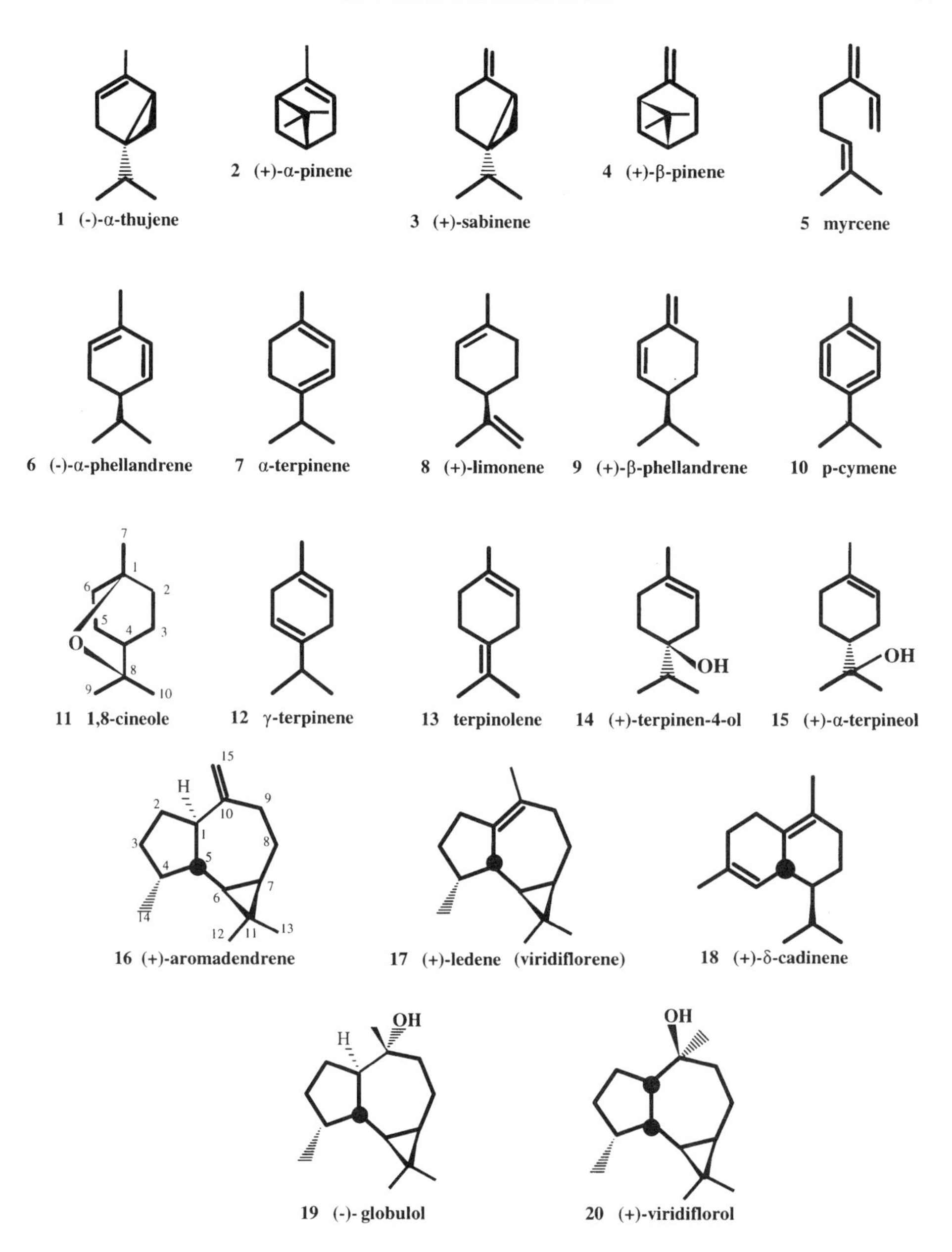

Figure 1 Chemical structures of the significant constituents in tea tree oil

The first gas chromatography–mass spectrometry (GCMS) investigation of tea tree oil (Swords and Hunter 1978) reported forty eight constituents of which only eight were unassigned. Assignments were based on GCMS data with preparative GC, liquid chromatography (LC) and infrared spectroscopy (IR) confirmation used for some identifications. It must be noted however that this oil was not a typical commercial oil as cineole content (16.5%) was above standard limits (Brophy *et al.* 1989b; Standards Association of Australia 1985; International Standards Organisation 1996) and *p*-cymene (11.4%), α-terpinene (2.7%) and γ-terpinene (11.5%) indicated substantial oxidation (Brophy *et al.* 1989b). Components identified in tea tree oil above trace levels (>0.1%) included α-gurjunene (0.23%), β-terpineol (0.24%), *allo*-aromadendrene (0.45%), α-muurolene (0.12%) 8-*p*-cymenol (0.13%) and viridiflorene (1.03%). The short, packed Carbowax 20 M GC column used for the GCMS analysis had obvious deficiencies. Even the 100 m Carbowax 20 M capillary column failed to resolve α-thujene from α-pinene, β-phellandrene from 1,8-cineole and had difficulty in resolving myrcene from α-phellandrene and 1,4-cineole and terpinen-4-ol from β-elemene.

These authors described the sesquiterpene hydrocarbon viridiflorene for the first time and proposed structure (21) (Figure 2). This structure was inconsistent with the dehydration product of viridiflorol (20) which should have structure (17) because of the absolute configuration assignments of Buchi *et al.* (1969) following the synthesis of (−)-aromadendrene (22), the enantiomer of naturally occurring (+)-aromadendrene (16). Viridiflorene must then have structure (17) which is also the structure of ledene, the dehydration product of ledol (23), the C10 enantiomer of (+)-viridiflorol (20). This assignment has been confirmed by nuclear magnetic resonance (NMR) by both Australian and French workers (Southwell and Tucker 1991; Faure *et al.* 1991). (+)-Ledene (17) has been well known as a dehydration product of viridiflorol (e.g. Birch *et al.* 1959) and also as a natural constituent of essential oils (e.g. Taskinen 1974). Consequently the name viridiflorene should be replaced by ledene in tea tree and other essential oil reports.

A more recent GCMS analysis of *M. alternifolia* oil corrected the viridiflorene structure, listed a total of 97 constituents and identified several new components (Brophy

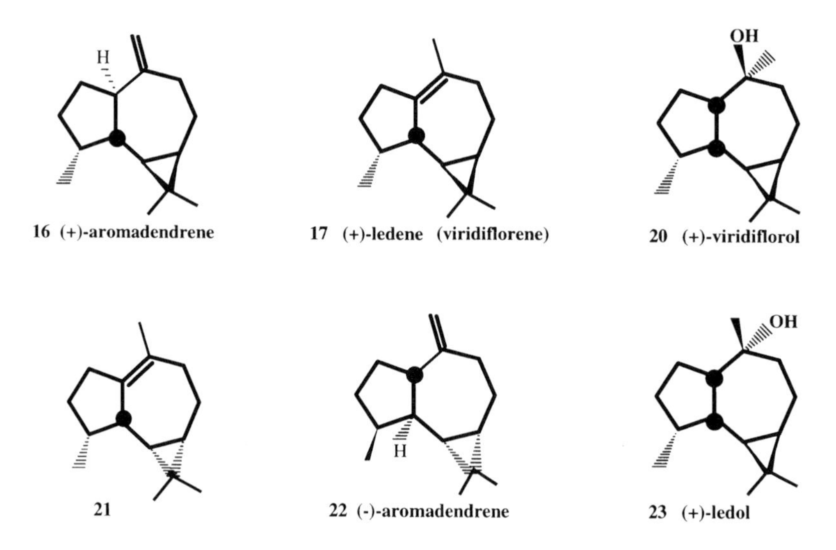

Figure 2 Chemical structures of sesquiterpenoids conformationally related to ledene

et al. 1989b). The most significant findings of this investigation included (a) determining suitable GC and column conditions for the optimal separation of tea tree oil constituents, b) the identification of the *cis* and *trans* alcohol pairs of sabinene hydrate, *p*-menth-2-en-1-ol and piperitol, (c) the composition of atypical and aged oils, (d) the affect of distillation time on oil composition, (e) the micro-scale analysis of a single leaf (6 mg) for quality determination and (f) the presence of oil precursors in the ethanolic extracts of tea tree flush growth.

This last finding was subsequently investigated more thoroughly (Southwell and Stiff 1989). Although individual mature leaves when immersed in ethanol gave an extract which accurately reflected the quality of the oil if the leaves were distilled, the same could not be said for the brighter green flush growth. This flush growth was found to be rich in *cis*-sabinene hydrate (24) which was replaced by γ-terpinene (12) and terpinen-4-ol (14) as the leaves matured. Very little of (24) and other minor precursors were detected in the distilled oil due to the lability of *cis*-sabinene hydrate (Erman 1985; Fischer *et al.* 1987, 1988).

Comprehensive GCMS analyses were performed on eight tea tree clones propagated at the University of California (Kawakami *et al.* 1990). These clones, originally claimed to be *M. alternifolia* stock, have since been shown to contain at least one *M. linariifolia* variety (Southwell *et al.* 1992). Oils were obtained by both simultaneous purging and extraction (SPE) to give headspace analysis and steam distillation and extraction (SDE) to give oil analyses. Only two of the distilled oils were from terpinen-4-ol type clones. The headspace (SPE) analysis concentrated the more volatile monoterpene hydrocarbons especially sabinene, α-thujene, α-terpinene and γ-terpinene at the expense of the less volatile oxygenated terpenoids especially terpinen-4-ol. These compositions were similar to the composition of the initial vapour cloud that emerges from the still condenser prior to condensation (Southwell 1988). Although this investigation reported for the first time a number of new sesquiterpenoids, none exceeded 0.6% of the total oil.

The two commercial terpinen-4-ol type tea tree oils sourced from *M. alternifolia* and *M. linariifolia* are distinguishable on the grounds of oil chemistry (Southwell and Stiff 1990). The former when analysed by GC gives a smaller peak for α-thujene than for α-pinene and the latter gives the converse. As these peaks are always the first two significant chromatographic peaks, a glance at the GC trace will tell which oil has been analysed as long as α-pinene and α-thujene are resolved. The mean α-thujene : α-pinene ratio was 0.33 (n = 521) for *M. alternifolia* and 1.49 (n = 180) for *M. linariifolia.* Non-polar or intermediate polarity stationary phases are best for this analysis as resolution is greater. The α-thujene : α-pinene ratio is upset as the percentage of cineole in the oil increases. As α-thujene is associated with the terpinen-4-ol biogenetic pathway, higher cineole means less terpinen-4-ol and a corresponding lower α-thujene : α-pinene ratio. Hence the test is only suitable for the terpinen-4-ol chemical varieties. A similar way of distinguishing the two species is to measure the *cis*-sabinene hydrate (24) to *trans*-sabinene hydrate (25) ratio in the ethanolic extract of the flush growth of both species. The mean *cis*-sabinene hydrate : *trans*-sabinene hydrate ratio was 7.1 : 1 (62 extracts from 11 trees) for *M. alternifolia* and 0.7 : 1 (29 extracts from 6 trees) for *M. linariifolia* (Southwell and Stiff 1990).

Terpinolene varieties of *Melaleuca* have, from time to time, received passing mention in the literature (Laakso 1966; Brophy *et al.* 1989a; Kawakami *et al.* 1990). Their existence was formally acknowledged and data documented by Southwell *et al.* (1992) and Butcher *et al.* (1994). Both *M. alternifolia* and *M. trichostachya* were found to have varieties with from 10–57% terpinolene, high proportions of cineole (13–56%) and insufficient terpinen-4-ol (1–20%) for the tea tree oil market.

With the advent of chiral GC columns, the enantiomeric ratios of tea tree oil constituents were determined (Russell *et al.* 1997; Leach *et al.* 1993; Cornwell *et al.* 1995). Seven monoterpenes were resolved in this way and their percentages and enantiomeric ratios shown in Table 2. These ratios provide valuable criteria for checking the authenticity of tea tree oil especially blends with (−)-terpinen-4-ol from *Eucalyptus dives* that have been detected in the past. Similar chiral resolutions were achieved using lanthanide shift reagents on the individual tea tree oil constituents terpinen-4-ol and α-terpineol (Leach *et al.* 1993). The complexity was however too great for these shifts to be meaningful for an entire oil. The 58% enantiomeric excess obtained for a standard (99%) sample of terpinen-4-ol ($[\alpha]_D + 29°$) was consistent with both the 30% enantiomeric excess for terpinen-4-ol ($[\alpha]_D + 16\%$) fractionated from tea tree oil (Russell *et al.* 1997) and the maximum rotation values ($+47° – +48°$) reported on enantiomerically pure samples (Naves and Tullen 1960; Ohloff and Uhde 1965; Verghese 1966).

Because of the efforts of these investigators over the years, the chemistry of tea tree oil is now, well established. There is a need to clarify some of the minor and trace constituent assignments and to establish the enantiomeric composition of the sesquiterpenoids. Then efforts can concentrate on which of these constituents are beneficial or detrimental (to the commercial uses of the oil) and how to maximise or minimise their contribution to commercial oils.

Table 2 The enantiomeric composition of the seven tea tree oil constituents resolved by chiral gas chromatography on β-cyclodextrin

Constituent	*KI*	%
(−)-α-pinene	1007	0.18
(+)-α-pinene	1017	1.68
(−)-α-phellandrene	1058	a
(+)-α-phellandrene	1058	a
(−)-limonene	1079	a
(+)-limonene	1082	0.51
(±)-β-phellandrene	1092	0.22
(±)-β-phellandrene	1095	0.36
(±)-linalool	1254	tr
(±)-linalool	1257	tr
(+)-terpinen-4-ol	1334	24.73
(−)-terpinen-4-ol	1340	13.13
(−)-α-terpineol	1381	0.69
(+)-α-terpineol	1385	2.03

a, insufficient resolution to allow accurate quantitation; tr, trace.

Molecular Perspective

The components in the accompanying Table (Table 3) have been reported as volatile constituents from the leaves of the terpinen-4-ol variety of *Melaleuca alternifolia*. The list is based on the comprehensive list of Brophy *et al.* (1989b) supplemented with the additional components by Swords and Hunter (1978), Kawakami *et al.* (1990) and Leach *et al.* (1993). Correlation of entries from the different sources is confounded by the use of different GC column lengths and stationary phases. Entry sequence is based on the non-polar BP1 FSOT column of Brophy *et al.* (1989b) with entries from the other sources cited. The assignments for minor and trace constituents may not be reliable where mass spectra alone were used for these identifications. Indeed specialists in this field have suggested that "the injection of an essential oil into a GCMS instrument with a fully automatic library search was not a vigorous scientific exercise worthy of publication" (Stevens 1996) and indicate that such identifications be supported by retention indices comparisons on two columns of different polarity. Furthermore, one regulatory body now insists that identification be confirmed by at least two methods (Liener 1966). Consequently some of the minor and trace component identifications listed in Table 3 must be considered tentative especially where supporting evidence is not available.

EXTRACT CONSTITUENTS

Although steam distillation is the preferred method for the isolation of essential oils, some commercial products are obtained by alternative processes. For example, citrus oils are isolated from the peels of citrus by cold pressing and many perfumery products (e.g. jasmin, boronia, acacia) by solvent extraction. This latter method is usually preferred for low-volume high-value products. More recently carbon dioxide and supercritical fluids have been replacing conventional solvents especially for important flavour constituents (Kerrola 1995). Little is known of the extractive constituents of *M. cajuputi* and *M. quinquenervia*. Some workers have investigated the alcoholic extraction of *M. alternifolia*, *M. linariifolia*, *M. lanceolata* and *M. uncinata* (Brophy *et al.* 1989b; Southwell and Stiff 1989; Southwell and Stiff 1990; Jones *et al.* 1987).

Solvent extraction has never been seriously considered as an alternative procedure for obtaining tea tree oil. The cost of the process, the scale of the operation and the problems that would be associated with marketing a chemically different product have been deterrents for the industry.

On a laboratory scale however, these alternative procedures have been found to be very useful analytical techniques which have also contributed to our knowledge of leaf chemistry.

Volatile Constituents

As solvent extraction removes both volatile and non-volatile constituents from leaf, the volatile constituents are those which can be readily analysed, without derivatisation, by

Table 3 Volatile constituents of *Melaleuca alternifolia* leaf

No.	*Constituent*	*Retention index*[A]			%[B]	*Reference*[C]	*ID*[D]
		Polar[b]	*Non-polar*[b]	*Chiral*[d]			
1	α-thujene	1039	926	976	0.4	b,c,*d*	MS,RT
2	(−)-α-pinene	1034	933	1007	0.2	a,b,c,*d*	MS,RT,COGC
3	(+)-α-pinene	un	un	1017	1.7	a,b,c,*d*	MS,RT
4	camphene	1075	943	—	tr	a,*b*	MS,RT
5	sabinene	1128	969	—	0.2	a,*b*,c	MS,RT,COGC
6	β-pinene	1114	973	—	0.3	a,*b*,c	MS,RT
7	myrcene	1172	984	1025	0.5	a,b,c,*d*	MS,RT,COGC
8	(−)-α-phellandrene	1172	998	1058	0.3	a,*b*,*d*	MS,RT,COGC
9	(+)-α-phellandrene	un	un	1058	un	a,*b*,*d*	MS
10	1, 4-cineole	un	1006	1085	tr	a,*b*,*d*	MS,RT
11	α-terpinene	1187	1010	1062	9.9	a,b,c,*d*	MS,RT,COGC
12	*p*-cymene	1285	1013	1079	5.0	a,b,c,*d*	MS,RT,COGC
13	(−)-limonene	1206	1021	1079	un	a,b,c,d	MS,RT
14	(+)-limonene	un	un	1082	0.5	a,b,c,*d*	MS,RT,COGC
15	(±)-β-phellandrene	1216	1022	1092	0.2	b,c,*d*	MS,RT
16	(±)-β-phellandrene	un	un	1095	0.4	b,c,*d*	MS
17	1, 8-cineole	1216	1022	1135	3.0	a,b,c,*d*	MS,RT,COGC
18	*trans*-ocimene	un	1040	—	tr	*b*	MS
19	γ-terpinene	1257	1050	1100	21.2	a,b,c,*d*	MS,RT,COGC
20	*trans*-sabinene hydrate	un	1058	—	tr	*b*,c	MS,RT,COGC,NMR
21	*p*-α-dimethylstyrene	1468	1075	—	0.2	a,b,*d*	MS,RT
22	terpinolene	1298	1081	—	3.2	a,b,c,*d*	MS,RT
23	(±)-linalool	1570	1084	1254	tr	a,b,c,*d*	MS,RT,COGC
24	(±)-linalool	un	un	1257	tr	a,b,c,*d*	MS
25	*cis*-sabinene hydrate	un	1089	1301	tr	*b*,c,*d*	MS,RT,COGC,NMR,IR
26	*trans*-menth-2-en-1-ol	1584	1112	—	0.2	*b*,c	MS,RT,COGC
27	*cis*-menth-2-en-1-ol	1652	1129	—	0.1	*b*,c	MS,RT,COGC
28	(+)-terpinen-4-ol	1626	1166	1334	24.7	a,b,c,*d*	MS,RT,COGC,NMR,IR
29	(−)-terpinen-4-ol	un	un	1340	13.1	a,b,c,*d*	MS,RT
30	(−)-α-terpineol	1722	1177	1381	0.7	a,b,c,*d*	MS,RT
31	(+)-α-terpineol	un	un	1385	2.0	a,b,c,*d*	MS,RT,COGC
32	*trans*-piperitol	1705	1185	—	tr	*b*,c	MS,RT,COGC
33	*cis*-piperitol	1771	1195	—	tr	a,*b*,c	MS,RT,COGC
34	hexanol	—	—	—	tr	*a*,b	MS
35	allyl hexanoate	—	—	—	tr	*a*,b	MS
36	camphor	—	—	—	tr	*a*,b	MS
37	1-terpineol	—	—	—	tr	*a*,b	MS
38	β-terpineol	—	—	—	0.2	*a*,b	MS
39	piperitone	—	—	—	tr	*a*,b	MS
40	$C_{15}H_{24}$	—	—	—	tr	*a*,b	MS

Table 3 *(Continued)*

No.	*Constituent*	*Retention index*[A]			%[B]	*Reference*[C]	*ID*[D]
		Polar[b]	*Non-polar*[b]	*Chiral*[d]			
41	α-cubebene	1479	—	1394	tr	*a,b,d*	MS,RT
42	α-ylangene	1485	—	—	tr	*b*	MS,RT
43	unidentified	1490	—	—	tr	*a,b*	MS
44	$C_{15}H_{24}$	1496	—	—	tr	*b*	MS
45	methyl eugenol	2042	1377	1538	tr	*b,d*	MS,RT,COGC
46	α-copaene	1504	1379	1389	0.2	a,b,*d*	MS,RT
47	unidentified (β-cubebene?)	1506	1381	1408	tr	*b*,c,d	MS
48	unidentified	1541	—	1415	tr	*b*,d	MS
49	α-gurjunene	1544	1414	1442	0.5	a,b,c,*d*	MS,RT,IR
50	$C_{15}H_{24}$	1558	—	1462	tr	*a,b*,d	MS
51	unidentified	1560	—	1473	tr	*a,b*,d	MS
52	β-elemene	—	—	—	0.1	a,*b*	MS,RT
53	β-caryophyllene	1619	1423	1483	0.4	a,b,c,*d*	MS,RT,COGC
54	α-bulnesene	—	—	1489	0.2	b,*d*	MS
55	β-gurjunene	—	—	1452	tr	*a,b,d*	MS,RT
56	β-maaliene	—	—	1456	tr	*d*	MS
57	aromadendrene	1623	1444	1501	1.6	a,b,*d*	MS,RT,COGC,IR
58	unidentified	1654	—	—	tr	*b*	MS
59	γ-gurjunene	—	—	1510	0.2	*d*	MS
60	$C_{15}H_{24}$	1658	—	—	tr	*b*	MS
61	*allo*-aromadendrene	1663	1465	1519	0.7	a,b,c,*d*	MS,RT,IR
62	$C_{15}H_{24}$	1673	—	1522	0.5	b,*d*	MS
63	$C_{15}H_{24}$	1677	1474	—	0.2	a,*b*,c	MS
64	humulene	1688	1457	—	tr	*a,b*	MS,RT,COGC
65	unidentified	—	—	—	tr	*a*	MS
66	γ-muurolene	—	1710	—	tr	*a,b*	MS
67	1,2,4-trihydroxymenthane	—	1470	—		*b*	MS,RT,COGC, NMR,IR
68	ledene (viridiflorene)	1713	1497	1555	1.8	a,b,c,*d*	MS,RT,COGC, NMR,IR
69	$C_{15}H_{24}$	—	—	1542	tr	*b*,d	MS
70	$C_{15}H_{24}$	1734	—	1559	tr	*b*,d	MS
71	$C_{15}H_{24}$	1739	—	1577	tr	*b*,d	MS
72	α-muurolene	1743	—	1536	0.3	a,b,*d*	MS,RT,IR
73	α-amorphene	1744	—	1546	0.3	b,*d*	MS,RT
74	bicyclogermacrene	1754	—	1550	0.1	b,*d*	MS,RT
75	unidentified	—	—	—	tr	a,b,c,*d*	MS
76	calamenene	1757	1515	1596	0.3	a,b,c,*d*	MS,RT,IR
77	δ-cadinene	1777	1520	1582	1.9	a,b,c,*d*	MS,RT,IR
78	unidentified	—	1523	—	tr	*b,*	MS
79	cadina-1,4-diene	1802	1530	1607	0.3	a,b,c,*d*	MS,RT
80	nerol	1807	—	—	tr	*a,b*	MS

Table 3 *(Continued)*

No.	*Constituent*	*Retention index*[A]			*%*[B]	*Reference*[C]	*ID*[D]
		Polar[b]	*Non-Polar*[b]	*Chiral*[d]			
81	$C_{15}H_{24}$	1812	—	1610	tr	b,d	MS
82	*p*-cymen-8-ol	1852	—	1397	0.1	a,b,*c*,d	MS
83	unidentified	1880	—	1615	tr	*b,d*	MS
84	$C_{15}H_{24}$	1944	—	1631	tr	*b,d*	MS
85	unidentified	1956	—	1673	tr	*b,d*	MS
86	palustrol	1980	—	1689	tr	*b,d*	MS,RT
87	$C_{15}H_{26}O$	2027	—	—	tr	*b*	MS
88	unidentified	2033	—	—	tr	*b*,d	MS
89	$C_{15}H_{26}O$	2038	1550	1717	0.2	b,*d*	MS
90	ledol	2057	1604	—	tr	*b*	MS,RT,COGC
91	cubenol	2080	—	1759	0.5	b,*d*	MS,RT
92	unidentified	2084	—	1725	tr	*b*,d	MS
93	$C_{15}H_{26}O$	2098	—	1738	0.1	*b,d*	MS
94	globulol	2103	1585	1755	0.8	b,*d*	MS,RT,COGC
95	viridiflorol	2113	1593	1764	0.6	b,*d*	MS,RT,COGC
96	unidentified	2123	—	—	tr	*b*	MS
97	spathulenol	2152	1573	1750	0.2	b,*d*	MS,RT,COGC
98	rosifoliol	2133	1599	—	tr	*b*	MS,RT,COGC
99	unidentified	2142	1625	—	tr	*b*	MS
100	unidentified	2175	—	—	tr	*b*	MS
101	unidentified	2193	—	—	tr	*b*	MS
102	unidentified	2200	—	—	tr	*b*	MS
103	$C_{15}H_{26}O$	2210	—	—	tr	*b*	MS
104	unidentified	2217	—	—	tr	*b*	MS
105	unidentified	2222	—	—	tr	*b*	MS
106	$C_{15}H_{26}O$	2228	—	—	tr	*b*	MS
107	unidentified	2258	—	—	tr	*b*	MS
108	unidentified	2292	—	—	tr	*b*	MS
109	unidentified	2309	—	—	tr	*b*	MS
110	epicubenol	2001[c]	—	—	0.4	*c*	MS,RT
111	*β*-eudesmol	2052[c]	—	—	0.3	*c*	MS,RT
112	cadinol T	2126[c]	—	—	0.1	*c*	MS,RT
113	*δ*-cadinol	2149[c]	—	—	0.2	*c*	MS,RT

[A] Retention index from reference superscripted.

[B] Yield (%) from reference in italics e.g. *a*.

[C] a, Swords and Hunter (1978); b, Brophy *et al.* (1989); c, Kawakami *et al.* (1990); d, Leach *et al.* (1993); tr, trace < 0.1%; un, unresolved.

[D] Identification (ID) by means of comparative mass spectrum (MS), retention time (RT), GC co-injection (COGC), Nuclear Magnetic Resonance Spectrometry (NMR) and Infrared Spectroscopy (IR).

gas chromatography (GC). Ethanolic extraction of a single tea tree leaf (1–10 mg) in a 0.1 ml GC vial insert gave a solution suitable for GC analysis in the usual way (Southwell and Stiff 1989; Brophy *et al.* 1989b). The 30 hour room temperature extraction was reduced to one hour following 10 seconds of microwave irradiation (Russell *et al.* 1997). The monoterpene region of the resultant trace (Figure 3) accurately reflected the quality of the oil obtained by steam distillation of the same leaf material when mature leaf was extracted.

This microextraction method revealed numerous interesting facts about the chemistry of tea tree which led to many significant analytical uses.

The first of these was developed in response to the increasing production of and demand for tea tree oil. Prior to the 1980s, supplies had been adequately sourced from natural stands growing in specific areas in the New South Wales northern rivers region. Production could only be increased by either seeking good quality stands growing further afield or establishing plantations. Both approaches present quality control problems because of the abundance of the undesirable high cineole chemical variety. The microextraction analytical procedure provided an easy way to check both the quality of distant natural stands and the genetic quality of parent trees providing propagation material for plantation establishment (Brophy *et al.* 1989b). This extraction method then provides an excellent way of determining whether the tree under investigation has acceptable levels of terpinen-4-ol and 1,8-cineole without having to carry out the resource-consuming steam distillation procedure.

When the brighter green flush growth was extracted, γ-terpinene (12) and terpinen-4-ol (14) were found to be present in their precursor sabinene hydrate forms. These thujane precursors (Figure 4) convert to their more stable end products either during steam distillation by artifact formation or *in vivo* as the leaf matures (Southwell and Stiff 1989). Hence accurate figures from microextraction quality checks are best obtained using the dark green mature tea tree leaves.

This difference between flush growth and mature leaf also led to many significant observations. For example, the along-the-branch variation of key constituents *cis*-sabinene hydrate (24), *trans*-sabinene hydrate (25), sabinene (3), terpinen-4-ol (14) and γ-terpinene (12)

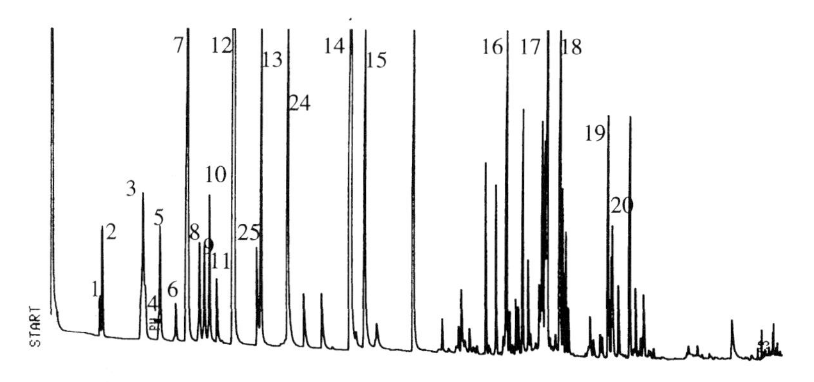

Figure 3 Gas chromatographic trace of *M. alternifolia* single leaf extract on a 60 m AT35 FSOT column. Peak numbers are consistent structures shown in Figures 1 and 4

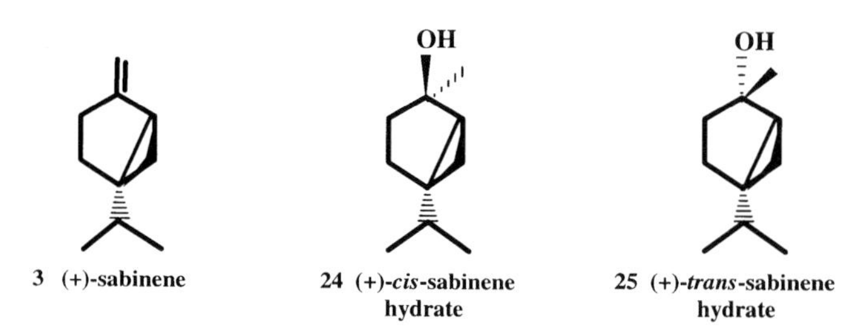

Figure 4 Chemical structures of the precursor thujanes from *Melaleuca*, terpinen-4-ol type flush growth extracts

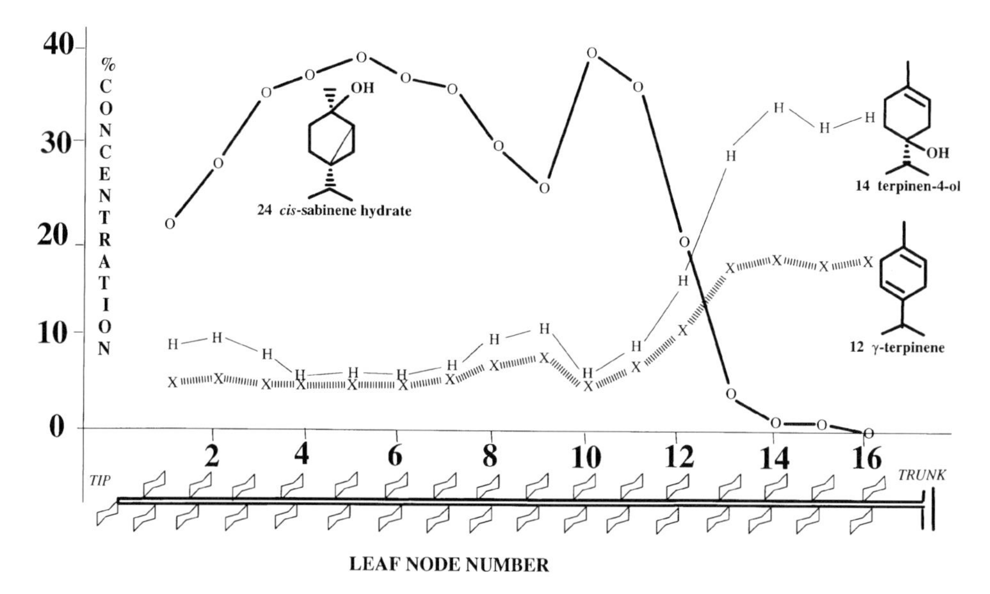

Figure 5 Along-the-branch variation in the concentrations of *cis*-sabinene hydrate, terpinen-4-ol and γ-terpinene in individual leaves of *M. alternifolia*

was monitored from the tip to the main stem by the single leaf microextraction method (Southwell and Stiff 1989). The high *cis*-sabinene hydrate levels in the flush growth gave way abruptly to high levels of γ-terpinene and terpinen-4-ol as the leaf matured (Figure 5). The point of inflection occurred at exactly the stage where flush growth ceased and mature leaf began irrespective of the percentage flush on the branch. Searches for similar relationships in the closely related *Melaleuca bracteata* (black tea tree) and *Leptospermum petersonii* (lemon-scented tea tree) did not reveal significant along-the-branch differences (Southwell 1989) other than a minor inverse relationship between citral and citronellal with the latter species.

This extraction method also provided another way of distinguishing between *M. alternifolia* and the closely related *M. linariifolia* (see above) (Southwell and Stiff 1990). The ratio of the precursors *cis*-sabinene hydrate: *trans*-sabinene hydrate was approximately 7.1 : 1 for the flush growth ethanolic leaf extract of *M. alternifolia* and approximately 0.7 : 1 for *M. linariifolia*. Measuring this ratio in the flush extract provides an alternative method for distinguishing the two species to measuring the α-thujene/α-pinene ratio in the oil (Table 4).

Microextraction also proved ideal for determining the quality of seedlings before planting out (Russell *et al.* 1997). With the existence of a number of chemical varieties of *M. alternifolia* and *M. linariifolia* and the likelihood of cross pollination (Butcher *et al.* 1992), plantation establishment using the right genetic material is essential. Freshly emerging leaves from propagation material including seedling dicotyledon leaves are ideally suited for single leaf extraction analysis. A preliminary analysis of supposedly good quality terpinen-4-ol type tea tree seedlings showed higher than expected concentrations of α-pinene (approx. 12%) and terpinolene (approximately 12%) in early stages of development. Terpinen-4-ol and the precursor sabinene hydrates were not present in the dicotyledon leaves. Single leaf analysis, on a day by day basis as each successive leaf pair formed showed that, by the time leaf set 10 was 10 weeks old, terpinolene concentrations had dropped to normal (3%) and

Table 4 The minimum, maximum, mean and standard deviation (s.d.) *cis* : *trans* sabinene hydrate (A) and *α*-thujene : *α*-pinene (B) ratios for *M. alternifolia* and *M. linariifolia* flush growth leaf extracts from the terpinen-4-ol variety

	Ratio	*Species*	*Min*	*Max*	*Mean*	*s.d.*
A	*cis* : *trans*	*M. alternifolia*	4.2	13.4	7.11	1.65
	sabinene hydrate	*M. linariifolia*	0.5	0.9	0.65	0.13
B	α-thujene : α-pinene	*M. alternifolia*	0.1	0.7	0.33	0.09
		M. linariifolia	0.6	2.3	1.49	0.37

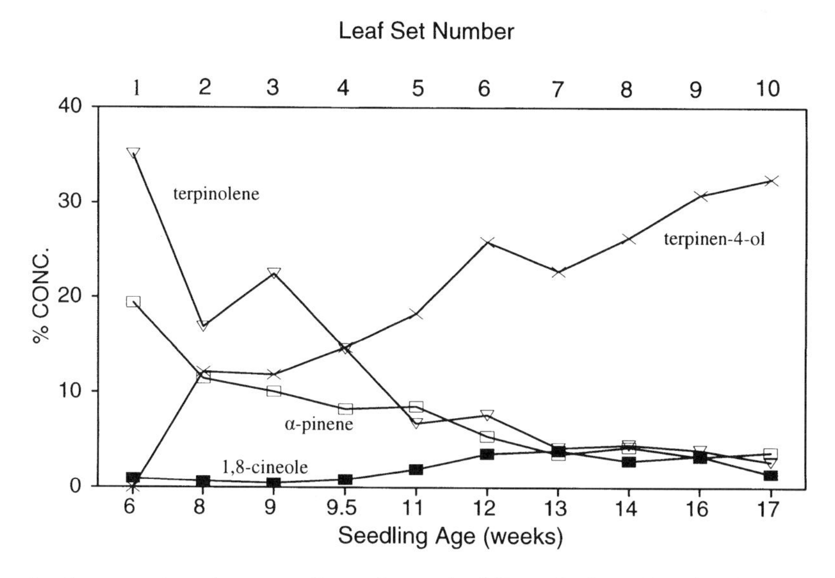

Figure 6 Concentration of tea tree oil constituents in different leaf sets 6 weeks after emergence

Table 5 The concentrations (%) of key components in the emerging 2-week-old dicotyledon (normal type) and ten-week-old leaf-set-ten (parentheses) leaves of seedlings of the three main chemotypes of tea tree

Component	*α-Pinene*	*1,8-Cineole*	*Terpinolene*	*Terpinen-4-ol*
Chemotype				
Terpinen-4-ol	20(3)	0(2)	32(3)	0(36)
Cineole	15(6)	40(71)	10(1)	0(1)
Terpinolene	0(1.5)	12(18)	25(57)	0(1)

terpinen-4-ol increased to 36% (Figure 6). As a result of this microextraction method, recommendations were made to the industry concerning the best times to analyse tea tree seedlings for quality before planting out. Table 5 shows the early changes in oil constituent concentrations from emerging to 10 week old leaf set 10 leaves. Consequently leaf set 10 at 6 weeks of age will give an accurate estimate of final oil quality. Dicotyledon analysis will predict the chemotype propagated.

Supercritical and dense carbon dioxide extractions of tea tree oil have been attempted on laboratory scale. In one example, investigations using a variety of supercritical carbon dioxide extraction conditions showed that although oils chemically identical to the distilled oil can be obtained, precursor components such as *cis*-sabinene hydrate are likely to extract without substantial conversion to the end products terpinen-4-ol and γ-terpinene (Wong 1997). As it is difficult to harvest material devoid of flush growth, this extraction method could give a product with low terpinen-4-ol and high precursor contents unless temperatures approaching those of a steam distillation are used. Thus differing compositions for the product and the higher costs and specialised equipment associated with such an extraction may discourage further investigation.

Leaf Waxes and Proline Analogues

In contrast with the volatile constituents, the non-volatile components in *Melaleuca*, especially the *M. alternifolia* group have received little attention. Some such isolates are shown in Figure 7.

One published analysis of leaf waxes (Courtney *et al.* 1983) reported the isolation of the triterpenoid ursolic acid (26) from the leaf wax of *M. quinquenervia*. Studies of the leaf waxes of the *M. alternifolia* group have not been reported in the literature.

L-proline (27) and its analogues N-methyl-L proline (28), 4-hydroxy-N-methylproline (29) and 4-hydroxy-N, N-dimethylproline (30) have been isolated from various *Melaleuca* species (Naidu *et al.* 1987; Jones *et al.* 1987). The structures of these constituents were supported by X-ray crystallographic studies. Concentrations were found to increase when the plants were subjected to water or salinity stress under laboratory conditions. The best sources of these compounds were found to be *M. uncinata*, *M. lanceolata, M. cuticularis, M. populiflora* and *M. viridiflora*. These nitrogenous constituents are under investigation as potential stress alleviating constituents and are increasing the biomass yields of pastures and crop species. Investigations aimed at commercialising these compounds are underway with field trials in several locations on the east coast of Australia (Naidu 1997).

COOH

HO

26 ursolic acid

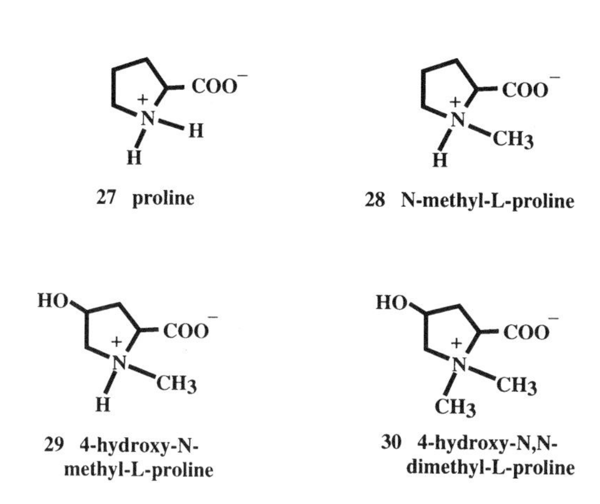

Figure 7 Chemical structures of significant non-volatile constituents in *Melaleuca* leaf

CHEMICAL VARIATION

The species known to yield the terpinen-4-ol type of tea tree oil are *M. alternifolia, M. linariifolia* and *M. dissitiflora* (Brophy *et al.* 1989b). Other species may also be suitable sources of the commercial oil. *M. uncinata* R. Br. *sensu lato*, for example has a similar terpinen-4-ol type oil which only fails to meet standards because of a high *p*-cymene concentration (17.6%) (Brophy *et al.* 1990; Brophy and Lassak 1992). In addition, *M. parviflora* from Vietnam has been reported to yield a similar oil (Dung *et al.* 1994). The identification of this species has now been revised to *M. alternifolia* (Dung 1996).

The above species also yield other types of tea tree oil. The most abundant is the 1,8-cineole chemical variety (Brophy *et al.* 1989b). *M. alternifolia* also has a terpinolene-rich chemical variety (Southwell *et al.* 1992).

Penfold *et al.* (1948), before the advent of gas chromatography used the *o*-cresol method for the determination of 1,8-cineole (Cocking 1920) to separate three chemical varieties of *M. alternifolia* based on 1,8-cineole content. The "Type" variety contained 6–14 percent, "Variety A" 31–45 percent and "Variety B" 54–64 percent. The first of these, the Type form, was the only one recommended for medicinal use (Penfold *et al.* 1948; Guenther 1948; Lassak and McCarthy 1983). More recent investigations (Brophy *et al.*

1989b) have noted the high 1,8 cineole Variety B form but suggest a gradation to the Type form rather than the existence of a clear intermediate Variety A form. On the other hand Butcher *et al.* (1994) propose a total of five chemical varieties: the initial three proposed by Penfold *et al.* (1948) and two varieties of the terpinolene type named Variety C (15–20% terpinen-4-ol; 30–36% 1,8-cineole; 10–18% terpinolene) and Variety D (1–2% terpinen-4-ol; 17–34% 1,8-cineole; 28–57% terpinolene). More thorough surveys of individual trees including greater numbers of samples are giving data falling between these proposed varieties suggesting a merger of these proposed varieties as has happened with *Eucalyptus punctata* (Southwell 1973).

The existence of these chemical varieties is of economic importance for nursery suppliers or plantation owners collecting seed and raising their own seedlings. The quality of the seed bearing parent is not a reliable guide to the quality of the seed as cross pollination with a parent of a different variety is sometimes possible. The analysis of seedling leaf, even as early as the dicotyledon leaf stage, is a reliable method for determining tree quality when the onset of different biogenetic pathways is understood. Tea tree breeding programs are seeking to establish seed orchards of high yielding, high quality varieties of the terpinen-4-ol variety of *M. alternifolia* (Chapter 7, this volume). Although this species is the one at present preferred by the industry, the Australian and International Standards indicate that the oil may be sourced from *M. alternifolia*, *M. linariifolia* and *M. dissitiflora* and other species of *Melaleuca* yielding a comparable oil. Trials involving the production of *M. linariifolia* (Southwell and Stiff 1990) and *M. dissitiflora* oil (Williams and Lusunzi 1994) have not yet led to large scale plantings of these species.

QUALITY ASSURANCE AND STANDARDS

Oil Quality

The existence of chemical varieties other than the commercial terpinen-4-ol type has meant that these have been harvested from the wild or sometimes cultivated, resulting in substandard oils being found in the marketplace. Hence oil quality is critical. As with other essential oils, quality before the advent of gas chromatography was defined by measuring the physical constants of refractive index, optical rotation, specific gravity and solubility in alcohol. Measurement of these constants is still mandatory for compliance with International, National and Pharmacopoeia Standards because of its usefulness defining physical constant ranges for the unique combination of constituents in any one essential oil.

The early standards for tea tree oil such as the BPC (British Pharmaceutical Codex 1949) and the Australian Standard K-175 (Standards Association of Australia 1967) were based on these physical constants (Table 6). The BPC also included an alcohol determination based on the measurement of ester numbers before and after acetylation. As this method is not appropriate for tertiary alcohols (Guenther 1948), reliable tea tree oil alcohol determinations were not achieved until oils were analysed by gas chromatography. The revised Australian Standard 2782–1985 (Standards Association of Australia

Table 6 Tea tree oil standards and monographs

Standard	*Year*	*Title*	*Content*
British Pharmaceutical Codex	1949	Oleum *Melaleucae*	Definition, Description, Solubility, Density, Refractive Index, Optical Rotation, Ester Value, Constituents
Australian Standard K-175	1967	Oil of *Melaleuca alternifolia*	*As above except for Constituents, Ester Value*
Martindale, the Extra Pharmacopoeia	1972–1993	*Melaleuca* Oil	Synonyms, Description, Solubility, Density, Composition (partial), Storage, Uses
Australian Standard 2782	1985	Oil of *Melaleuca*, Terpinen-4-ol Type	*As for AS K-175 with cineole (max) and terpinen-4-ol (min) percentages*
French Standard T75–358	1991	Huile essentielle de *Melaleuca* type *terpinene-4-ol*	*As for AS 2782 with GC profile table (13 components), typical traces (2) and flash point*
International Standard 4730	1996	Oil of *Melaleuca*, Terpinen-4-ol Type	*As for T 75–358 with 14 component profile and 3 traces*
Deutscher Arzneimittel Codex Draft	1996	Teebaumol	*As for ISO 4730 with Acid Number, TLC Method and extra analyte (Δ^3-carene)*

1985) for Oil of *Melaleuca,* Terpinen-4-ol Type set a maximum level (15%) for 1,8-cineole and a minimum concentration (30%) for terpinen-4-ol as determined by gas chromatography (GC). Standards Australia submitted a further revision of this standard to the International Standards Organisation (ISO) who subsequently published ISO 4730 (International Standards Organisation 1996). This submission included a Chromatographic Profile Table listing the required ranges for fourteen constituents and was used as a basis for French Standard T75-358 by AFNOR (Association Française de Normalisation 1996) and the updated Australian Standard AS 2782-1997 (Standards Australia 1997). Ledene (viridiflorene), sometimes hidden under the α-terpineol peak on polar GC stationary phases, was a notable omission. In addition, the upper limit for sabinene (3.5%) was not high enough to cover all *M. dissitiflora* terpinen-4-ol type oils (Brophy and Doran 1996). Furthermore, the upper limit for sesquiterpene and sesquiterpene alcohol components was based on a few extraordinary analytical results. This profile Table was then included in the ISO Standard which also contains three typical chromatograms (for information only) on polar (BP 20), non-polar (OV 101) and intermediate polarity (AT 35) stationary phases. Analysis on an appropriate intermediate polarity phase (Figure 3) is preferable because polar phases sometimes do not resolve α-thujene (1) from α-pinene (2) and, β-phellandrene (9) from 1,8-cineole (11) and non-polar phases, although separating the former peaks, do not resolve the latter.

The tea tree oil industry is seeking to have Oil of *Melaleuca* relisted as a monograph in pharmacopoeias. The German Pharmaceutical Codex has published a monograph (Deutscher Arzneimittel-Codex 1996) which outlines a TLC and GLC method for oil analysis. The latter is based on the published literature and includes the determination of Δ^3-carene which has not been reported in tea tree but is viewed with suspicion because of the eczematous properties of its auto-oxidation products (Tisserand and Balacs 1995).

In the market place, most tea tree oil is bought and sold on 1,8-cineole and terpinen-4-ol gas chromatographic area percent figures alone. This ensures that the oil is sourced from the low-cineole–high terpinen-4-ol chemical variety. A technique involving the addition of a known weight of internal standard (usually *n*-tridecane, *n*-tetradecane or *n*-pentadecane) or the measurement of the four physical constants (optical rotation, refractive index, relative density and solubility in alcohol) ensure that the oil has not been diluted by either solvent or non-volatile materials.

More sophisticated means of quality control have been examined. These include gas chromatography–mass spectrometry (GCMS), nuclear magnetic resonance (NMR), infrared spectroscopy (IR) and chiral column gas chromatography.

GCMS has been used to help identify the components in the oil (Swords and Hunter 1978; Brophy *et al.* 1989b; Kawakami *et al.* 1990) but is unlikely to be used for routine quality control.

Similarly chiral stationary phases have been used to determine the enantiomeric excess of the chiral constituents in tea tree oil (Leach *et al.* 1993). Now that these ratios have been determined, there is little need to use chiral stationary phases routinely. In special circumstances they may however be useful for detecting adulteration. For example, some years ago, a blend of tea tree oil and a eucalyptus oil fraction reached the international market. Although the eucalyptus oil fraction was rich in terpinen-4-ol, such a source of this key terpene alcohol is characterised by a negative optical rotation. Consequently the ratio of (+)-terpinen-4-ol to (−)-terpinen-4-ol was much lower in the blend (1:2) than in the unadulterated (2:1) tea tree oil (Leach *et al.* 1993).

High field proton and carbon magnetic resonance spectroscopy (NMR) is being increasingly used in essential oil analysis (Formacek and Kubeczka 1982). For example, enantiomeric purities can be confirmed by the use of chiral shift reagents. Key components, terpinen-4-ol and α-terpineol were examined using ^{13}C and ^{1}H NMR DEPT, INEPT and ^{13}C-^{1}H HETCOR analyses prior to the addition of lanthanide shift reagents (Leach *et al.* 1993). Although the resolution of methyl signals of the enantiomers was possible in this way, application to the analysis of a complete tea tree oil is confounded by the complex nature of the oil and the abundance of signals in the appropriate regions.

Infrared spectroscopy (IR) although less informative than other techniques, is useful in indicating the degree of hydroxyl absorption and the presence of the multiple fingerprint region peaks for cineole in higher cineole oils.

Tree Quality

The tea tree oil industry has moved in the last decade from being an industry based on wild or bush harvesting to being based on commercial plantations. There is the requirement then to ensure that the trees involved, whether they be wild stands for harvest, seed trees for propagation material or trees in established plantations, are of the right chemical variety and contain acceptable concentrations of cineole and terpinen-4-ol. For the rapid determination of tree quality, the time consuming leaf distillation step needs to be bypassed.

To accelerate the quality control procedure, leaf extraction methods were developed (Brophy *et al.* 1989b; Southwell and Stiff 1989) (see above). Hence individual leaves on any tree can be assessed by GC in less than one hour by inserting them into a vial, adding ethanol, irradiating with microwaves and injecting the resultant ethanolic solution into the gas chromatograph. Fresh leaf was found to extract much more rapidly than dried leaf.

This extraction method is an ideal way of assessing the quality of a leaf, a tree or a number of trees by varying the vial size and solvent volume according to the number of leaves examined. The method can also be used quantitatively by adding a known quantity of internal standard to a known weight of leaf before chromatography. The chromatographic behaviour of the internal standard varies with the GC stationary phase with the n-alkanes tridecane, tetradecane and pentadecane eluting close to terpinen-4-ol. This method has been used for determining oil quantity and quality for tea tree breeding programs (Chapter 7, this volume).

Seedling Quality

With the quality of trees cultivated in plantation being critical for the marketing of the resultant oil, then the quality of the propagation material becomes equally critical. Most plantation trees have so far been propagated from seed as the higher costs and massive numbers of trees planted out makes the logistics of propagation by tissue culture or cuttings unmanageable. The latter methods have been tried successfully on a small scale and do guarantee more consistent quality. With propagation by seed, an oil quality check on the mother tree will not necessarily determine the oil quality of the progeny as fertilisation from a poor quality father may have occurred. Hence unless seed is collected from a region where cross pollination from poor quality father trees is impossible, there is a chance the seed could be of poor quality.

In addition, seed has often been bought from merchants unaware of the chemical varieties available within individual *Melaleuca* species. This has been seen with attempts to establish tea tree plantations in at least two overseas countries. In one of these, of the eight clones described (Kawakami *et al.* 1990) only one (Clone II) met the requirements of the International Standard 4730. Clone I, although containing acceptable quantities of terpinen-4-ol and 1,8-cineole also contained excessive concentrations of sabinene which, along with the high α-thujene/α-pinene ratio (Southwell and Stiff 1990) suggested *M. linariifolia*. Clones III to VIII contained insufficient terpinen-4-ol with excessive concentrations of 1,8-cineole and Clones III and VIII contained excessive quantities of terpinolene as well.

Hence there was a need for the determination of oil quality in seedlings to prevent the large scale planting of substandard quality trees. Such a method was developed at the Wollongbar Agricultural Institute (Russell *et al.* 1997) by using the extraction analysis method described above.

With careful analysis of the first dicotyledon leaves to emerge from a seedling, it was possible to predict whether the variety in question was the terpinen-4-ol, cineole or terpinolene chemotype. Although oil yields were very low at this stage, ethanolic extraction with microwave irradiation followed by GC analysis, gave a sufficiently intense profile to type each variety. A misleading aspect of this method was that the commercial terpinen-4-ol variety contained substantial percentages of terpinolene (approx. 12%), α-pinene (12%) and β-pinene (14%) and no terpinen-4-ol. This early stage analysis gave the wrong idea that the terpinen-4-ol variety was actually the terpinolene chemovar because of high terpinolene concentrations. The cineole variety contained a similar proportion (approx. 10%) while the terpinolene chemovar had much more (approx. 25%) terpinolene (Table 5).

The key component for measurement is 1,8-cineole. In the commercial terpinen-4-ol variety, cineole is usually low (0.5–5%). In the terpinolene variety it rises to approximately 12% whereas in the cineole variety it reaches values as high as 40–70%. Although the measurement of terpinen-4-ol means nothing in the dicotyledon leaves (all varieties read zero), values increase as the seedling ages. With, for example, a ten week old leaf set ten analysis (i.e. ethanolic extract analysis of the tenth leaf pair to emerge when tested ten weeks after emergence) terpinen-4-ol proportions are approximately 36% with the commercial variety and only about 1% with the other chemical varieties. Terpinen-4-ol first appears as *cis*- and *trans*-sabinene hydrate (ratio approximately 7:1 with *M. alternifolia* and 0.7:1 with *M. linariifolia*) (Southwell and Stiff 1989, 1990) in bright green flush seedling growth as it does with the flush growth on mature plants. These investigations of the ontogenetic variation in oil constituent percentages (Figure 8) indicate that different biogenetic pathways (e.g. the pinene, the cineole/limonene/α-terpineol, the terpinolene and the α-thujene/sabinene/sabinene hydrate/γ-terpinene/terpinen-4-ol pathways are initiated at different stages of the plant's ontogeny. Clearly the cineole, pinene and terpinolene pathways appear at earlier stages of development in seedling growth than the sabinene hydrate/terpinen-4-ol pathway. Consequently producers can reliably assess the quality of their seedlings at both early and late stages of the seedlings development.

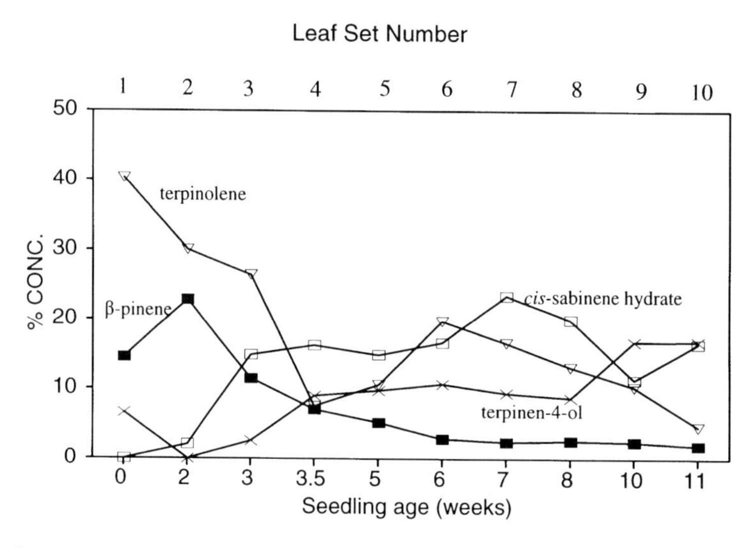

Figure 8 The concentration of key consitituents in *M. alternifolia* seedling leaves at emergence

BIOGENESIS

Although little work has been done specifically on *Melaleuca* oil biogenesis, the pathways in other genera producing similar monoterpenoids have been investigated. For example, Croteau (1987) summarised the current thinking concerning the metabolism from acetyl-CoA through mevalonic acid pyrophosphate and isopentenyl pyrophosphate to geranyl pyrophosphate. Recent investigations now suggest that isopentenyl pyrophosphate is formed, not from mevalonic acid but via the alternative triose phosphate/pyruvate pathway (Lichtenthaler *et al.* 1997; Eisenreich *et al.* 1997). Croteau's review (1987) continued to examine the cyclisation reactions of geranyl pyrophosphate (34,35) and linalyl pyrophosphate (36,37) that produce cyclic monoterpenoids such as limonene, α-terpinene, γ-terpinene, sabinene, 1,8-cineole, α-pinene, terpinen-4-ol, α-terpineol etc. Cyclase enzymology is complex in that different cyclases can produce the same product and an individual cyclase can produce multiple cyclic products. Definitive investigations involving partially purified enzymes, cell-free extracts, isotopic labelling and substrate substitution are adding gradually to our knowledge of these pathways. Some conclusions can also be drawn by studying the co-occurrence and concentration variation of significant metabolites at various stages of ontogeny in different chemical varieties.

With the high 1,8-cineole variety of *Melaleuca* species, the 1,8-cineole concentration increases concomitantly with limonene and α-terpineol suggesting that all three cyclic products are derived from a linalyl pyrophosphate derived, enzyme bound moiety containing (31) (Figure 9). Conversely, the terpinen-4-ol variety is richer in congeners α-terpinene, γ-terpinene, terpinolene and terpinen-4-ol, all derived from a moiety such as (32) which is easily obtained from (31) by 1,2-hydride shift. Evidence for such a hydride shift has been provided following investigations on the formation of *cis* and *trans* sabinene hydrate in marjoram (Hallahan and Croteau 1989). The product of this shift is now available for further cyclisation to form the cyclopropane moiety (33) which has been implicated in pathways to the thujanes in both marjoram (Hallahan and Croteau 1988) and tea tree (Southwell and Stiff 1989). Tea tree flush growth contains similar monoterpenes to the sabinene hydrate variety of sweet marjoram, *Majorana hortensis* (Southwell and Stiff 1989). The difference is that in marjoram, the sabinene hydrates remain and do not appear as terpinen-4-ol and γ-terpinene in the mature leaf. The fact that steam distillation causes the hydrates to convert to terpinen-4-ol and γ-terpinene has been

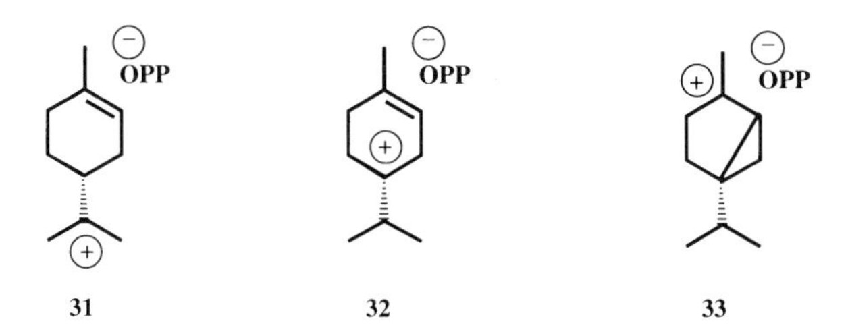

Figure 9 Chemical structures of likely intermediates in the biogenetic pathways to *Melaleuca* oil constituents

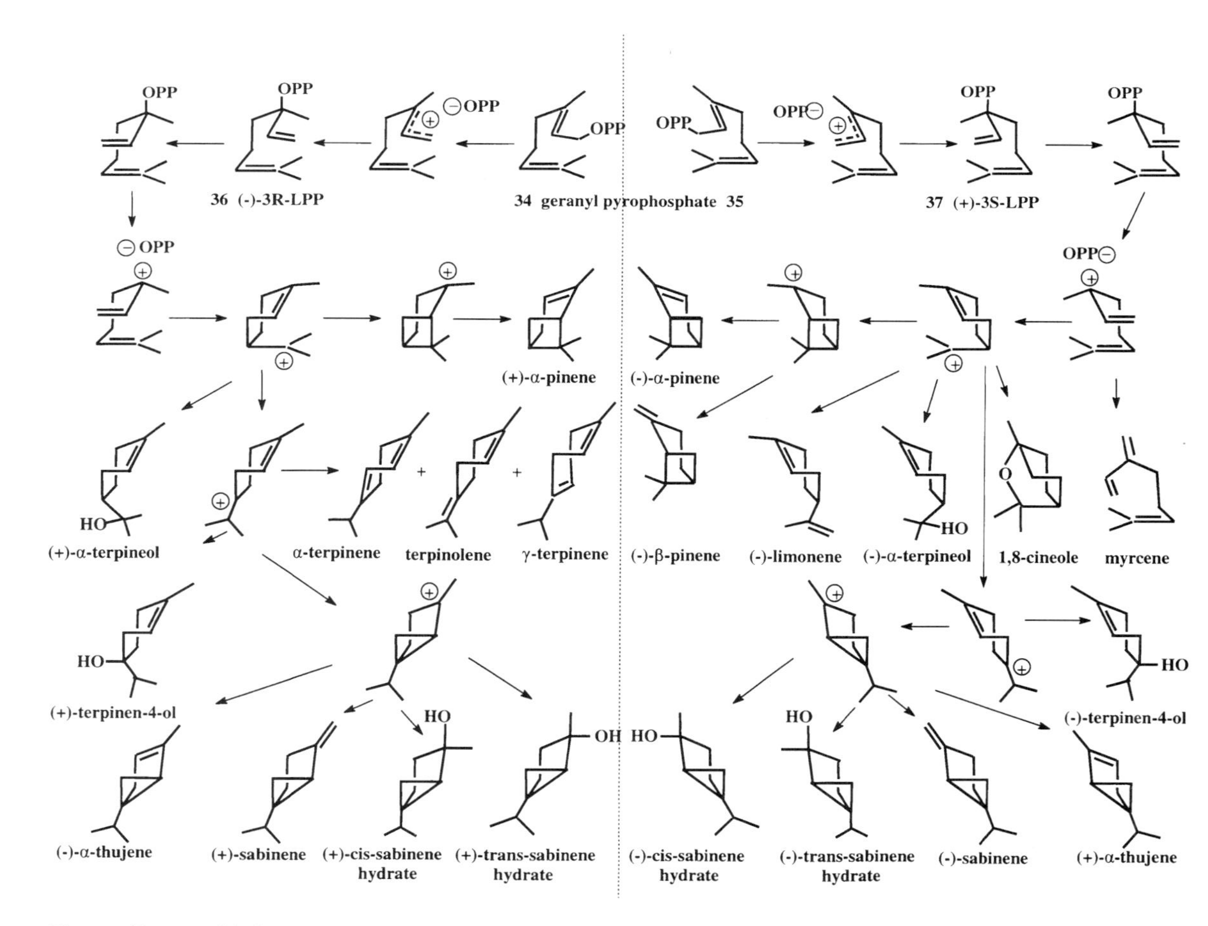

Figure 10 Possible biogenetic pathways for the formation of *Melaleuca* terpinen-4-ol type monoterpenene constituents

advantageous for the tea tree oil industry. Distilling flush growth does not give a different quality oil as the precursor sabinene hydrates convert to the end product terpinen-4-ol and γ-terpinene during the distillation in a similar way to the change that takes place as the leaf matures. The inability of marjoram to carry out this conversion as the leaf matures has meant that the quality of marjoram products varies according to the processing method. Consequently extraction methods favour the hydrates and distillation methods terpinen-4-ol (Fischer *et al.* 1987). The biosynthetic capacity of marjoram seems confined to the synthesis of the sabinene hydrate skeleton in contrast to that of *Melaleuca* which converts the thujane skeleton to the menthane skeleton as the leaf matures. Whether this change in tea tree is a secondary enzymic transformation of thujanes to terpinenes or the closing down of one biogenetic pathway and the concomitant initiation of the other is yet to be confirmed. The comparative concentration of the metabolites however suggests the former.

Cornwell *et al.* (1995) distilled young leaves of *M. alternifolia* in ^{18}O-labelled water to obtain label in the resultant terpinen-4-ol and α-terpineol. This indicated that if terpinen-4-ol (14) was formed directly from *cis*-sabinene hydrate (24), the process involved hydration of the terpinen-4-yl cation (32) rather than a 1,4-hydroxyl-shift in *cis*-sabinene hydrate (24) which does not account for the non-oxygenated byproducts. In addition the ratio of these products is similar to the ratio of the same products obtained by the acid catalysed rearrangement of sabinene hydrate. Hence these authors conclude that the sabinene hydrate degradation during ontogenesis is a purely chemical breakdown rather than enzymic secondary metabolism as suggested by Southwell and Stiff (1989). This does not however, address the comparison with marjoram where the sabinene hydrates are retained even in the mature leaf (Fischer *et al.* 1987, 1988).

By applying these findings to the commercial terpinen-4-ol chemical variety of tea tree oil, biogenetic pathways like those outlined in Figure 10 are likely to be appropriate for monoterpenoid formation.

In addition, *Melaleuca* oil contains up to approximately 12% of sesquiterpenoids, mostly sesquiterpene hydrocarbons. The flush growth extracts also indicated the presence of sesquiterpene precursors which convert to aromadendrene (16), ledene (17), δ-cadinene (18) etc. as either the leaf matures or the leaf is distilled. Although these precursors have not been isolated and positively identified, a biogenetic precursor like bicyclogermacrene (38) as suggested by Taskinen (1974) for marjoram is plausible (Figure 11) (Brophy *et al.* 1989b; Ghisalberti *et al.* 1994).

The terpinen-4-ol chemical varieties of both *M. linariifolia* and *M. dissitiflora* would be expected to display similar biogenetic pathways to *M. alternifolia*. The GC difference in the α-thujene/α-pinene ratios (see above) for the oils of *M. alternifolia* and *M. linariifolia* suggests that the thujane pathways differ because of the greater concentration of α-thujene in the latter. A similar difference was also seen in the GC traces of the ethanolic extracts of the flush growth of both species. In addition to the α-thujene/α-pinene ratio, *M. alternifolia* gave a *cis*:*trans* sabinene hydrate ratio of approximately 7:1 whereas in *M. linariifolia* the ratio was approximately 0.7:1. These ratios are very similar to those obtained by Hallahan and Croteau (1989) for the sabinene hydrate ratios when both the "natural" (−)-(3R)-linalyl pyrophosphate (36) and the "unnatural" (+)-(3S)-linalyl pyrophosphate (37) precursors respectively were used as substrates for sweet

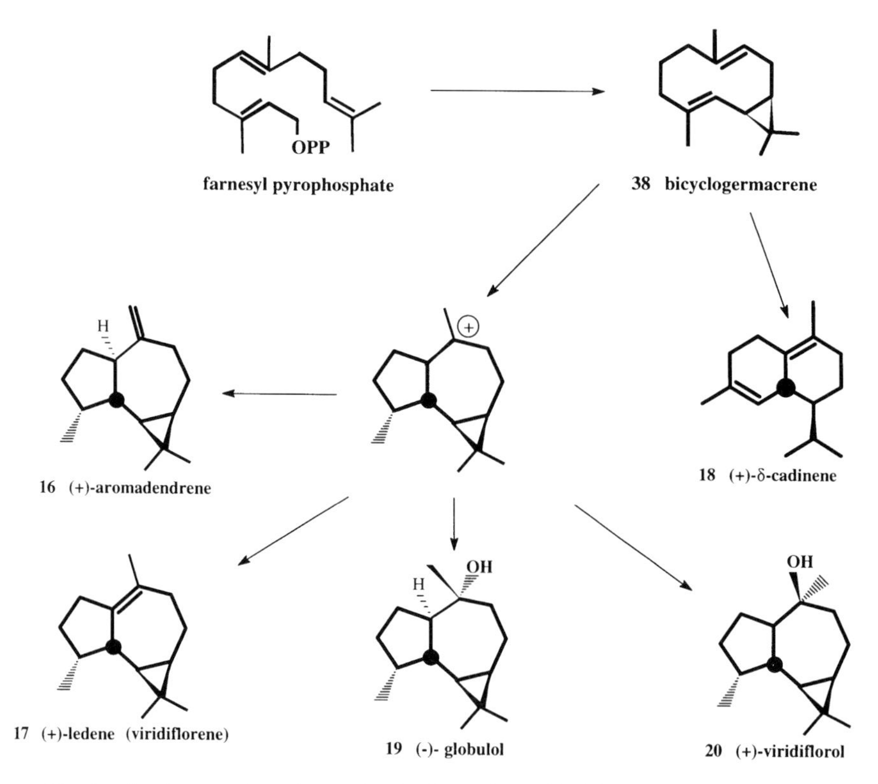

Figure 11 Possible biogenetic pathways for the formation of *Melaleuca* terpinen-4-ol type sesquiterpene constituents

marjoram (*Majorana hortensis*) partially purified cyclase. This suggests that the difference between the two species is significant at the enzyme level where for *M. linariifolia* the "unnatural" (3S)-linalyl pyrophosphate is better accommodated even though there is minimal difference at the end of the biogenetic pathways especially with the steam distilled oil constituents (Southwell and Stiff 1990). Thus the sabinene hydrate cyclase from *M. linariifolia* should bind preferentially to the right handed screw form of geranyl pyrophosphate (35) which isomerises to the bound (+)-(3S)-linalyl pyrophosphate (37) in contrast to the *M. alternifolia* cyclase which should bind to the left handed screw form of geranyl pyrophosphate (34) isomerising to the bound (−)-(3R)-linalyl pyrophosphate (36) (Figure 10).

Determination of oil quality in tea tree seedlings by extraction has thrown some light on the progressive initiation of biogenetic pathways in *Melaleuca.* At the dicotyledon leaf stage of development, the pinene, cineole and terpinolene pathways are evident with no initiation of the thujane sabinene and sabinene hydrate or the terpinene terpinen-4-ol and γ-terpinene pathways (Russell *et al.* 1997). α- and β-Pinene appear to be the products of one cyclase. Croteau (1987) suggested a possible association with myrcene which is not supported by *Melaleuca* seedling ontogeny where myrcene concentrations increase only after the formation of α- and β-pinene. 1,8-Cineole is also formed immediately and varies little in the course of development of the seedling. The other products which, like cineole, are possibly derived from moiety (31) (Figure 9) are α-terpineol (15), terpinolene (13) and limonene (8). Terpinolene, which can also be derived from moiety (32) (Figure 9), is present in the dicotyledon leaves in higher proportions than in the mature leaves or the distilled oil. This decreasing proportionality is in contrast to the constant proportion of cineole suggesting some differences in formation pathways. On the other hand, along-the-branch analysis of the terpinolene variety shows an inverse proportion relationship between cineole and terpinolene suggesting that, in this variety, the two are similarly derived (Southwell *et al.* 1992). Limonene (8) and α-terpineol (15), known to increase in concentration as cineole increases, only appear later in seedling ontogeny (Figure 8).

The most remarkable concentration changes in seedling development occur, as in mature tree along-the-branch analysis, with the thujanes sabinene, *cis*- and *trans*-sabinene hydrate and the terpinenes α-terpinene, γ-terpinene, terpinolene, and terpinen-4-ol. These biogenetic pathways seem to be initiated in earnest at a seedling age of around three weeks when the third leaf set begins to emerge (Figure 8). From three weeks to ten weeks both *cis*-sabinene hydrate and terpinen-4-ol concentrations increase. After ten weeks, the more mature leaves have greater concentrations of terpinen-4-ol than *cis*-sabinene hydrate (Figure 6).

MELALEUCA OIL STABILITY

Health authorities require shelf life measurements to be made so that "use by..." dates can be printed on health care product labels. For these purposes an understanding of the chemical changes taking place as an oil ages is essential. The first and most obvious chemical change in oil composition to be noted is an increase in *p*-cymene concentration as the menthadienes α-terpinene (7) γ-terpinene (12) and terpinolene (13) oxidise (Figure 12). The rate at which this change occurs is variable as some oils are remarkably stable over ten years whereas others can oxidise after two years if storage conditions are poor (Table 7) (Southwell 1988; Brophy *et al.* 1989b). Storage in cool, dark, dry, inert atmosphere (or minimum surface area/volume ratio) and inert containers (stainless steel or tinted glass) ensures optimum stability.

Some aged oils have been reported to be deep yellow in colour before depositing small amounts of oil insoluble crystals. These crystals were found to be 1S,2S,4S-trihydroxy-*p*-menthane (39) from both *M. linariifolia* (Jones and Oakes 1940; Davenport *et al.* 1949) and *M. alternifolia* (Brophy *et al.* 1989b) terpinen-4-ol type oils. The formation of this oxidation product occurs concurrently with but by no means as extensively as, the formation of *p*-cymene from the terpinenes (Figure 12).

Shelf-life trials conducted in our laboratories have shown that the antimicrobial activity of aged oils is not decreased as long as terpinen-4-ol levels are maintained. Indeed some researchers have reported increased activity with aged oils in which terpinen-4-ol concentrations have been maintained (Markham 1996). Consequently the antimicrobial stability of aging oils can be assessed by measuring terpinen-4-ol content.

Table 7 Monoterpenoid composition comparison of aged oils of *M. alternifolia*

Sample no.	1	2	3	4	5
Age, years	10	10	5	2	1
Rel. deterioration rate	slow	rapid	rapid	rapid	moderate
Composition, %					
α-thujene	0.6			0.2	0.8
α-pinene	2.2	3.2	tr	2.0	2.5
sabinene	0.1		tr	tr	
β-pinene	0.6	0.3	tr	0.4	0.7
myrcene	0.5	0.2	tr	0.1	0.7
α-phellandrene	0.2	tr		tr	0.4
α-terpinene	5.8	0.2		0.1	6.6
p-cymene	4.3	32.0	21.7	35.3	8.0
β-phellandrene + cineole + limonene	0.4	7.3	2.1	3.1	4.0
γ-terpinene	15.0	tr	tr	tr	17.6
terpinolene	2.7	tr	tr	tr	3.1
terpinen-4-ol	41.6	31.5	45.9	23.8	37.3
α-terpineol	3.7	6.4	9.6	8.2	2.9
1,2,4-trihydroxymenthane	tr	4.6	2.5	3.6	tr
ISO Standard Status	pass	fail	fail	fail	fail

The overall chemical stability however, is best assessed by measuring α-terpinene, γ-terpinene and *p*-cymene in addition to terpinen-4-ol. Chemical stability trials at the Wollongbar Agricultural Institute indicated that well-stored oils, 9–13 years old, only show a little oxidation and actually increase in terpinen-4-ol content. Poorly stored oils however, can oxidise rapidly (even in 12 months) and one 47 year old oil contained 37.7% *p*-cymene and 3.2% 1S,2S,4S-trihydroxy-*p*-menthane (39) while still retaining 26.7% terpinen-4-ol (Table 8). 1,8-Cineole content remains remarkably constant during aging as it does during the different stages of plant ontogeny.

PROCESS MODIFIED OILS

Typical *Melaleuca* oil compositions are based on steam or hydro-distillations of several hours duration for either laboratory or commercial scale operations. It has already been seen that this composition changes for solvent extraction (especially when flush growth or early seedling leaf is included) (Southwell and Stiff 1989, 1990), for supercritical fluid extraction (Wong 1997), and for headspace analysis (Southwell 1988; Kawakami *et al.* 1990). The compositions of these process-modified oils are shown in Table 9.

Of commercial significance for oil quality is the compositional variation that can occur using different distillation times. For example Brophy *et al.* (1989b) found that the first 30 minutes of a laboratory scale-distillation gave a 55.9% terpinen-4-ol oil whereas the remaining 60 minutes gave only 25.1% terpinen-4-ol (Table 9). This trend also exists in commercial-scale distillations albeit within a reduced time-frame (Russell *et al.*

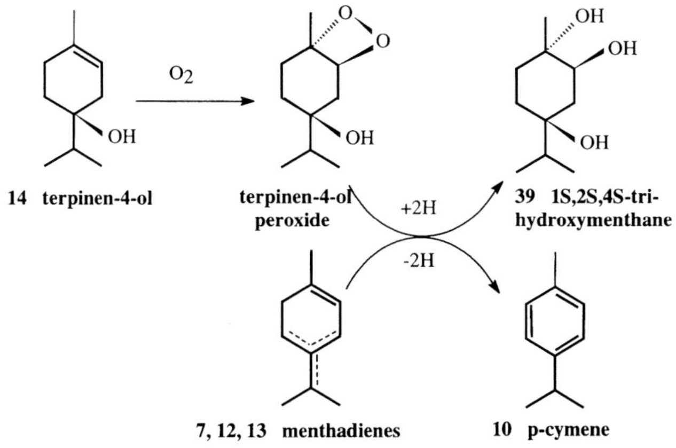

Figure 12 The oxidation of oil of *Melalecua*, terpinen-4-ol type constituents

Table 8 Shelf-life tests for tea tree oil chemical stability (25–35°C) at the Wollongbar Agricultural Institute (*italics* indicate composition of freshly distilled oil)

Age (years)	*Percentage composition*					
	terpinen-4-ol	*cineole*	*p-cymene*	*α-terpinene*	*γ-terpinene*	*1S,2S,4S-trihydroxy-p-menthane*
47	26.7	2.2	37.7	0	0	3.2
13	38.7	1.2	5.0	7.4	18.4	0
	34.5	*1.3*	*3.7*	*9.8*	*21.2*	*0*
11	34.6	8.9	5.7	7.1	19.3	0
	31.8	*9.3*	*3.4*	*9.8*	*22.5*	*0*
9	40.4	3.5	7.5	5.2	13.4	0.1
	39.6	*3.1*	*1.6*	*10.4*	*20.1*	*0*
1	32.0	15.7	11.1	1.9	7.1	0.3
	29.0	*14.8*	*5.8*	*6.1*	*14.3*	*0*

1997). This variation in constituent percentages showed that the more polar constituents distil over before the less polar, lower boiling components suggesting that hydrodiffusion acts to extract more polar components first (Koedam, 1987). More detailed studies of a similar nature were reported by Stiff (1996) for hydrodistillations and by Johns *et al.* (1992) for steam distillations. Consequently, distillation for reduced periods of time will enhance terpinen-4-ol content and reduce the percentage of terpene hydrocarbons that occur in a completely distilled oil, the former enhancing the concentration of the active ingredient in *Melaleuca* and the latter reducing the concentration of some possible allergenic fractions (Southwell *et al.* 1997).

Table 9 Percentage composition of process modified *M. alternifolia* products compared with a typical terpinen-4-ol type oil

	Typical oil[a]	*Head space*[a]	*Flush extract*[b]	*Mature extract*[c]	*Early dist.*[d]	*Late dist.*[d]	*Interface fraction*[c]	*Water soluble*[c]
α-pinene	2.1	4.2	1.5	3.0	1.4	3.5	2.0	tr
sabinene	1.6	12.3	5.6	1.5	0.2	0.1	0.2	tr
α-terpinene	10.4	19.8	0.3	9.5	7.8	14.0	4.7	tr
limonene	0.6	2.8	0.5	1.9	un	un	un	tr
p-cymene	3.8	10.0	0.3	2.8	1.3	1.4	4.7	tr
1,8-cineole	0.5	8.8	4.3	2.2	5.7	4.1	3.8	1.6
γ-terpinene	19.4	24.6	6.3	26.4	15.6	29.1	9.9	tr
terpinolene	4.1	4.7	1.4	3.7	2.6	4.8	2.2	tr
terpinen-4-ol	42.9	3.2	7.6	32.9	55.9	25.1	19.2	76.5
α-terpineol	3.9	1.2	0.6	2.3	3.8	2.1	4.3	13.1
aromadendrene	0.2	tr	tr	1.6	0.3	1.2	7.8	tr
ledene	0.2	tr	tr	1.5	0.5	1.5	6.6	tr
δ-cadinene	0.5	tr	tr	1.6	0.3	1.2	6.0	tr
trans-sabinene hydrate	tr	tr	7.6	—	—	—	—	—
cis-sabinene hydrate	tr	tr	39.0	tr	—	—	—	—

[a] Kawakami *et al.* (1990), [b] Southwell and Stiff (1989), [c] Russell *et al.* (1997), [d] Brophy *et al.* (1989b).
un, unresolved; tr, trace.

Some of the more polar components of an essential oil will always dissolve in the condensate. For terpinen-4-ol type *Melaleuca* oils, these are the more polar 1,8-cineole, terpinen-4-ol and α-terpineol components. Extraction of several batches of condensate with diethyl ether has shown that about 0.07% of oil can be recovered from the water containing the dissolved oil. This oil was made up of approximately 73.7% terpinen-4-ol, 13.2% α-terpineol, 1.6% 1,8-cineole and traces ($<1\%$) of the other alcohols (*cis*-, and *trans*-menth-2-en-1-ol, *cis*-, and *trans*-piperitol) (Russell *et al.* 1997). Although distillation condensate is dilute, it may well provide a suitable source of terpinen-4-ol for spray application (e.g. horticultural plant pathogens as investigated by Bishop (1995) and Bishop and Thornton (1997)).

In addition, the interface of the oil and aqueous layers of the condensate sometimes contain distillation debris and is kept separate from the commercial oil. Analysis of this interface oil showed enhanced sesquiterpenes and reduced terpinen-4-ol typical of the "dregs" of a distillation (Russell *et al.* 1997). Extreme conditions of distillation, especially time and pressure, can "overcook" an oil and give similarly substandard results (Table 9).

As with any distilled oil, composition can be modified by further fractional distillation. This can be the method of choice for preparing 98% pure terpinen-4-ol, for enhancing the terpinen-4-ol content of an oil by removing outlying volatile constituents and for removing the potentially allergenic sesquiterpenoid fraction. 1,8-Cineole however can not be removed in this way without adversely effecting the composition of the oil by removing quantities of desirable constituents as well (Russell *et al.* 1997).

METABOLISM OF TEA TREE OIL

Little is known about the metabolism of tea tree oil in mammals including humans. The metabolism of the oil by the tea tree plantation pest Pyrgo beetle, *Paropsisterna tigrina*, has been investigated by examining the frass volatiles from larvae and adults (Southwell *et al.* 1995). At first, no obvious metabolites were observed when the beetles fed on commercial terpinen-4-ol type *Melaleuca* plantation tea tree. When higher 1,8-cineole tea tree leaf was the sole diet, (+)-2β-hydroxycineole (40) was isolated as the principal metabolite (Figure 13). Reinvestigation of the frass dropped when the commercial terpinen-4-ol variety was fed, showed that traces of (+)-2β-hydroxycineole occurred in the frass. Other paropsine beetles when fed high cineole diets also metabolised cineole but to other isomers of hydroxycineole. *Chrysophtharta bimaculata* hydroxylated cineole in the 3α (41), 9 (42) and 2α (43) positions, *Faex nigroconspersa* larvae in the same positions in different proportions and *F. nigroconspersa* adults in different proportions again (Table 10, Figure 13). Investigations now need to establish whether these seemingly species-specific metabolites are being used as pheromones for insect communication.

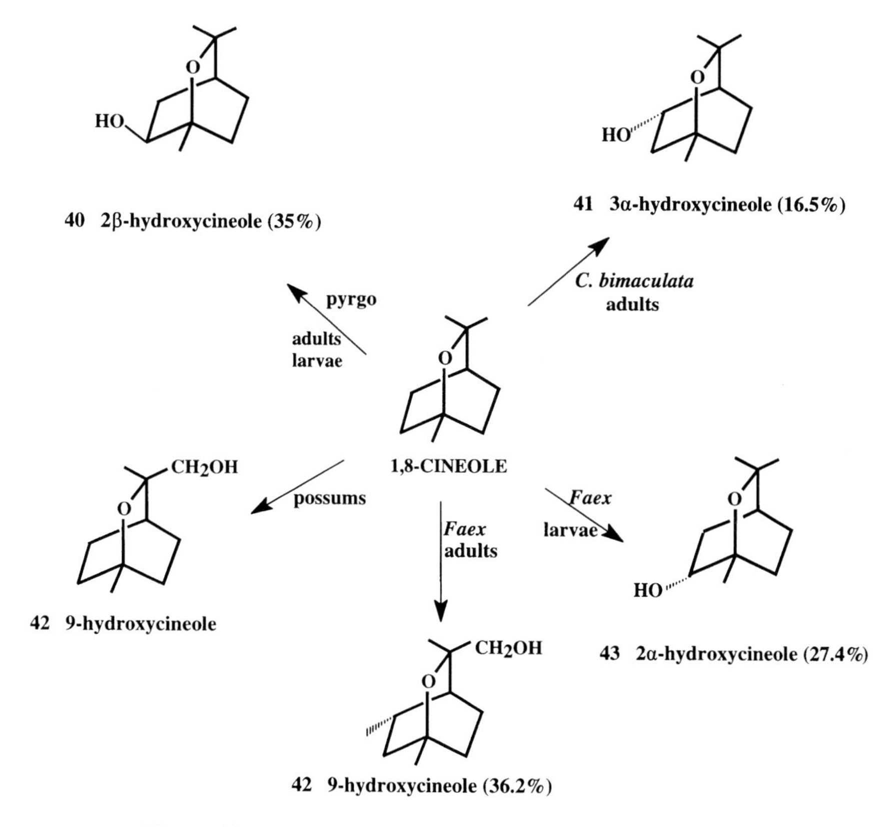

Figure 13 The chemical structures of metabolites of 1,8-cineole

Table 10 The proportions (%) of 2α (43), 2β (40), 3α (41) and 9 (42) hydroxycineoles detected in the frass volatiles of paropsine beetles feeding on high cineole *Melaleuca* leaf

	Paropsisterna tigrina adults	*P. tigrina* larvae	*Faeax nigro-conspersa* adults	*F. nigro-conspersa* larvae	*Chrysophtharta bimaculata* adults
2α-hydroxycineole (43)	1.7	2.0	11.4	27.4	0.6
2β-hydroxycineole (40)	38.7	39.8	0.5	0.5	0.8
3α-hydroxycineole(?) (41)	1.6	1.9	6.7	4.3	16.5
9-hydroxycineole (42)	1.7	0.7	36.2	5.2	3.4

OTHER USES OF TEA TREE

General Uses

Most of the other additional uses of *Melaleucas* are based on the physical or aesthetic rather than the chemical properties of the wood (used for construction), branches (for broom fences), bark (for bark paintings, art) or entire tree (in windbreak, landscaping, swamp reclamation) (Wrigley and Fagg 1993).

Proline Analogues

The discovery of substantial quantities of proline analogues (Figure 7) in the leaves of some *Melaleuca* species highlights a potentially commercial source of industrial chemical compounds. Coupled with the recent interest in biochemical changes in plants suffering water and salinity stress, is the ability of drought resistant species to accumulate proline and proline analogues including glycinebetaine. Investigations have shown that, when subject to water or salinity stress under glass house or laboratory conditions, proline (27) and *trans*-4-hydroxy-N-methyl-L-proline (29) in *M. lanceolata* and *trans*-4-hydroxy-N-methyl-L-proline (29) and N,N'-dimethyl-*trans*-4-hydroxy-L-proline (30) levels in *M. uncinata* increased (Naidu *et al.* 1987).

Further investigations are now translating these laboratory findings into field situations as trial plantings of *M. bracteata, M. uncinata, M. lanceolata, M. populiflora* and *M. viridiflora* are being established in eastern and north eastern Australia (Naidu 1997). The best species for the commercial production of proline analogues will be chosen for the production of natural agricultural chemicals which will alleviate stress and increase biomass yields in pastures and crops.

REFERENCES

Association Française de Normalisation (1996) Huile essentielle de *Melaleuca* type terpinène-4-ol. *Huiles essentielles.* Tome 2 (5e édition), AFNOR, Paris La Défense, pp. 529–536.

Baker, R.T. and Smith, H.G. (1906) Essential oil of *Melaleuca thymifolia. J. Proc. Roy. Soc. NSW*, **40**, 60.

Baker, R.T. and Smith, H.G. (1907) Essential oil of *Melaleuca uncinata* and *M. nodosa. J. Proc. Roy. Soc. NSW*. **41**, 196.

Baker, R.T. and Smith, H.G. (1910) Essential oil of *Melaleuca genistifolia* Sm. (*M. bracteata*). *J. Proc. Roy. Soc. NSW*, **44**, 592.

Baker, R.T. and Smith, H.G. (1911) Essential oil of *Melaleuca genistifolia*, *M. gibbosa* and *M. pauciflora*. *J. Proc. Roy. Soc. NSW*, **45**, 365.

Baker, R.T. and Smith, H.G. (1913) Essential oil of *Melaleuca leucadendron*. *J. Proc. Roy. Soc. NSW*, **47**, 196.

Birch, A.J., Grimshaw, J., Speake, R.N., Gascoigne, R.M. and Hellyer, R.O. (1959) *Tetrahedron Lett.*, **3**, 15–18.

Bishop, C.D. (1995) Antiviral Activity of the Essential Oil of *Melaleuca alternifolia* (Maiden & Betche) Cheel (Tea Tree) Against Tobacco Mosaic Virus. *J. Essent. Oil Res.*, **7**, 641–644.

Bishop, C.D. and Thornton, I.B. (1997) Evaluation of the Antifungal Activity of the Essential Oils of *Monarda citriodora var. citriodora* and *Melaleuca alternifolia* on Post-Harvest Pathogens. *J. Essent. Oil Res.*, **9**, 77–82.

British Pharmaceutical Codex (1949) Oleum Melaleucae, General Medical Council, London, pp. 597–598.

Brophy, J.J., Boland, D.J. and Lassak, E.V. (1989a) Leaf Essential Oils of *Melaleuca* and *Leptospermum* species from tropical Australia. In D.J. Boland (ed.). *Trees for the Tropics*, ACIAR, Canberra, pp. 193–203.

Brophy, J.J., Davies, N.W., Southwell, I.A., Stiff, I.A. and Williams, L.R. (1989b) Gas Chromatographic Quality Control for Oil of *Melaleuca* Terpinen-4-ol Type (Australian Tea Tree). *J. Agric. Food Chem.* **37**, 1330–1335.

Brophy, J.J. and Doran, J.C. (1996) Essential Oils of Tropical *Asteromyrtus*, *Callistemon* and *Melaleuca* Species. ACIAR, Canberra pp. 76–77.

Brophy, J.J., Lassak, E.V. and Boland, D.J. (1990) Steam volatile leaf oil of *Melaleuca globifera* R. Br., *M. lateriflora* Benth., *M. symphyocarpa* F. Muell. and *M. uncinata* R. *Flav. Frag. J.*, **5**, 43–48.

Brophy, J.J. and Lassak, E.V. (1992) Steam volatile leaf oils of some *Melaleuca* species from Western Australia. *Flav. Frag. J.* **7**, 27–31.

Buchi, G., Hofheinz, W. and Paukstelis, J.V. (1969) The synthesis of (−)-aromadendrene and related sesquiterpenes. *J. Am. Chem. Soc.*, **91**, 6473–6478.

Butcher, P.A., Bell, J.C. and Moran, G.F. (1992) Patterns of genetic diversity and nature of the breeding system in *Melaleuca alternifolia* (Myrtaceae). *Austral. J. Bot.*, **40**, 365–375.

Butcher, P.A., Doran, J.C. and Slee, M.U. (1994) Intraspecific variation in leaf oils of *Melaleuca alternifolia* (Myrtaceae). *Biochemical Systematics and Ecology*, **42**, 419–430.

Cheel, E. (1924) Notes on *Melaleuca*, with descriptions of two new species and a new variety. *J. Proc. Roy. Soc. NSW*, **58**, 89–197.

Cocking, T.A. (1920) New method for the estimation of cineole in *Eucalyptus* oil. *Pharm. J.*, **105**, 81–83.

Cornwell, C.P., Leach, D.N. and Wyllie, S.G. (1995) Incorporation of oxygen-18 into terpinen-4-ol from the $H_2{}^{18}O$ steam distillates of *Melaleuca alternifolia* (Tea Tree). *J. Essent. Oil Res.*, **7**, 613–620.

Courtney, J.L., Lassak, E.V. and Speirs, G.B. (1983) Leaf wax constituents of some *Myrtaceous species*. *Phytochemistry*, **22**, 947–949.

Croteau, R. (1987) Biosynthesis and catabolism of monoterpenoids. *Chem. Rev.*, **87**, 929–954.

Davenport, J.B., Jones, T.G.H. and Sutherland, M.D. (1949) The essential oils of the Queensland flora. Part XXIII. A re-examination of the essential oil of *Melaleuca linariifolia*. *Univ. Queensl. Pap., Dep. Chem.*, **1**(36), 1–12.

Deutscher Arzneimittel-Codex 1986 (1996) Teebaumol, Melaleucae aetheroleum, 8, Erganzung, T-019, 1-3, Govi-Verlag Pharmazeutischer Verlag GMBH Frankfurt Am Main/Eschborn.

Dung, N.X., Bao, P.Q. and Leclercq, P.A. (1994) Chemical composition of the leaf oil of *Melaleuca parviflora* Link from Vietnam. *J. Essent. Oil Res.* **6**, 419–420.

Dung, N.X. (1996) Center for Education and Development of Chromatography (EDC) of Vietnam. Personal communication.

Eisenreich, W., Sagner, S., Zenk, M.H. and Bacher, A. (1997) Monoterpenoid essential oils are not of mevalonoid origin. *Tetrahedron Lett.*, **38**, 3889–3892.

Erman, W.F. (1985) Chemistry of the Monoterpenes, Marcel Dekker, New York, pp. 826–827.

Faure, R., Ramanoelina, A.R.P., Rakotoniraing, O., Bianchini, J.P. and Gaydou, E.M. (1991) Two-dimensional nuclear magnetic resonance of sesquiterpenes. 4. Application to complete assignment of ^{1}H and ^{13}C NMR spectra of some aromadendrene derivatives. *Magn. Reson. Chem.*, **29**, 969–971.

Fischer, N., Nitz, S. and Drawert, F. (1987) Original flavour compounds and the essential oil composition of majoram (*Majorana hortensis* Monech). *Flav. Frag. J.* **2**, 55–61.

Fischer, N., Nitz, S. and Drawert, F. (1988) Original composition of majoram flavour and its changes during processing. *J. Agric. Food Chem.*, **36**, 996–1003.

Formacek, V. and Kubeczka, K.H. (1982) *Essential Oils Analysis by Capillary Gas Chromatography and Carbon-13 NMR Spectroscopy*, Wiley, Chichester.

Ghisalberti, E.L., Patalinghug, W.C., Skelton, B.W. and White, A.H. (1994) Structural studies of aromadendrane-1,2-diols. *Aust. J. Chem.*, **47**, 943–949.

Guenther, E. (1948) *The Essential Oils*, Van Nostrand, New York.

Guenther, E. (1968) Australian tea tree oils. Report of a field survey. *Perfum. Essent. Oil Rec.*, **59**, 642–44.

Hallahan, T.W. and Croteau, R. (1988) Monoterpene biosynthesis: Demonstration of a geranyl pyrophosphate:sabinene hydrate cyclase in soluble enzyme preparations from sweet marjoram (*Majorana hortensis*). *Arch. Biochem. Biophys.*, **264**, 618–631.

Hallahan, T.W. and Croteau, R. (1989) Monoterpene biosynthesis: Mechanism and stereochemistry of the enzymic cyclisation of geranyl pyrophosphosate to (+)-*cis*- and (+)-*trans*-sabinene hydrate. *Arch. Biochem. Biophys.*, **269**, 313–326.

International Standards Organisation (1996) Oil of *Melaleuca*, Terpinen-4-ol Type, 4730, ISO, Geneva.

Johns, M.R., Johns, J.E. and Rudolph, V. (1992) Steam distillation of tea tree (*Melaleuca alternifolia*) oil. *J. Sci. Food Agric.* **58**, 49–53.

Jones, G.P., Naidu, B.P., Paleg, L.G., Tiekink, E.R.T. and Snow, M.R. (1987) 4-Hydroxy-N methylproline analogues in *Melaleuca spp. Phytochem.*, **26**, 3343–3344.

Jones, T.G.H. (1936) Essential oils from the Queensland flora. Part X. *Melaleuca linariifolia. Proc. Roy. Soc. Qld.*, **48**, 48–50.

Jones, T.G.H. and Oakes, H.C. (1940) The crystalline solid formed in the oil of *Melaleuca linariifolia. Univ. Queensl. Pap., Dep. Chem.*, **1**(18), 1–3.

Kawakami, M., Sachs, R.M. and Shibamoto, T. (1990) Volatile constituents of essential oils obtained from newly developed tea tree (*Melaleuca alternifolia*) clones. *J. Agric. Food Chem.*, **38**, 1657–1661.

Kerrola, K. (1995) Literature review: Isolation of essential oils and flavour compounds by dense carbon dioxide. *Food Rev. Int.*, **11**, 547–573.

Koedam, A. (1987) Some aspects of essential oil preparation. In P. Sandra and C. Bicchi, (eds) *Capillary Gas Chromatography in Essential Oil Analysis*, Huethig, Heidelberg, pp. 13–27.

Laakso, P.V. (1966) Fractionation of tea tree oil (*Melaleuca alternifolia*). In Hanc, O. and Hubik, J. (eds) Scientiae Pharmaceuticae *Proceedings of the 25th Congress of Pharmaceutical Sciences*, Prague, 1965. Czechoslovakian Medical Press: Prague, 1, pp. 485–492.

Lassak, E.V. and McCarthy, T. (1983) Australian Medicinal Plants, Methuen, Sydney, p. 97.

Leach, D.N., Wyllie, S.G., Hall, J.G. and Kyratzis, I. (1993) The enantiomeric composition of the principal components of the oil of *Melaleuca alternifolia. J. Agric. Food Chem.*, **41**, 1627–1632.

Lichtenthaler, H.K., Schwender, J., Disch, A. and Rohmer, M. (1997) Biosynthesis of isoprenoids in higher plant chloroplasts proceeds via a mevalonate independent pathway. *FEBS Lett.*, **400**, 271–274.

Liener, I.E. (1996) Editorial. *J. Agric. Food Chem.*, **44**, 1.

Markham, J. (1996) Antimicrobial effectiveness of tea tree oil. In Conference Program and Papers, *Tea Tree Oil into the 21st Century—The Proof and the Promise.* Austteam. Lismore, pp. 30–37.

Naidu, B.P. (1997) Commonwealth Scientific and Industrial Research Organisation. Personal communication.

Naidu, B.P., Jones, G.P., Paleg, L.G. and Poljakoff-Mayber, A. (1987) Proline analogues in *Melaleuca* species: Response of *Melaleuca lanceolata* and *M. uncinata* to water stress and salinity. *Aust. J. Plant Physiol.*, **14**, 669–677.

Naves, Y.R. and Tullen, P. (1960) Die bisher beobachtete Hochstdrehung von terpinenol-4 betragt 48.3. *Bull. Soc. Chim. France*, 2123.

Ohloff, von G. and Uhde, G. (1965) Abosolute Konfiguration von Terpinenol-(4), *Helv. Chim. Acta*, **48**, 10–28.

Penfold, A.R. (1925) The Essential Oils of *Melaleuca linariifolia* (Smith) and *M. alternifolia* (Cheel). *J. Proc. Roy. Soc. NSW*, **59**, 306–325.

Penfold, A.R., Morrison, F.R. and McKern, H.H.G. (1948) Studies in the physiological forms of the Myrtaceae, Part II. The occurrence of physiological forms in *Melaleuca alternifolia* Cheel. *Researches on the Essential Oils of the Australian Flora*; Museum of Technology and Applied Science: Sydney **1**, 18–19.

Russell, M., Southwell, I.A. and Stiff, I.A. (1997) NSW Agriculture. Unpublished results.

Southwell, I.A. (1973) Variation in the leaf oil of *Eucalyptus punctata. Phytochem.*, **12**, 1341–1343.

Southwell, I.A. (1988) Australian tea tree: Oil of *Melaleuca*, terpinen-4-ol type. *Chem. in Aust.*, **55**, 400–402.

Southwell, I.A. (1989) Terpenoid pathways in Australian tea trees. In S.C. Bhattacharyya, N. Sen and K.L. Sethi (eds) 11th International Congress of Essential Oils Fragrances and Flavours. New Delhi, India, 12–16 November, 1989. Oxford. New Delhi.

Southwell, I.A. and Stiff, I.A. (1989) Ontogenetical changes in monoterpenoids of *Melaleuca alternifolia* leaf. *Phytochemistry*, **28**, 1047–1051.

Southwell, I.A. and Stiff, I.A. (1990) Differentiation between *Melaleuca alternifolia* and *M. linariifolia* by monoterpenoid comparison. *Phytochemistry*, **29**, 3529–3533.

Southwell, I.A., Freeman, S. and Rubel, D. (1997) Skin irritancy of tea tree oil. *J. Essent. Oil Res.*, **9**, 47–52.

Southwell, I.A. and Tucker, D.J. (1991) NSW Agriculture, University of New England. Unpublished results.

Southwell, I.A., Stiff, I.A. and Brophy, J.J. (1992) Terpinolene varieties of *Melaleuca. J. Essent. Oil Res.*, **4**, 363–367.

Southwell, I.A., Maddox, C.D.A. and Zalucki, M.P. (1995) Metabolism of 1,8-cineole in tea tree (*Melaleuca alternifolia* and *M. linariifolia*) by pyrgo beetle (*Paropsisterna tigrina*) *J. Chem. Ecol.*, **21**, 439–453.

Standards Association of Australia. (1967) *Oil of Melaleuca alternifolia*; Standards Australia: Sydney, K175.

Standards Association of Australia. (1985) Essential Oils—*Oil of Melaleuca, Terpinen-4-ol Type*; Standards Association of Australia: North Sydney, AS 2782-1985.

Standards Australia. (1997) *Oil of Melaleuca, terpinen-4-ol type (Tea Tree oil)* Standards Australia: Sydney, AS 2782-1997.

Stevens, R. (1996) Editorial. *Flav. Frag. J.*, **11**, 1.

Stiff, I.A. (1996) Ph.D. Thesis. Macquarie University, Sydney.

Swords, G. and Hunter, G.L.K. (1978) Composition of Australian Tea Tree Oil (*Melaleuca alternifolia*). *J. Agric. Food Chem.*, **26**, 734–737.

Taskinen, J. (1974) Composition of the essential oil of sweet marjoram obtained by distillation with steam and by extraction and distillation with alcohol-water mixture. *Acta Chem. Scand.*, **B28**, 1121.

Tisserand, R. and Balacs, T. (1995) *Essential Oil Safety*, Churchill Livingston, Edinburgh, p. 186.

Verghese, J. (1966) Terpinen-4-ol, *Perfumery and Essential Oil Record*, 694–498.

Williams, L.R. and Lusunzi, I. (1994) Essential oil from *Melaleuca dissitiflora*: a potential source of high quality tea tree oil. *Indust. Crops Prod.*, **2**, 211–217.

Wong, V. (1997) Supercritical Fluid Extraction (SFE) of Monoterpenes from Australian Tea Tree Leaves (*Melaleuca alternifolia*). B. Appl. Sci. (Hons) Thesis. University of Western Sydney, Hawkesbury, Richmond.

Wrigley, J.W. and Fagg, M. (1993) Bottlebrushes, Paperbarks and Tea Trees and all other Plants in the *Leptospermum* Alliance. Angus and Robertson, Sydney pp. 5–17.

3. CULTIVATION OF TEA TREE

ROBERT T. COLTON AND G. JOHN MURTAGH

NSW Agriculture, Orange, Australia;
Agricultural Water Management, Goonellabah, Australia

INTRODUCTION

In its natural habitat, the tea tree plant grows as a small 5–8 metre tree. In commercial plantations however, it is grown as an intensive row crop. It is planted at high densities to maximise leaf and oil production per hectare and is harvested every six to eighteen months by cutting the whole plant close to ground level. Plantations are currently established using wild seed, with some screening for chemotype, but there is considerable plant to plant variation. The variation will be less with the new varieties under development as discussed in Chapter 7.

SITE REQUIREMENTS

Tea trees are hardy perennial plants, adapted to a wide range of soil types. They are drought tolerant and can survive flooding and fire (Murtagh 1996a). However, while the trees are very adaptable, the production of consistent high yields of oil requires fairly specific climate and soil conditions.

Tea tree grows naturally on damp soil within humid, sub-tropical areas of the north coast of New South Wales (NSW) (Anderson 1956). Sites which mimic these conditions have proven to be highly productive for commercial plantations as have warmer districts in Queensland. However, other coastal areas with poorer soils and lower rainfall supplemented with irrigation are also suitable sites for productive plantations.

CLIMATE

As a native of wet sub-tropical areas, tea tree responds to warm temperatures and a continuous supply of moisture. In the northern rivers area of NSW where most tea tree oil is produced, annual average rainfall generally ranges from 1000–1600 mm. Rainfall has a strong summer incidence with the driest conditions occurring during July–September (Bureau of Meteorology 1988).

Although tea tree needs ample moisture to grow, it can survive in very dry conditions. In extreme drought, it will shed its leaves and will re-shoot after rain (Small 1981a).

High relative humidity appears to be important in achieving high oil yields, but the effect has not been fully quantified. In a glasshouse study, Murtagh and Lowe (1997) found that oil concentration was significantly greater when relative humidity was high (81% min) than when it was low (40%). Oil concentration in the leaf is known to fluctuate seasonally as well as from day to day (Murtagh and Etherington 1990). The reasons for this are complex and are not fully understood but relative humidity levels seem to be a factor in the generally lower levels of oil in the leaf during the dry winter-spring period. (Drinnan 1997b).

The peak temperature for tea tree growth exceeds 30°C (Curtis 1996). In this controlled environment experiment growth increased with increasing temperature up to the highest level tested of 35°C day/30°C night temperature. This represents an unusual tolerance to high temperatures. In north Queensland, annual growth rates in plantations are about 50% higher than in northern NSW (Drinnan 1997b). In the northern rivers area, winter temperatures restrict growth so that crops are dormant or near dormant for 3–4 months. Sustained growth starts in September when mean air temperature reaches 17°C and finishes about the end of May (15.5°C) (Murtagh 1996b). Allowing for differences between soil and air temperatures in spring and autumn, 16°C is considered the lower limit for sustained growth.

Temperature affects tree growth rates and it also appears to affect the concentration of oil in the leaf. In NSW, oil concentration typically increases from 40–50 mg/g of dry leaf in September to 60–90 mg/g in January. It begins to decline during March and falls most rapidly during winter. The lower September figure occurs further away from the coast where frosts are common (this volume, Chapter 6). However, warmer night temperatures such as those experienced in the tropics do not adversely affect oil concentration as was once feared (Southwell and Wilson 1993).

Frost will severely damage young growth but the height of older trees offers a degree of protection and damage is normally minimal (Colton and Murtagh 1990).

Plantations are currently located along the sub-tropical and tropical coastal strip from Port Macquarie to far north Queensland. This area provides a long warm to hot growing season, moderate to high rainfall sometimes supplemented with irrigation and high to very high humidity. In this environment, well managed plantations produce average oil yields of 150–250 kg/ha/yr with the higher yields more common in the more tropical areas.

Experience with tea tree in cooler, drier inland areas is very limited. Trees will undoubtedly grow and produce oil in many areas but yields will be lower and the economics more questionable.

Trees grow best in sheltered areas where they are protected from the drying effects of wind.

NATURAL HAZARDS

Several studies have shown that the genus *Melaleuca* is very tolerant of flooding and waterlogging (Barlow 1988; van der Mozel *et al.* 1991). Established *M. alternifolia* plants can survive total immersion for short periods, up to seven days, without plant loss.

Partial immersion for periods of 1–3 months will not kill established plants but will reduce their rate of growth (Colton and Murtagh 1990).

Gomes and Kozlowski (1980) found that seedlings of *M. quinquenervia* were very tolerant of flooding for up to 90 days. Experience with *M. alternifolia* is that seedlings will not survive flooding when they are completely immersed.

Tea tree can tolerate and recover from bushfires with new shoots emerging from charred stems, similar to the way eucalypts recover after fire.

SOILS

In its natural environment, tea tree is found on alluvial soils in swampy areas and along drainage lines of the coastal river flood plains, but it will grow well on a range of soils provided there is adequate moisture (Weiss 1997). However, it is less well suited to shallow light sandy soils, heavy clay soils and acid sulfate soils.

Trees grow best on deep sandy loams or friable loams which retain deeper moisture over the dry season or have irrigation available. Light, sandy soils, having little organic matter, do not retain their moisture or nutrients and require regular irrigation over the dry season to maintain growth. Lighter soils are much less risky if they have a permanent watertable within reach of the tree roots to provide a buffer against dry spells. These sites are more likely to be found on a valley floor than on ridges and hillsides.

Heavy clay soils will grow tea tree, but growth is often impeded by the high bulk density and lack of oxygen in these soils. Adequate surface drainage to allow water to get away after heavy rain or flooding is essential on heavy textured soils and those in low lying situations. Poor drainage can limit access to plantations during the wet season, preventing timely harvesting and management operations. Soil compaction is also a common problem following the use of harvesting and other heavy equipment during wet periods.

Good moisture holding capacity is an important consideration where production relies on rainfall and some compromise is often necessary between the heavier, moisture retentive soils and the lighter, high growth soils which may tend to dry out.

Acid sulfate soils are found in the wetlands of many of the coastal rivers. These soils contain iron pyrite or iron sulphide. While they are covered by water and thus insulated from the atmosphere, this pyritic layer is innocuous. When the land is drained and oxygen reaches the iron sulphide, it is oxidised to become sulphuric acid (Creagh 1993). The combination of sulphuric acid and aluminium which is also released is very deadly to both vegetation and most soil and water life forms. It is better to avoid shallow acid sulfate soils than to consider managing and treating them.

Tea tree grows naturally on acid soils pH 4.5–5.5 (measured in water) but also grows well on soils with neutral reaction (Weiss 1997). It will not thrive on alkaline soils. It is intolerant of saline soils, particularly if they are also prone to waterlogging (van der Moezel *et al.* 1991).

Plants grown on fertile soils will carry a much heavier canopy than those where soil fertility is low (Colton and Murtagh 1990). Fertiliser can substitute for low fertility on some soil types (Drinnan 1997b).

CHARACTERISTICS OF MAJOR GROWING AREAS

The tropical and sub-tropical east coast of Australia experiences high rainfall and humidity in summer followed by dry winters and springs. Towards the north, temperatures are higher, particularly in winter, and rainfall becomes increasingly summer dominant. There is a much more pronounced dry season in winter and spring. These differences can be seen in Tables 1–3.

Rainfall in the northern rivers area of NSW is significantly higher than in the Mareeba-Dimbulah area of north Queensland, and is much less summer dominant (Table 1).

Daily maximum temperatures for the northern rivers are similar to those at Mareeba in January, but January minima and winter maxima and minima are much lower at the southern centres (Table 2).

Table 1 Rainfall incidence and seasonal distribution

	December–May	*June–November*	*Year*
	mm	mm	mm
Mareeba	825	86	914
Dimbulah	632	96	718
Broadwater	947	534	1481
Casino	731	377	1108
Grafton	691	383	1074

Table 2 Mid summer and mid winter temperatures

	January		*July*	
	Mean daily maximum (°C)	*Mean daily minimum* (°C)	*Mean daily maximum* (°C)	*Mean daily minimum* (°C)
Mareeba	31.2	20.8	25.3	11.2
Broadwater	28.2	19.5	19.2	7.4
Casino	31.4	18.7	21.3	6.3
Bellingen	29.5	17.9	19.9	4.6

Table 3 Relative humidity

	February		*October*	
	9.00 am %	3.00 pm %	9.00 am %	3.00 pm %
Mareeba	79	66	58	44
Casino	81	58	68	50
Grafton	79	61	67	51

February relative humidity levels are similar in the northern rivers and at Mareeba, but October levels tend to be lower in Mareeba. Humidity is highest in February–March and lowest in the September to November period in both areas (Table 3).

The Mareeba-Dimbulah area is on the edge of the Atherton Tableland, west of Cairns at an elevation of 400–500 metres. Being west of the coastal ranges, rainfall is considerably lower than on the coast and falls away sharply from east to west. Although the area is located in the tropics, its altitude can lead to light frosts, with an occasional heavy frost. Most tea tree is grown on light, sandy, tobacco-growing soils which need regular fertiliser applications and irrigation. These soils were selected because tea tree is grown as a substitute for tobacco, and do not necessarily represent the best soil type for tea tree.

In the northern rivers area, tea tree is grown largely within its natural habitat on the coastal river flood plains. Soils are more likely to be medium to heavy textured and summer waterlogging often occurs. Light frosts occur in most winters.

PLANT PROPAGATION AND ESTABLISHMENT

Nursery Seedling Production

Since tea tree seed is particularly small and seedlings have a poor competitive ability, plantations are established from seedlings raised in nurseries or seed beds (Plate 6) (Murtagh 1991a). Seed (Plate 14) is collected from high yielding individual trees (Plate 2, 13) (List *et al.* 1996) which have been screened for chemotype and are known to yield well. Seed with higher yield and oil concentrations is expected to be available from the tea tree breeding program from 1998.

A single productive tree will yield 50–400 grams of seed and a gram of seed will usually contain 8000–12000 viable seeds (Colton and Murtagh 1990). Tea trees flower (Plate 3) in spring and seed is normally mature after fifteen to eighteen months. After maturity, seed remains viable in the capsule and can be collected at any time of the year. If kept in a warm dry place after collection, capsules open and release the seed.

Small plantations (a few hectares) are usually established with seedlings purchased from a commercial nursery. Larger plantations find it more economical and convenient to set up their own nursery and produce seedlings to suit their own planting schedule.

Seedlings can be grown either in cell type containers or as open rooted seedlings. Containerised seedlings are easier to grow, are well adapted to mechanised transplanting and suffer less transplanting shock than open rooted seedlings. They can be planted over most of the year and will suffer fewer losses even when weather conditions are less than ideal provided irrigation is available. Open rooted seedlings are much cheaper to produce and can be quite reliable for those with appropriate skills and experience.

Seedlings usually spend 6–8 weeks in a plastic tunnel and a further 6–8 weeks hardening off under shade cloth. They are then usually 15 cm high and ready to plant out.

Vegetative Propagation

Clonal propagation using cuttings or tissue culture is rare in plantation establishment. It has not been cost competitive with large scale rapid seedling production systems from seed. Its big advantage of providing a uniform stand of plants will become more important once elite, high yielding, high quality selections begin to be released from the breeding program. A successful micro propagation protocol has been developed using coppice shoots (List *et al.* 1996).

Plant Population and Row Spacing

Leaf yield is strongly influenced by plant density, with the highest yield per hectare being obtained at the densest planting (Small 1981b). Small's highest density was 23,000 trees/ha but subsequent work showed that the trend to higher oil yields continued to at least 36,000 trees/ha (Peak 1982). Industry experience in both NSW and Queensland supports a population of about 35,000 as being optimum. These high plant populations achieve full ground cover more quickly after harvest and compete better with weeds by shading.

Row spacing and planting layout are largely determined by the size of the tractors, mowers, cultivators, sprayers and harvesters likely to be used in the plantation. Single rows at 75–100 cm intervals or twin rows on 1.5 m beds, giving elevated rows with an effective 75 cm spacing, are common and can accommodate most machinery. In-row plant spacings to achieve 35,000 plants/ha for these row spacings are: 75 cm rows–38 cm apart, 100 cm rows–28 cm apart, 1.5 m beds–38 cm apart.

Planting

Cell grown seedlings can be transplanted at almost any time of the year provided soil moisture is adequate or irrigation is available. In practice, spring is preferable as it avoids the cold weather and frost risk of winter and the very wet soils and possible heat waves of summer and autumn.

Because spring is normally very dry, irrigation is essential to ensure good establishment. This is particularly so with open rooted seedlings as the risk of plant losses is higher, especially if periods of drying winds and high temperatures occur. (Colton and Murtagh 1990).

Seedlings are planted with a normal seedling transplanting machine (Plate 7) which can place a band of fertiliser under the plants and can apply a quantity of water to each plant. Pre-emergence herbicide is usually applied immediately after transplanting followed by irrigation (Plate 8). Seedling establishment is enhanced if the soil is kept moist for the first 4–6 weeks or until rain falls.

Field Preparation

The amount and type of land preparation needed will depend on previous land use, soil type and topography. It may involve drainage, levelling and the break up of compacted layers in addition to normal weed control and topsoil preparation for transplanting (Colton and Murtagh 1990).

A good surface drainage system is important, particularly on low lying heavy soils to allow timely access to fields for harvesting and other operations following periods of heavy rainfall or flooding. Levelling land to a uniform grade will facilitate drainage. It is also essential if flood irrigation is being considered. A system of drains big enough to remove the excess water from a 200 mm rainfall event within two days is warranted, in higher rainfall coastal areas.

Compacted layers are quite common, particularly on heavier soils which have been cropped or have had heavy machinery over them while they were wet. These hard layers inhibit root development and restrict moisture infiltration and will reduce the growth rates of tea tree. Deep ripping to the bottom of the compacted zone will overcome the problem, but it can re-occur quickly if heavy equipment is used when the soil is saturated.

On sites where major drainage work and levelling is required, preparation should begin up to one year before the anticipated transplanting time. Where less preparation is necessary, the lead time can be much shorter but a rushed preparation can be counter productive.

When fields are ready to plant, the top 10 cm of the soil should be well prepared, moist, free of clods and trash and weed free. In this condition, pre-emergence herbicides can work effectively and seedlings will get away to a good start. Tea tree seedlings do not compete well with weeds and rely on pre-emergence herbicides or other forms of weed control for 2–3 months after planting.

IRRIGATION

Irrigation is normally necessary in the period after planting to ensure that plants establish quickly and to minimise losses, particularly if soil and weather are less than ideal. Open rooted seedlings are unlikely to become established unless they are irrigated regularly for the first few weeks or there is regular rainfall.

Once plants are well established, the need for irrigation will depend on location, soil type, depth to watertable, seasonal rainfall incidence and the frequency of below average rainfall years.

Tea tree is a drought resistant plant and can survive in very dry conditions (Murtagh 1996a). Day to day growth is fed from its root mass which is found in the top 40–50 cm of the soil but survival depends on the deep sinker roots when the top layers of the soil dry out (Drinnan 1997b). Where the sinker roots reach ground water and the average rainfall exceeds 1,000 mm/year irrigation may not be economic.

Moisture stress causes cessation of growth and if severe and prolonged, it will cause leaves to drop. This represents a loss of oil and a loss of income.

Drinnan (1997b) showed that in the light sandy soils at Mareeba, water stress for periods of more than 2–4 weeks reduced oil concentration. Concentrations were highest when soils were continuously moist but not too wet. Soils which were heavier, higher in organic matter or mulched stayed moist and kept oil concentrations higher. Frequent irrigation achieved the same results. Murtagh (1991b) found that on the heavier, moisture retentive soils in the Lismore-Casino area irrigation had no effect on oil concentration.

Drinnan (1997b) found that in North Queensland trees were using 5–9 mm of water per day, which is about 0.8–1.0 times the pan evaporation rate, in the period leading up to harvest. In that environment annual moisture requirement was about 1,400 mm per year, which equates to 14 ML/ha. Allowing for adequate rainfall for three months actual irrigation requirement is about 10.5 ML/ha.

In the light sandy soils of the Mareeba-Dimbulah area irrigation is essential during most of the year. Evapo-transpiration on fully grown trees ranges from about 4 mm/day in June–July to 8–9 mm/day in October–November. Peak demand in spring requires about 60 mm of water per week, normally applied by sprinkler in 3–4 applications. When trees are half grown (12 weeks after harvest) water requirements are about half those of fully grown trees.

In northern NSW, where tea tree is grown along the high rainfall coast on moisture retentive soils the need for irrigation is more tactical and its economics need to be considered on a case by case basis.

In spite of the relatively favourable water relations on many plantations evapo-transpiration typically exceeds median monthly rainfall in 5–12 months of the year, and trees must survive or grow on deep subsoil moisture if it is available.

Murtagh (1991b), in a three year study, identified three growth phases in tea tree during the dry spring period, each with a different response to irrigation. The trees had access to deep subsoil moisture. Early in the growing season (August–September), growth was slow and the absolute response to irrigation was small. Cool temperatures appeared to restrict growth and to reduce the availability of water by increasing the internal resistance of plants to water uptake.

A strong flush of growth commenced about mid-October and over the next month growth was rapid, with little regard to the total water potential in the shoots or in soil water content. Although an irrigation response was obtained during this phase, it was reduced by the relatively small response in shoot water potential to irrigation. During the remainder of the dry season however growth rates were strongly related to the water potential in flush shoots and there was a strong response to irrigation during December to February, when air humidity was high.

The spring flush of growth occurs during the normal dry period and this coincidence is the most important factor which reduces the need for irrigation under NSW north coast conditions. Response to irrigation will be site and soil type specific with drier sites being more responsive. Responses are unlikely where subsoil is kept moist by groundwater.

The economics must be considered carefully as irrigation infrastructure and operating costs are expensive. For example, Reilly (1988) estimated the cost of installing spray irrigation would add 25% to the capital cost of plantation establishment, excluding land.

NUTRITION

There has been very little research into the nutritional requirements of plantation tea tree and there is no published information on the subject.

This lack of information is not surprising considering that plantations were first established only 15 years ago and most are less than 10 years old. Many on the NSW north coast are planted on alluvial soils which are reasonably fertile. Early trial work by growers and research workers found tea tree to be generally unresponsive to added fertiliser on these soils. A typical result was achieved on two soils in the Casino district in 1994. On a flood plain alluvial and a peat soil, Clarke (1995) found no response in either biomass production or oil concentration when rates of up to 150 kg/ha of nitrogen, with generous basal dressings of phosphorus and potassium, was applied to young trees.

Australian natives are regarded as being generally unresponsive to fertilisers. Weiss (1997) reported the application of fertilisers to established eucalypts to be unprofitable and indicated that the few results from fertiliser trials had shown no significant increase in either oil or foliage yield. Diatloff (1990) found that three species of *Leptospermum* were not responsive to nitrogen fertilisation in either plant growth or oil concentration.

However, there is plenty of evidence of eucalypts responding to applied nutrients, particularly when grown on lower fertility soils and where there was adequate soil moisture (Cromer 1996).

As NSW tea tree plantations age, particularly on the less fertile soils, tree growth is slowing and fertiliser responses are becoming more common. The infertile coastal podzols (red and yellow podzolics) and the peat soils are among the first to respond. These soils have very low cation exchange capacities, very low phosphorus levels and often potassium levels are not adequate for high yielding field crops. Peat soils have excellent physical properties but contain very low levels of plant nutrients, especially potassium, and are commonly low in copper. (Clarke 1997).

Many of these soils also suffer from nutrient imbalances as a result of low pH, low calcium or salinity problems. Using soil tests to identify these problems and correcting them with lime, dolomite or gypsum is proving to be a critical first step in developing a good nutrition program.

On these soils, trees are responding to annual rates per hectare of nitrogen 70–100 kg, phosphorus 15–20 kg, and potassium 35–45 kg as well as a range of trace elements. On the relatively fertile river and floodplain alluvials, where trace elements are not normally deficient, growers are starting to apply NPK fertilisers at moderate rates of: N 60–80, P 10, K 20 kg/ha, to maintain soil fertility (Clarke 1997). Application occurs after harvest and in several other split dressings during the main growth period of late October to March. Industry experience is that tea trees respond much better to small amounts of fertiliser applied regularly. Some growers are reporting growth responses from regular applications of commercial foliar nutrient preparations.

It is desirable that nitrogen be in a slowly available form if possible, so that it is less prone to leaching and denitrification losses during the wet season. Nutrient fortified, composted spent leaf is an ideal slow release fertiliser which also maintains organic levels of soils. (Clarke 1997).

In the Mareeba-Dimbulah area of north Queensland, where tea tree is being grown in light, infertile, sandy soils formerly used to grow tobacco, plantation yields are uneconomic if trees were not fertilised regularly. Compared to unfertilised trees on these fully irrigated plantations, adequate nutrition lifts biomass production by about 50% and lifts oil concentration from 5% to 7% in dry leaf. (Drinnan 1997b).

Commercial practice in these plantations is to apply a complete fertiliser in two applications in each regrowth cycle, with additional nitrogen and potash applied every 4–6 weeks. Fertiliser is broadcast and watered-in early in the cycle, while trees are small and is applied through the sprinkler irrigation once regrowth becomes tall. In each regrowth cycle of 8–9 months typical applications (kg/ha) are: N 150–200, P 20–25, K 150–200 (Drinnan 1997b).

Leaf Tissue Nutrient Levels

There are few published data on tea tree leaf tissue nutrient levels. Limited studies in north Queensland (Drinnan 1997b) and northern NSW (Virtue, Murtagh and Lowe 1997) show that tea tree leaf tissue levels are similar to other important native species, at least for the major nutrients (Table 4). Drinnan (1997a) also showed that unfertilised trees growing in infertile sandy soil and exhibiting nutrient deficiency symptoms, had N and K levels in leaves which were only half those found in healthy, productive trees. P levels were similar whether trees were fertilised or not but several trace elements particularly copper (Cu), zinc (Zn) and iron (Fe) were much lower in unfertilised trees (Table 5).

Mycorrhizal Associations

Mycorrhizae are beneficial, soil borne fungi which develop symbiotic associations with plants. The fungal hyphae of mycorrhizae extend the root surface area of plants to

Table 4 Optimum/normal nutrient levels in leaves of four native tree species

Nutrient	*M. alternifolia*[1]	*Leptospermum* spp[2]	*Eucalyptus* spp[3]	Macadamia[4]
N%	1.5–2.0	1.1–1.8	0.8–1.5	1.3–1.8
P%	0.1–0.2	0.1–0.8	0.05–0.1	0.08–0.1
K%	0.9–1.5	0.7–1.6	0.5–0.7	0.5–0.8
Ca%	0.4–1.0	0.8–1.0	0.4–1.0	0.5–0.9
S%	0.2	—	—	0.18–0.25
Mg%	0.1–0.4	0.1–0.3	0.2–0.4	0.08–0.10
Ca (mg/kg)	10	—	4–10	5–12
Zn (mg/kg)	40	—	15–40	15–50
Fe (mg/kg)	30–100	—	50–100	25–200
B (mg/kg)	40–80	—	20–50	20–75

Sources: 1. Drinnan *et al.* (1997); 2. Reuter *et al.* (1986); 3. Judd *et al.* (1996); 4. Incitec Analysis Systems. Interpretation Chart No. 256. October 1996.

Table 5 Leaf nutrient levels in fertilised and unfertilised tea tree at Mareeba

Nutrient	*Optimum/normal*	*Unfertilised/deficient*
N%	1.8–20	0.9–1.0
P%	0.1	0.1
K%	1.5	0.7–0.8
Ca (mg/kg)	10	1–5
Zn (mg/kg)	40	1–5
Fe (mg/kg)	30–100	10

increase water and nutrient uptake, particularly the uptake of phosphorus (Mengel and Kirkby 1987).

The fungus enhances the uptake of phosphorus, nitrogen and trace elements in forest trees and is most effective in increasing the growth of eucalypts in soils that are deficient in phosphorus and nitrogen. Forest trees take up nutrients mostly from the surface soil where fine roots and fungal hyphae are most abundant and nutrients and organic matter are concentrated (Grove *et al.* 1996).

Mycorrhizal associations have been observed in a number of myrtaceous species including *Melaleuca* spp. (Khan 1993). They have also been observed in *M. alternifolia* in northern NSW (Virtue 1997).

Mycorrhizae are likely to be active in most northern NSW plantations as these occur, for the most part, within areas where tea tree occurs naturally.

Effluent Nutrient Removal

Tea tree is an attractive candidate for effluent reuse on agricultural crops because of its wide adaptability in warm environments, perenniality, tolerance of wet conditions, annual "cut and carry" harvest, and high economic return (Murtagh 1996c). Its ability to accumulate high levels of phosphorus (P) in leaves without any reduction in growth can provide an economic method of stripping P from effluent and waste water.

Bolton and Greenway (1995) found that tea tree thrived when irrigated with standard secondary treated effluent, containing 5 mg/L of nitrogen and 6 mg/L of phosphorus. They found that accumulated P levels in leaves were higher (up to 11 mg P/g DW) than in leaves of effluent-fed *Phragmites australis* (1.7 mg P/g DW), a species regularly used as a constructed wetland macrophyte. The regular harvesting of tea tree for oil provides a renewable and economical sink for the removal of phosphorus from effluent waters.

YIELD CHARACTERISTICS

Plant Composition at Harvest

Trees will be from 1–2 m high at harvest, depending on growing conditions and the interval from last harvest. This height will be achieved in north Queensland every 8–9 months but will take about 12 months in northern NSW. At the first harvest trees will have a single, thicker stem and will not be so easily cut as at subsequent harvests, when they will have several thinner stems.

Harvested biomass contains 40–45% dry matter but can range from 30–55%. The dry harvested plant is about one third oil bearing leaf in average sized trees but ranges from 27% in big trees (which are more woody) to 40% in smaller trees (Colton and Murtagh 1990).

In plantations having an optimum plant population of about 35,000 plants/hectare harvested biomass could average about 20–25 t/ha fresh weight under good growing conditions, but could range from 10 t/ha in poor growing conditions to 35 t/ha under excellent growing conditions (Table 6). This yield range equates to 4–14 t/ha dry biomass and 2–5 t/ha of dry leaf.

Oil concentration can also vary. In freshly harvested biomass, which contains stems, twigs and leaf, it can range from 5–13 mg/g. In dry leaf the range is 30–80 mg/g (3–8%) with 50–60 mg/g being fairly typical (Table 7).

Oil Yield

A medium oil concentration of 55 mg/g and an average biomass production of 4 t/ha of dry leaf will result in an oil yield of about 200 kg/ha per harvest (Table 8). On the NSW north coast, where crops are harvested annually this yield becomes the annual yield. In north Queensland where crops are harvested every 8–9 months, this per harvest figure becomes 260–300 kg/ha per annum.

In practice, both NSW and Queensland growers are averaging about 180–200 kg/ha/harvest, with some achieving up to 250 kg. Combinations of above average management and more productive sites are resulting in oil yields of 250–300 kg/ha/harvest. However, restricted growth caused by weeds, poor soil or dry conditions can easily reduce oil yields to 100–150 kg/ha/harvest.

In new plantations, full yields are not achieved until year 3 (Small 1986). Oil concentration is normally lower at the first harvest (Drinnan *et al.* 1997).

Table 6 Typical biomass yield (t/ha) under different growing conditions

	Growing conditions		
	Poor	*Good*	*Excellent*
Total harvested yield (fresh)	9	22	34
Total harvested yield (dry)	4	9	14
Leaf yield (dry)	2	4	5

Table 7 Typical oil concentrations in different plant fractions

	Oil concentration (mg/g)		
	Low	*Medium*	*High*
Whole harvested plant (fresh)	5	9	13
Leaf (dry)	30	55	80

Table 8 Oil yield with different combinations of growth rate and oil concentration

Oil concentration in leaf (mg/g)		*Oil yield* (kg/ha/harvest)		
		Poor growth	*Good growth*	*Excellent growth*
Low	30	50	110	155
Medium	55	95	200	280
High	80	135	290	390

HARVESTING

When to Harvest

The aim is to harvest when trees have reached their maximum leaf yield and leaves contain their maximum oil concentration. However these two factors do not necessarily reach their peaks at the same time of the year.

In a four year, month of harvest study Murtagh (1996b) showed that in NSW biomass yields are highest when annual harvests occur between July and September and lowest when harvests occur in May. The yield advantage of winter—early spring harvests results from the fact that the most efficient regrowth period (4–6 months after harvest) coincides with the best growing months (January–March) when temperature, moisture and humidity are closer to optimum.

Regrowth is always slower in the three months following harvest because light interception, and hence photosynthesis, are limited by the plants' small leaf area. (Murtagh 1996b). With July–September harvests, this period of slower growth mainly coincides with the cooler, drier spring period.

Murtagh and Smith (1996) found that oil concentration in northern NSW varies significantly from month to month and between years but tends to be lowest in the July–September period. Drinnan (1997b) supports these findings, reporting the highest oil concentrations over summer and the lowest in late winter–spring in north Queensland. Harvesting at different times has very little effect on oil quality or composition (Murtagh and Smith 1996; Drinnan 1997a).

When total biomass and oil concentration are combined they tend to balance out, leaving no clear cut advantage for any one month of harvest. However, in years when oil concentration is particularly low during July–September harvesting should be delayed. In frost prone areas April–June harvests are not advisable as new coppice growth is readily damaged. On heavier soils prone to waterlogging summer harvesting may be inadvisable, to avoid soil compaction.

A crop is ready to harvest once its canopy is fully developed and trees have reached maximum leaf yield (Plates 9–11). Once trees have reached this stage they tend to start losing lower leaves and stems begin to thicken (Colton and Murtagh 1990). Delaying harvest beyond this point increases harvest costs and reduces distillation efficiency for no additional oil yield. Tree size will vary with growing conditions but 1.5–2 m is a common height for trees ready to harvest in well grown plantations.

This stage is reached after 8–9 months regrowth in north Queensland, about 12 months in the NSW Northern Rivers, and may take up to 15–18 months in cooler, slower growing sites further south. The interval from planting to the initial harvest is likely to be from 3–6 months longer than the interval between subsequent harvests. At the initial harvest, trees will have single stems which will thicken if harvest is delayed unduly and can quickly exceed the cutting capacity of the harvester.

Harvesting Equipment

Harvesting involves cutting the whole tree about 150 mm from the ground, chopping the material into 10–30 mm lengths, and blowing it into field bins for transport to the distillery (Plate 11).

Single or multi-row heavy duty forage harvesters are commonly used. They must be robust enough to handle the woody stems, have a high throughput capacity and be operated by a powerful tractor, given the bulk of material commonly handled (Colton and Murtagh 1990).

Field bins are usually tipping trailer bins with a capacity of up to 3 tonnes. They can also be cartridges which are transported to the distillery on a trailer and lowered onto a fixed base. Once distillation is complete spent leaf is tipped out and the bins are returned to the field for refilling.

Coppicing

Tea tree has a strong coppicing ability (Plate 9) (Small 1981; Colton and Murtagh 1990) and vigorous coppicing is vital for rapid regrowth. Harvest initiates coppice regrowth, with new shoots beginning to appear within a few weeks. Blake (1983) found that coppicing vigour is influenced by a range of factors including month of harvest. He quoted a number of species which gave the best coppicing after cutting in winter or early spring, but this was not a universal finding. While Murtagh (1996b) observed few differences in coppicing vigour following different months of harvest, anecdotal evidence in NSW supports slower early growth following late autumn harvests, probably due to the onset of winter dormancy.

Major plant losses have been recorded where plants were harvested under high water and temperature regimes in the glasshouse (Murtagh and Lowe 1997). Following harvest, guttation water accumulated on cut stems and provided favourable conditions over several days for pathogen invasion. Cut surfaces subsequently blackened and plants died. Drinnan (1997a) reports similar experiences following field harvesting during the wet season in north Queensland. Losses only occurred when summer harvests occurred on very wet soil and plants continued to pump water through the cut surface for some time after harvest. Plants develop new shoots as normal but symptoms begin to appear 3–4 weeks after harvest and plants die. Plant surfaces above and below ground become blackened as plants die. Queensland Department of Primary Industry pathologists have identified the soil borne pathogen as charcoal root disease (*Macrophomina spp*). Plants from some seed sources appear more susceptible than others. Losses, which have been as high as 60%, have been greater from first harvests in new plantations but have also occurred following coppice harvests.

REFERENCES

Anderson, R.H. (1956) The Trees of New South Wales (*Government Printer: Sydney*).

Barlow, B.A. (1988) Patterns of differentiation in tropical species of Melaleuca L. (Myrtaceae). *Proceedings of the Ecological Society of Australia*, **V15**, 239–247.

Blake, T.J. (1983) Coppice systems for short rotation intensive forestry: the influence of cultural, seasonal and plant factors. *Australian Forest Research*, **13**, 279–291.

Bolton, K.G.E. and Greenway, M. (1995) Growth characteristics and leaf phosphorus concentrations of three *Melaleuca* species sand cultured in different effluent concentrations. *Proc. National Conference on Wetlands for Water Quality Control*, Townsville, September 1995.

Bureau of Meteorology (1988) Meteorological Summary, July 1988. In *Climatic Averages Australia.* Australian Government Publishing Service, Canberra.

Clarke, B. (1995) NSW Agriculture. Unpublished results.

Clarke, B. (1997) NSW Agriculture. Personal communication.

Colton, R.T. and Murtagh, G.J. (1990) Tea-tree oil—plantation production. *NSW Agriculture and Fisheries, Agfact P6.4.6.*

Creagh, C. (1993) Working together with acid sulphate soils. *Ecos 77 CSIRO 1993.*

Cromer, R.N. (1996) Silviculture of eucalypt plantations in Australia. In P.M. Attiwil and M.A. Adams (eds.), *Nutrition of Eucalypts*, CSIRO, Australia, pp. 259–273.

Curtis, A. (1996) Growth and Essential Oil Production of Australian Tea Tree (*Melaleuca alternifolia* (Maiden and Betche) Cheel), Master of Agricultural Science Thesis, The University of Queensland.

Diatloff, E. (1990) Effects of applied nitrogen fertiliser on the chemical composition of the essential oil of three *Leptospermum spp. Australian Journal of Experimental Agriculture*, **30**, 681–685.

Drinnan, J.E. (1997a) Queensland Department of Primary Industries. Personal communication.

Drinnan, J.E. (1997b) Development of the North Queensland Tea Tree Industry. Final report for project DAQ-184A. Rural Industries Research and Development Corporation.

Drinnan, J.E., Virtue, J.G., Murtagh, G.M. and Lowe, R.F. (1997) Queensland Department of Primary Industries—NSW Agriculture. Personal communication.

Gomes, A.R.S. and Kozlowski, T.T. (1980) Response of *Melaleuca quenquenervia* seedlings to flooding. *Physiologia Plantarum*, **49**, 373–377.

Grove, T.S., Thomson, B.D. and Malazcjuk, N. (1996) Nutritional physiology of Eucalypts: Uptake, Distribution and Utilisation. In P.M. Attiwil and M.A. Adams (eds.), *Nutrition of Eucalypts*, CSIRO, Australia, pp. 77–108.

Judd, T.S., Attiwil, P.M. and Adams, M.A. (1996) Nutrient concentrations in Eucalyptus: a synthesis in relation to differences between taxa, sites and components. In P.M. Attiwil and M.A. Adams (eds.), *Nutrition of Eucalypts*, CSIRO, Australia, pp. 123–153.

Khan, A.G. (1993) Occurrence and importance of mycorrhizae in aquatic trees of NSW, Australia. *Mycorrhizae*, **3**, 31–38.

List, S.E., Brown, P.H., Low, C.S. and Walsh, K.B. (1996) A micropropagation protocol for *Melaleuca alternifolia (tea tree). Australian Journal of Experimental Agriculture*, **36**, 755–760.

Mengel, K. and Kirkby, E.A. (1987) Principles of Plant Nutrition. International Potash Institute: Wonblaufen-Bern, Switzerland. 4th Edn. Ch. 2, pp. 25–112.

Murtagh, G.J. (1991a) *Tea tree oil. In New Crops* (eds. R.J. Jessop and R.L. Wright), Inland Press, Melbourne, pp. 166–174.

Murtagh, G.J. (1991b) Irrigation as a management tool for production of tea tree oil. Final report for project DAN-19A. Rural Industries Research and Development Corporation.

Murtagh, G.J. (1996a) Challenges facing Australian Agriculture—Tea tree oil. *Proc. Aust. Institute of Valuers and Land Economists Rural Conf.* Kooralbyn, July 1996.

Murtagh, G.J. (1996b) Month of harvest and yield components of tea tree. I. Biomass. *Australian Journal of Agricultural Research*, **47**, 801–815.

Murtagh, G.J. (1996c) The irrigation response by tea tree and implications for effluent reuse. In P.J. Polglase and W.M. Tunningley (eds.), *Some Application of Wastes in Australia and New Zealand: Research and Practice*, CSIRO Forestry and Forest Products, Canberra, pp. 137–141.

Murtagh, G.J. and Etherington, R.J. (1990) Variation in oil concentration and economic return from tea tree (Melaleuca alternifolia Cheel) oil. *Australian Journal of Experimental Agriculture*, **30**, 675–679.

Murtagh, G.J. and Lowe, R.F. (1997) NSW Agriculture. Unpublished results.

Murtagh, G.J. and Smith, G.R. (1996) Month of harvest and yield components of tea tree. II Oil concentration, composition and yield. *Australian Journal of Agricultural Research*, **47**, 817–827.

Peak, C.M. (1982) Essential oil production from *Melaleuca alternifolia.* Wollongbar Agricultural Research Centre, Biennial Report, p. 53.

Reilly, T. (1988) Investment in tea tree plantations. Reports of the 1st Tea Tree Oil Seminar, Lismore, November 1988, pp. 42–47.

Reuter, D.J. and Robinson, J.B. (eds.) (1986) '*Plant Analysis*' Inkata Press, Melbourne.

Small, B.E.J. (1981a) Department of Agriculture, NSW. Personal communication.

Small, B.E.J. (1981b) Effects of plant spacing and season on growth of *Melaleuca alternifolia* and yield of tea tree oil. *Australian Journal of Experimental Agriculture*, **21**, 439–442.

Small, B.E.J. (1986) Tea tree oil. Department of Agriculture, NSW *Agfact P6.2.1*, 2nd edition.

Southwell, I.A. and Wilson, R.W. (1993) The Potential for tea tree oil production in northern Australia. *Acta Horticulture*, **331**, 223–227.

Van der Moezel, P.G., Pearse-Pinto, G.V.N. and Bell, D.T. (1991) Screening for salt and waterlogging tolerance in Eucalyptus and Melaleuca species. *Forest Ecology and Management*, **40**, 27–37.

Virtue, J.G., Murtagh, G.M. and Lowe, R.F. (1997) NSW Agriculture. Personal communication.

Virtue, J.G. (1997) Weed interference in the annual regrowth cycle of plantation tea tree (*Melaleuca alternifolia*). PhD Thesis. The University of Sydney.

Weiss, E.A. (1997) *Essential Oil Crops*, CAB International, Oxford, pp. 302–311.

Plate 6 Cell grown *Melaleuca alternifolia* seedlings in the nursery (R. Colton)

Plate 7 A four-row planter adds fertiliser and water to the seedlings as they are planted (R. Colton)

Plate 8 Herbicides keep seedlings weed free for up to 12 weeks (R. Colton)

Plate 9 Trees coppice vigorously after harvest (R. Colton)

4. WEED MANAGEMENT IN TEA TREE PLANTATIONS

JOHN G. VIRTUE

Animal & Plant Control Commission, Primary Industries, SA, Australia

INTRODUCTION

Weed management integrates preventative and control techniques to minimise crop yield loss due to weed interference. Preventative techniques include quarantine to avoid new weeds, and providing conditions unsuitable for weed establishment. Control techniques include chemical, physical and biological methods of weed destruction, and manipulation of crop competitiveness. Weed interference slows crop growth by competition for resources of sunlight, soil water and soil nutrients, and by production of growth-inhibiting chemicals. Weed interference reduces tea tree yield during both the initial establishment phase and in the annual regrowth cycles. This chapter describes the nature of the weed problem in tea tree plantations and the effect of weeds on tea tree yield. An understanding of the mechanisms of weed interference in tea tree plantations then provides the basis for discussing effective and sustainable weed management techniques. The chapter is based on research conducted on the north coast of New South Wales (NSW), Australia, where tea tree was first established in plantations.

WEEDS IN TEA TREE PLANTATIONS

A simple definition for a weed is an unwanted plant. In cropping systems weeds are primarily considered as plants which interfere with crop growth and consequently reduce crop yield. Table 1 lists those weeds which were regionally widespread in tea tree plantations in a 1992–3 survey. The predominant weeds were herbaceous annuals and perennials, and included broadleaf, grass and sedge weeds. Annual weeds go through a rapid succession of germination, growth, flowering, seed set and death. Annual weeds readily colonise bare ground, as occurs after a soil cultivation. Perennial weeds have a life cycle that alternates between growth and flowering phases over many years. Perennial weeds are often favoured by agricultural systems which have minimal soil disturbance. Some perennials reproduce vegetatively from root fragments or bulbs, and soil cultivation aids their spread within a field.

Environmental conditions in a typical tea tree plantation favour strong weed growth. In a 1992 survey of 28 Australian plantations (Virtue 1997) the majority were located on the upper north coast of NSW. The subtropical climate is characterised by relatively high annual rainfall (ranging from 1,000–1,400 mm), wettest in February–March (late summer) and driest in August–September (early spring). Winters are relatively mild and summers warm and humid. Such a climate allows both winter and summer-growing annual weeds, and favours warm-season perennial grass weeds. Most plantations in the

Table 1 Weeds which were widespread over north-east NSW tea tree plantations in a 1992–3 field survey of 25 plantations (McMillan and Cook 1995)

Common within a field		*Scattered within a field*	
carpet grass	*Axonopus affinis*	unidentified grasses	Family *Poaceae*
couch grass	*Cynodon dactylon*	smartweed	*Persicaria* spp.
paspalum grass	*Paspalum dilatatum*	Paddy's lucerne	*Sida rhombifolia*
sedges	*Cyperus* spp.	slender celery	*Apium leptophyllum*
kidney weed	*Dichondra repens*	purple top	*Verbena bonariensis*
fleabane	*Conyza* spp.	button burrweed	*Soliva anthemifolia*
flatweed	*Hypochoeris radiculata*	starwort	*Aster subulatus*
		sowthistle	*Sonchus oleraceus*
		fireweed	*Senecio madagascariensis*
		pinrushes	*Juncus* spp.

survey were established on alluvial soils and were prone to flooding and waterlogging. The high fertility and moisture levels of such soils favour vigorous growth of herbaceous weeds.

THE EFFECTS OF WEEDS IN TEA TREE PLANTATIONS

In a 1992 survey (Virtue 1997), approximately 80% of tea tree growers with plantations larger than ten hectares considered weeds to be a major limit to production. The main effect of weeds in tea tree plantations is a reduction in tree growth due to weed interference, and thus a decrease in oil production. Other indirect effects of oil contamination and harvest inefficiency are minimal. Weeds do have some beneficial effects on soil health.

Weed Interference and Tea Tree Oil Yield

Weed interference encompasses the two mechanisms by which weeds reduce crop yields; competition and allelopathy. Weed competition is the use of sunlight, soil water and/or soil nutrients at the expense of the crop. Allelopathy is the production by weeds of chemicals which can inhibit crop growth. Tea tree oil yield quantity ($kg\,ha^{-1}$) is the product of leaf biomass and leaf oil concentration. Yield quality is determined by the oil's chemical composition. Experiments have shown weed interference reduces leaf biomass yield of both seedling and regrowth tea tree. No effects of weed interference on leaf oil concentration and chemical composition have been detected.

Weed interference strongly reduces growth of seedling tea tree. McMillan and Cook (1995) observed a 97% reduction in tree mass where annual grasses were uncontrolled in the period between 31 and 180 days after planting (Figure 1). Even in the short period of 31 to 55 days after planting, annual grasses reduced tree mass by 56%. Such high yield reductions due to herbaceous weed interference are typical for tree seedlings (Table 2). Sustained weed interference can even kill tree seedlings, leading to replanting costs.

Leaf yield reductions due to weed interference in the annual regrowth cycle are much less than at tea tree seedling establishment, but integrated over successive cycles they are a substantial long-term cost. Measured reductions in mature leaf biomass yield due to weeds have ranged from 9 to 44% on a per tree basis (Table 3). The reductions are due to slowed tea tree growth, and no significant effect of weed interference on the rate of tea tree leaf fall has been detected. The magnitude of leaf yield loss increases with weed biomass, which in turn is affected by seasonal conditions (temperature and soil moisture), tree density and weed species. Tree density is important in calculating economic loss on a per hectare basis. For example consider two tree densities, 15,000 and 30,000 trees ha^{-1}, each with a mature harvest oil yield of 8 g $tree^{-1}$. If weed interference reduces yield per tree by 40% for the lower density and 25% for the higher density then the economic costs are $2,160 and $2,700 respectively (assuming an oil price of $Aus45 kg^{-1}). Thus the greater potential yield of high density plantations means that weeds can cause greater economic losses. However if weed management costs per hectare are equivalent at low and high density, then the profits from weed control are greater at the higher tree density.

Indirect Effects

Indirect effects of weeds on tea tree yield are likely to be minor. Native pests and diseases of tea tree are unlikely to use exotic herbaceous weeds as an alternate host plant

Figure 1 Decline in tea tree seedling biomass with increasing periods of weed interference from planting (from McMillan and Cook 1995)

Table 2 Percentage yield reductions of tree seedlings due to high herbaceous weed interference. Reductions are relative to low weed interference or weed-free treatments

Species	*Time from planting (months)*	*Yield reduction (%)*	*Biomass measurement*	*Reference*
Eucalyptus delegatensis (alpine ash)	3.5	88	foliage	Ellis *et al.* (1985)
Liriodendron tulipifera (yellow-poplar)	12	99	total shoots	Kolb and Steiner (1990)
Malus domestica (apple)	24	53	total shoots	Goode and Hyrycz (1976)
Pinus elliotii (slash pine)	6	55	total shoots	Smethurst *et al.* (1993)
Pinus taeda (loblolly pine)	12	76	foliage	Britt *et al.* (1990)
Quercus rubra (northern red oak)	12	61	total shoots	Kolb and Steiner (1990)

Table 3 Measured reductions in mature regrowth leaf yield of tea tree due to weed interference

Row system	*Tree density (trees* ha^{-1})	*Row spacing* (m)	*Leaf yield reduction* (%)
4-row hedge	25,000	0.4	26 (year 1) 9 (year 2)
2-row hedge	42,000	0.5	27
single row	13,000	3.0	44 (year 1) 33 (year 2)
single row	18,000	1.5	40

in their life cycle. Weed biomass in harvested material should be minimal and thus not waste space in distillation bins. Tea trees are unlikely to be harvested until they have formed a canopy, shading then causing weed senescence. Where a plantation has wide row spacing then harvesters are directed along the tree rows, avoiding weeds between rows. Any weed essential oils collected during distillation would be very dilute in the tea tree oil.

Benefits of Weeds

Weeds have some advantages. They form a groundcover that protects the soil from erosion and insulates roots from high summer temperatures. Decaying weeds add to soil organic matter, improving soil structure and promoting biological nutrient cycling. Weeds act as a reservoir for nutrients, reducing losses due to leaching below the root zone. Weeds can also act as food sources for beneficial insect predators and parasites.

MECHANISMS OF WEED INTERFERENCE

An understanding of the mechanisms, relative importance and timing of weed interference in seedling and regrowth tea tree is important for informed weed management decisions.

Planting to First Harvest

Newly planted tea tree seedlings have a very restricted root system of only 2.5 cm diameter, and a shoot height of 10–15 cm. Their small size makes them poor competitors against rapidly growing weeds. A range of weed interference mechanisms is possible, with early shading and moisture competition being of most importance.

Shading

Plants require light for photosynthesis, to produce chemical energy for growth. Tea tree shoot growth declines quickly with increased shading. Weed shading is most likely in the first three months after planting, when tea tree seedlings and weeds are of similar height. This potential shading period is longer for autumn plantings as falling temperatures slow tea tree growth (Murtagh 1996) whilst cool season annual weeds actively grow. Older tea tree seedlings of approximately 0.5 m in height are still at risk of shading by erect, broadleaf weeds and climbing weeds.

Moisture Competition

Plant roots extract water from the soil for use in photosynthesis, nutrient transport, chemical processes and plant turgidity. Tea tree will be prone to weed competition for water where the root systems interact and soil water is insufficient to meet both tree and weed needs. This is particularly likely in soils without a high water table, with low water-holding capacities (e.g. sandy loams) and/or which receive low rainfall. The majority of tree and weed roots occur in the surface 30 cm of soil (Bowen 1985), so competition between these roots can occur. At this soil depth herbaceous weeds can have root densities 10–100 times greater than trees (Bowen 1985), giving the weeds a competitive advantage for water uptake. Thus young tea tree seedlings with their shallow root systems are very susceptible to moisture competition, and water stress results. Such a mechanism of weed interference is very important for tree seedlings and extreme water stress can cause deaths (Nambiar and Zed 1980; Sands and Nambiar 1984). Moisture competition decreases as trees age and larger root systems access soil water at depth.

Nutrient Competition

Plant roots extract soil nutrients for growth and chemical processes. Major soil nutrients are nitrogen, phosphorus and potassium. Competition for soil nutrients occurs near the soil surface where nutrients are concentrated (Nambiar and Sands 1993). As discussed above, dense weed root systems have a competitive advantage in this soil layer. This remains regardless of tea tree age. Competition between weeds and tea tree is strongest for nitrogen as it is the most mobile soil nutrient in both wet and dry soils (Nambiar and Sands 1993).

Interactions

Interactions between competition for light, water and nutrients will occur. Weed competition for surface soil moisture will also cause competition for nitrogen. Reduced tea tree seedling shoot growth due to moisture competition increases the likelihood of weed shading.

Allelopathy

Many plants release chemicals which are inhibitory to the growth of neighbouring plants. Such chemicals can slow water and nutrient uptake and inhibit photosynthesis. Allelochemical-producing weeds found in tea tree plantations include couch (*Cynodon dactylon*), Farmer's friend (*Bidens pilosa*), barnyard grass (*Echinochloa crus-galli*) and summer grass (*Digitaria sanguinalis*) (Putnam and Weston 1986). Allelopathy has not been investigated in tea tree plantations, however young tree seedlings would be most vulnerable due to their small root systems.

Annual Regrowth Cycles

After the first harvest tea tree begins its first regrowth cycle. This is achieved by tea tree's strong coppicing ability from cut stumps. Shoot regrowth then reaches harvestable maturity approximately every 12 months. Harvesting to near ground level is a major disturbance to the normal root/shoot balance in tea tree. Changes in root/shoot relations in the annual tea tree regrowth cycle are illustrated in Figure 2 for a typical summer harvest on the north coast of NSW.

Prior to harvest tea tree has a normal root/shoot balance, which represents an equilibrium between carbon uptake by leaves and nutrient (and water) uptake by roots (Cannell 1985). A typical myrtaceous root system consists of a lateral network of fine roots in the topsoil, and a taproot and several lateral sinker roots which reach deep into the subsoil (Lamont 1978). The shoot system of regrowth tea tree consists of five or less main stems to a height of 1.5–2.5 m. Where tree rows are spaced approximately 1 m or less apart canopies are merged.

The removal of all shoots at harvest suddenly gives a very high root/shoot ratio. New shoots grow from buds in the process of coppicing. This early regrowth is vigorous. Such vigour is attributed to the high root/shoot ratio enabling abundant uptake of soil water and nutrients, the utilisation of stem and root nutrient reserves, and a hormone balance favouring shoot growth (Blake 1982). The larger the tree prior to harvest then the greater this vigour. This vigour decreases with cooler temperatures, and average daily temperatures below approximately 16°C (Murtagh 1996) will strongly limit shoot emergence and growth. Thus on the north coast of NSW tea tree effectively has a winter dormancy.

The rapid coppice shoot growth hastens the return to the normal root/shoot balance. This return is also achieved by fine root death, due to a deficiency of photosynthate.

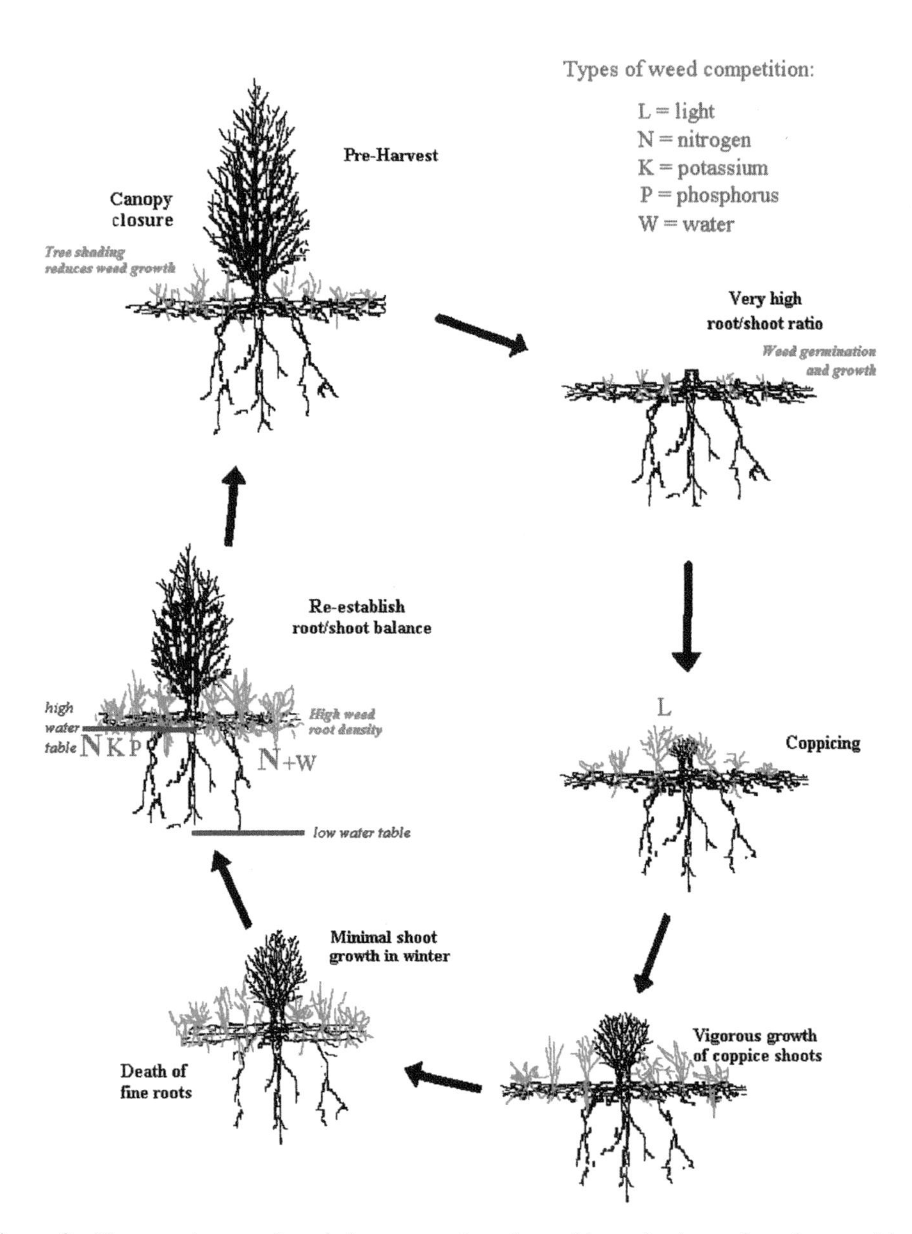

Figure 2 The annual regrowth cycle in tea tree plantations with mechanisms of weed competition indicated. A summer harvest is assumed

In spring, tea tree has regained its normal root/shoot balance, with increased demands on the root system for moisture and nutrients. Rising temperatures enable major shoot growth flushes in mid spring and mid summer. These are accompanied by new fine root growth. Tea tree root systems will not grow much larger with successive harvests as trees are only growing back to their original pre-harvest size.

The changes in root/shoot relations throughout the regrowth cycle influence the mechanisms and timing of weed interference. Harvest removes shading of the ground and so encourages weed germination and growth. The changes in root and shoot relations during the tea tree regrowth cycle present opportunities for different mechanisms of weed competition (Figure 2). Competition for soil nitrogen is the major mechanism of weed interference in the regrowth cycle of tea tree. Competition for light, moisture and other nutrients are of lesser importance. Allelopathy has not been investigated in the regrowth cycle and is not discussed.

Shading

Whilst early tea tree regrowth is of similar height to newly planted seedlings, the former has much more vigorous height growth. Shading is only likely early in the regrowth cycle where fast growing, climbing or upright weed species are present, and/or where the cool temperatures of a late autumn harvest slows tea tree regrowth.

Moisture Competition

Many tea tree plantations on the north coast of NSW are established on floodplains which have a high soil water table. Weed competition reducing surface soil moisture has not increased tea tree water stress in such plantations (Virtue 1997). It appears that the established root system of regrowth tea tree utilises soil water at depth to avoid moisture competition. A similar conclusion was reached by Sands and Nambiar (1984) and Woods *et al.* (1992) in young *Pinus radiata* plantations. Where a high soil water table is absent (e.g. hillslopes, drought years) then tea tree would compete with weeds for surface moisture from rainfall events. Such competition would be most intense during late regrowth, with the return to the normal root/shoot balance and with a larger leaf area increasing tree demands for water.

Nutrient Competition

Competition for soil nitrogen in the latter half of the tea tree regrowth cycle is a major mechanism of weed interference. Nitrogen supply is an important determinant of tea tree shoot growth, and weeds reduce both leaf nitrogen concentration and total shoot nitrogen uptake (Virtue 1997). The weed root density advantage in the nutrient-rich surface soil has been discussed previously. Nitrogen competition is greater in late regrowth as trees have lost the early advantage of the high root/shoot ratio. Competition for potassium and phosphorus has been occasionally observed in regrowth tea tree (Virtue 1997) but is of minor importance compared to nitrogen.

WEED MANAGEMENT IN TEA TREE PLANTATIONS

Weed management in tea tree plantations should include both preventative and control techniques. It is important to prevent the movement of new weeds into and within the plantation, and to provide conditions which are unsuitable for weed establishment. Direct weed control techniques such as herbicides are the main consideration in weed management, but control can also be achieved by increasing the competitiveness of tea tree.

Preventative Techniques

Preventing movement of weeds is much cheaper than the long term control costs of allowing their entry and spread within a plantation. Making conditions unsuitable for establishment of certain weed species also avoids control costs.

Quarantine

The movement of stock and machinery between plantations can spread new weeds. Stock to be used for grazing in plantations should be quarantined for two weeks. This ensures that the majority of weed seeds in the gut are defecated in a single area. Sheep should be recently shorn to limit externally-attached seeds of burr weeds. Mud attached to farm machinery can spread weed seed, bulbs and tubers. Such mud should be washed off before use in a new plantation.

Patch Management

New weeds tend to occur in discrete patches. Patches of those species that are strongly competitive, such as perennial grasses, tall broadleaves and vigorous climbers, should be mapped. One can then readily return to a patch for concerted direct control effort against the weed species. In addition, measures can be taken to limit spread of a patch and creation of new patches. This may involve fencing out stock, avoiding cultivation and/or restricting flood irrigation, depending on how the weed is dispersed.

Unsuitable Conditions for Weed Establishment

Long periods of soil waterlogging favours germination and establishment of sedges and rushes. These are difficult weeds to control. Laser-levelling prior to planting will allow a plantation to drain faster after rainfall. This reduces the incidence of these species and allows earlier machinery access to control all newly-germinating weeds.

Direct Control Techniques

The existing weed population needs to be managed to avoid periods of strong weed interference. Given the high susceptibility of seedling tea tree to many mechanisms of weed interference, it is vital to maximise weed kill throughout seedling establishment. In regrowth tea tree, weed control should particularly aim to avoid tree shading during early regrowth and competition for nitrogen during late regrowth. Good weed kill is especially important prior to and during late regrowth, as the greatest rate of leaf yield

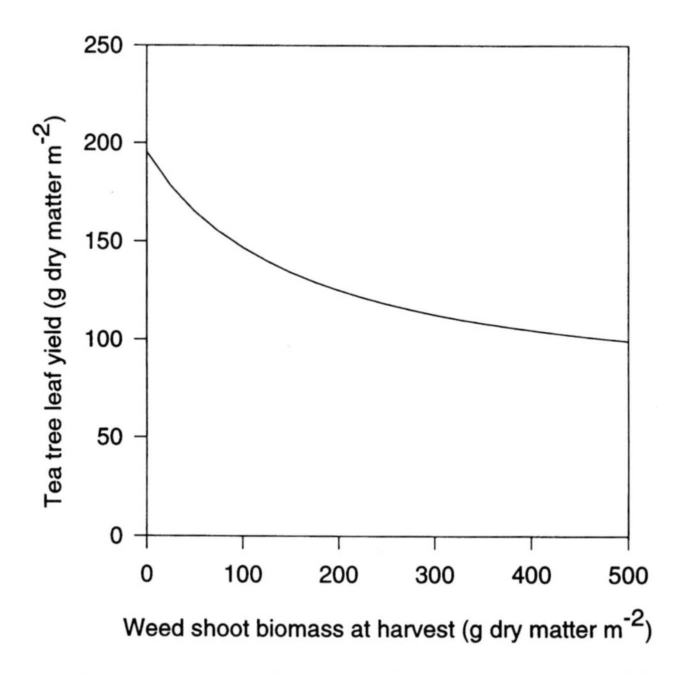

Figure 3 An example response curve of tea tree leaf yield to increasing weed biomass at mature harvest

loss occurs at low weed levels (Figure 3). Whilst a high degree of weed kill is not essential during early regrowth, access to weeds is easiest and weeds are smallest at this time. Thus it is an opportune time to particularly control perennial and long-season annual weeds to reduce later competition. The various direct weed control techniques differ in their ability to kill weeds and in their impacts on soil health.

Herbicides

Appropriately selected and applied herbicides are the most favourable direct weed control technique in tea tree in terms of weed kill, leaf yield and relative cost. Various herbicides have been screened for their safe use in tea tree (McMillan and Cook 1995). Table 4 summarises herbicides which can be used in NSW plantations. Specific recommendations regarding application rates and comments for use are given in Storrie *et al.* (1997). There are two general classes of herbicides used in tea tree plantations; pre-emergents and post-emergents. Pre-emergent herbicides are applied to the soil surface and kill germinating weeds. They are particularly suited to establishment of tea tree seedlings, and are best applied just prior to planting. It is best to select pre-emergent herbicide mixtures with broad spectrum action against a wide range of weed species. Post emergent herbicides are applied to already established weeds. As with pre-emergents there are some which are selective to certain families of weeds, such as grasses. Other post-emergents have a broad-spectrum action and need to be directed away from

Table 4 Herbicides for use in tea tree plantations in NSW. See Storrie *et al.* (1997) for details on rates and application methods

Timing	*Active ingredient*	*Trade names*	*Weeds controlled*
Pre-emergent	trifluralin 400 g L^{-1}	Treflan®, Tridan®	Annual grasses and some broadleaf weeds
	flowable simazine 500 g L^{-1}	various trade names	Annual broadleaf weeds
	metalachlor 720 g L^{-1}	Dual®	Annual grasses and some broadleaf weeds
	oryzalin 500 g L^{-1}	Surflan®	Annual grasses and some broadleaf weeds
Post-emergent	acifluorfen 224 g L^{-1}	Blazer®	Small broadleaf weeds
	clopyralid 300 g L^{-1}	Lontrel L®	Some annual broadleaf weeds
	fluazifop-p-butyl 212 g L^{-1}	Fusilade®	Annual and some perennial grasses
	haloxyfop 104 g L^{-1}	Verdict®	Annual and some perennial grasses
	2,2DPA 740 g kg^{-1}	various trade names	Grasses and certain broadleaves. Avoid spraying trees
	paraquat & diquat 135 + 115 g L^{-1}	Sprayseed®	Young annual grasses and broadleaf weeds. Avoid spraying trees

tea tree leaves to avoid damage. Care is required in selecting and applying herbicides to minimise effects on operator health, impacts on off-target organisms, and residues in any farm products. Populations of herbicide resistant weeds can develop if a single herbicide is relied upon as a sole means of control for that weed species. Herbicides are grouped according to their mode of action, and it is important to regularly switch between groups, to minimise development of resistance.

Soil Cultivation

Soil cultivation is a physical disturbance of the soil which controls weeds by uprooting, severing the root-shoot connection and/or weed burial. Cultivation is the most effective non-chemical technique to kill weeds in tea tree plantations. Small seedling weeds are killed more quickly, efficiently and cheaply than large weeds. The major limit to the effectiveness of cultivation is soil waterlogging. This may prevent machinery access and weed kill by desiccation is reduced. Other factors limiting cultivation's effectiveness are soil disturbance promoting new weed germination, selection for perennial weeds which reproduce vegetatively, and difficulty cultivating within tree rows and in closely-spaced hedges. Hand-hoeing should only be used for spot-weeding due to the high labour cost.

Aside from reducing weed interference, cultivation also favours tea tree growth in the short term through improved soil penetrability for root growth, increased rainfall infiltration and increased mineralisation of nutrients from soil aeration. In the long-term there may be negative impacts on tea tree growth through damage to soil and trees.

Excessive cultivations can damage soil structure so that rainfall infiltration and soil aeration is reduced. Cultivating wet soil can result in a compaction layer just below the tilled layer which then restricts root growth (Tisdall and Huett 1987). Cultivation lowers organic matter levels, causing a decline in soil fauna populations, reducing nutrient cycling and soil structural repair (Douglas 1987). Regular, deep cultivations will sever thick lateral roots of trees and eventually result in a deeper root system (Tamasi 1986), below the highest concentration of nutrients at the soil surface.

Mowing

Mowing is the cutting of weeds to near-ground level. Mowing reduces weed shading and seed set, but is ineffective for killing many weeds. In tea tree plantations mowing has given equivalent or lower yields than leaving weeds uncontrolled (Virtue 1997). Weeds with growing points below the cutting height regrow. Perennial grasses are particularly favoured by mowing and become increasingly dominant. The high root density of these grasses (Bowen 1985) and the regular promotion of vigorous regrowth with each mowing favours strong competition with tea tree for soil nitrogen.

Mowing does have some benefits. Soil organic matter levels are increased to improve nutrient cycling and soil structure. Maintenance of a groundcover also assists trafficability on wet soils, protects soil from erosion and keeps soil cool in summer.

The orchard weed control technique of a herbicide strip along the tree row and mowing in the inter-row is not appropriate for tea tree plantations. The mown sod improves machinery accessibility, but research has found that a strip width of 0.5 m either side of the tree row has been insufficient to avoid weed competition for soil nitrogen (Virtue 1997). For the average tea tree row spacing of 1.4 m for single row configurations (Virtue 1997), wider herbicide strips will overlap leaving no mown inter-row.

Grazing

Grazing animals can be used to control weeds. They require good management for effective weed control and to maintain animal health. Grazing has not been experimentally compared with other weed control techniques in tea tree, but one would expect only a marginal yield improvement over mowing. In the short term and with high stocking density weed biomass would be significantly reduced, increasing tea tree yield compared to no weed control. In the long term weeds which are non-palatable and/or recover well from grazing will increase and compete with tea tree. The attraction of grazing for weed control is the potential to minimise costs through sale of the animals. Sheep have become increasingly used in tea tree plantations for weed control. Compared to cattle they are less damaging to trees, graze closer to the ground and are less selective eaters. However they are poorly suited to the warm, humid and often waterlogged tea tree plantation environment, and health problems that may arise include worms, clostridial diseases, footrot, foot abscess, fleece rot, fly strike and lice.

Mulches

A mulch is a material that covers the soil surface. For weed control a mulch must stop weeds germinating and growing, by starving them of light. Mulches can be thin, impenetrable barriers such as woven plastic or paper. Alternatively mulches can be thick, loose layers such as spent tea tree litter or straw. Complete mulching of the soil surface provides excellent weed control but the cost of materials and application is very expensive. Post-distillation tea tree litter is a ready supply of mulch but given a retail purchase price of $Aus12–15 per cubic metre for landscaping, a 0.1 m thickness × 1.0 m width of mulch is worth $12,000–15,000 per hectare, before application costs. The mulch may only last several years, may have an allelopathic effect on the growing tea tree, and may not stop growth of perennial weeds. A thin layer of well-composted tea tree litter would be cheaper per hectare and still provide the benefits of organic mulches; protection from high summer temperatures, reduced soil erosion, greater rainfall infiltration and reduced soil evaporation, increased soil organic matter and nutrient addition (Firth 1991).

Legume Crops

Growing a legume grain crop in the year prior to planting enables early control of grass weeds with selective herbicides, and increases soil nitrogen for the seedling tea tree. Perennial cover crops are often used in orchards to provide a year-round groundcover to protect the soil and suppress weeds. Requirements of a perennial legume cover crop for densely planted tea tree are difficult to meet. The crop must establish quickly, persist under shaded conditions, remain prostrate and not shade tea tree, be shallow rooted to minimise moisture competition yet survive periods of low soil moisture, produce its own nitrogen requirements, and be tolerant of acidic, waterlogged soils. No species meets all of these requirements. Maku lotus, *Lotus pedunculatis*, is well-suited to acidic soils (Firth 1992) but twines up trees causing shading. Osciola white clover, *Trifolium repens*, has greater waterlogging tolerance than other cultivars but white clovers are shade intolerant (Firth 1992). Tropical legumes are strongly competitive with tree crops in warmer months (Firth 1992). The most appropriate role for annual and perennial legumes in tea tree plantations is to provide nutritious feed where grazing is the main weed control technique.

Summary

All weed control techniques have advantages and disadvantages in their use regarding extent of weed kill, relative cost, effect on soil health, tree damage, skills required and suitability in wet conditions. Weed kill should be maximised at minimum cost for greatest profit, but with long-term soil health in mind. A single technique cannot be solely relied upon as it will inevitably select for weeds that are tolerant to the technique. Rather, different techniques should be integrated within and between years, considering the stage of the regrowth cycle, weather conditions and the risk of yield loss due to weed competition.

Improving Tea Tree's Competitiveness

Plantation management decisions affecting tea tree growth have implications for tree competitiveness against weeds.

Harvest Time

Harvesting in late spring-summer provides warm temperatures for rapid coppice shoot emergence and growth. Tea trees are less likely to be shaded by weeds, will form a canopy sooner in the regrowth cycle to then shade weeds, and will suffer less fine root death.

Pre-harvest Tree Size

Larger tea trees at harvest result in faster regrowth due to a higher root/shoot ratio. The aim in the establishment phase of tea tree should be to grow a large, deep root system for vigorous regrowth in subsequent harvest cycles.

Plantation Configuration and Density

Choice of plantation layout is very important for weed management. Closely-spaced single row plantations have several advantages. They provide a high tree density to maximise yield per hectare, enabling increased expenditure on weed management per hectare. Close-spacings increase tea tree's competitiveness with canopy closure shading weeds. Single rows provide greater access to trees for cultivation and herbicide application than in hedge rows. Row crop equipment developed for other crops (e.g. vegetables, cotton) can be readily used. Tree spacing of 0.8–1.0 m between rows by 0.3–0.4 m within rows will give a tree density of approximately 30,000 trees ha^{-1}. Straight rows are essential for accurate weed control.

Nitrogen Nutrition

Nitrogen supply is a major determinant of tea tree leaf yield, and weed competition for soil nitrogen is a major mechanism of weed competition (Virtue 1997). Tea tree leaf nitrogen concentration needs to be greater than 1.5% dry mass (averaged over whole trees) to maximise yield (Virtue 1997). Nitrogen must be supplied to trees and not weeds. Research is required into tea tree uptake of nitrogen as no responses to fertiliser have been achieved (Virtue 1997). Foliar application of urea to trees (e.g. Coker *et al.* 1987) may provide a means of providing nitrogen whilst maintaining a weed groundcover for soil health.

High Water Table

To reduce moisture competition in regrowth tea tree, plantations should be established where there is a permanent and relatively high water table accessible.

Insect Management

Defoliation by insects during early regrowth has similar effects to cool temperatures, slowing shoot growth and increasing tree root death. Trees are set back for the rest of the regrowth cycle and will be more prone to late weed competition.

CONCLUSION

Herbaceous weeds are a major limit to leaf yield in tea tree plantations. Weed interference, if not controlled, will reduce the mature leaf yield of regrowth tea tree by an average of around 30%, and the growth of seedling tea tree by 60–90%. Leaf oil concentration and chemical composition are not significantly affected by weeds. Weeds compete for moisture, nutrients and light in seedling tea tree, but mainly for nitrogen in regrowth tea tree. This is because of the established root system of regrowth tea tree, which enables rapid early regrowth above weeds and access to water deep in the soil. Weed management is critical at seedling establishment, and weed kill should be maximised until first harvest. In subsequent regrowth cycles, weed control should be concentrated in the latter half of cycles, when weed competition is strongest. Weed management should aim to prevent weed establishment in plantations, integrate various direct weed control techniques, and promote tea tree's competitiveness against weeds.

REFERENCES

Blake, T.J. (1982) Coppice systems for short-rotation intensive forestry: The influence of cultural, seasonal and plant factors. *Australian Forestry Research*, **13**, 279–291.

Bowen, G.D. (1985) Roots as a component of tree productivity. In M.G.R. Cannell and J.E. Jackson (eds.), *Attributes of trees as crop plants*, Institute of Terrestrial Ecology, Titus Wilson & Son Ltd, United Kingdom, pp. 303–315.

Britt, J.R., Zutter, B.R., Mitchell, R.J., Gjerstad, D.H. and Dickson, J.F. (1990) Influence of herbaceous interference on growth and biomass partitioning in planted loblolly pine (*Pinus taeda*). *Weed Science*, **38**, 497–503.

Cannell, M.G.R. (1985) Dry matter partitioning in tree crops. In M.G.R. Cannell and J.E. Jackson (eds.), *Attributes of trees as crop plants*, Institute of Terrestrial Ecology, Titus Wilson & Son Ltd, United Kingdom, pp. 160–193.

Coker, A., Court, D. and Silvester, W.B. (1987) Evaluation of foliar urea applications in the presence and absence of surfactant on the nitrogen requirements of conditioned *Pinus radiata* seedlings. *New Zealand Journal of Forestry Science*, **17**(1), 51–66.

Douglas, L.A. (1987) Effects of cultivation and pesticide use on soil biology. In P.S. Cornish and J.E. Pratley (eds.), *Tillage: new directions in Australian agriculture*, Australian Society of Agronomy, Inkata Press, Melbourne, Australia, pp. 308–317.

Ellis, R.C., Webb, D.P., Graley, A.M. and Rout, A.F. (1985) The effect of weed competition and nitrogen nutrition on the growth of seedlings of *Eucalyptus delegatensis* in a highland area of Tasmania. *Australian Forestry Research*, **15**, 385–408.

Firth, D.J. (1991) The role of mulch and organic manures for sustainable horticulture in subtropical environments. *Agnote*, Reg1/038, NSW Agriculture.

Firth, D.J. (1992) Covercrops for subtropical orchards. *Agfact*, H6.3.10. NSW Agriculture.

Goode, J.E. and Hyrycz, K.J. (1976) The effect of nitrogen on young, newly-planted, apple rootstocks in the presence and absence of grass competition. *Journal of Horticultural Science*, **51**, 321–327.

Kolb, T.E. and Steiner, K.C. (1990) Growth and biomass partitioning of northern red oak and yellow-poplar seedlings: effects of shading and grass root competition. *Forest Science*, **36**(1), 34–44.

Lamont, B. (1978) Root systems of the Myrtaceae. *Australian Plants*, **10**(78), 74–78.

McMillan, M. and Cook, T. (1995) Final Report; DAN 74A—Herbicides for weed control in tea tree. Rural Industries Research and Development Corporation, Canberra, Australia.

Murtagh, G.J. (1996) Month of harvest and yield components of tea tree. I. Biomass. *Australian Journal of Agricultural Research*, **47**, 801–815.

Nambiar, E.K.S. and Zed, P.G. (1980) Influence of weeds on the water potential, nutrient content and growth of young radiata pine. *Australian Forestry Research*, **10**, 279–288.

Nambiar, E.K.S. and Sands, R. (1993) Competition for water and nutrients in forests. *Canadian Journal of Forestry Research*, **23**, 1955–1968.

Putnam, A.R. and Weston, L.A. (1986) Adverse impacts of allelopathy in agricultural systems. In A.R. Putnam and C.S. Tang (eds.), *The Science of Allelopathy*, Wiley, New York, U.S.A., pp. 43–56.

Sands, R. and Nambiar, E.K.S. (1984) Water relations of *Pinus radiata* in competition with weeds. *Canadian Journal of Forestry Research*, **14**, 233–237.

Smethurst, P.J., Comerford, N.B. and Neary, D.G. (1993) Weed effects on early K and P nutrition and growth of slash pine on a Spodsol. *Forest Ecology and Management*, **60**, 15–26.

Storrie, A., Cook, T., Virtue, J., Clarke, B. and McMillan, M. (1997) Weed management in tea tree plantations. NSW Agriculture.

Tamasi, J. (1986) Agrotechnical factors modifying the root system. In J. Tamasi (ed.), *Root Location of Fruit-trees and its Agrotechnical Consequences*, Akademiai Kiado, Budapest, Hungary, pp. 47–65.

Tisdall, J.M. and Huett, D.O. (1987) Tillage in Horticulture. In P.S. Cornish and J.E. Pratley (eds.), *Tillage: new directions in Australian agriculture*, Australian Society of Agronomy, Inkata Press, Melbourne, Australia, pp. 72–93.

Virtue, J.G. (1997) Weed interference in the annual regrowth cycle of plantation tea tree (*Melaleuca alternifolia*). PhD Thesis. The University of Sydney.

Woods, P.V., Nambiar, E.K.S. and Smethurst, P.J. (1992) Effect of annual weeds on water and nitrogen availability to *Pinus radiata* trees in a young plantation. *Forest Ecology and Management*, **48**, 145–163.

5. INSECT PESTS OF TEA TREE: CAN PLANTATION PESTS BE MANAGED?

A.J. CAMPBELL AND C.D.A. MADDOX

NSW Agriculture, Tropical Fruit Research Station, Alstonville, NSW, Australia

INTRODUCTION

Before 1980, the tea tree industry was largely an opportunistic cottage industry. Production of oil followed the harvesting and distillation of leaf from natural stands of *Melaleuca alternifolia* (Cheel) growing on the coastal flood plains of the Richmond and Clarence rivers of northern New South Wales (NSW). Natural populations were sparse and closely interspersed with *Eucalyptus*, *Acacia* and *Casuarina* species. Periodic defoliation by leaf chewing insects was common and pest management was nonexistent.

Since 1980 about 3,400 ha of plantation tea tree, of limited genetic diversity, have been established in NSW (Clarke 1996). Plantations are established with 30,000–40,000 plants per hectare (Colton and Murtagh 1991) and are mostly harvested annually.

Within plantations the young leaf available to pests has dramatically increased. At the same time no significant increase in the large wood or bark components observed in natural forests has occurred. This change in the ratio of the leaf to wood and bark components and the short cutting cycle has caused a shift in the composition of the insect fauna.

Moving to a plantation based industry has focused attention on the monitoring and control of insects damaging foliage. With further investigation and collecting under differing seasonal conditions, the number of pest species associated with tea tree will increase. Some currently recognised pests (e.g. *Paropsisterna tigrina* (Chapuis) commonly called Pyrgo beetle) may become less significant with improved pest control measures and monitoring strategies. Other invertebrate groups (e.g. mites and psyllids) may increase in significance, once their true impacts on oil yield are known.

This chapter contains basic information on the pests of tea tree in Australia based on the limited available knowledge. The pests vary seasonally, annually, and from plantation to plantation. The response of the NSW growers to a survey on pests and industry issues is summarised. Some producers have a poor knowledge of the pests and their control options. This poor knowledge base when combined with the inappropriate use of chemicals has led to the occasional residue in oil in the market place. The potential for residues will determine the direction of future plantation management practises for pest control. Tea tree pests are discussed according to the stage of growth and the plant component that they damage. A successful monitoring technique for *P. tigrina* is outlined and its value to the industry in terms of better pest management and more effective pesticide use is discussed.

INSECT FAUNA OF PLANTATIONS AND NATURAL STANDS

The initial entomological study by Treverrow (1992) on the north coast of NSW identified 66 phytophagous species, or species groups, found in tea tree. Of these, 35 were restricted to plantations, 21 were recovered from native areas and an additional 10 were common to both. The number of phytophagous species is now over 100 based on the collections made by the authors. Leaf beetles of the family Chrysomelidae (12 species each from a different genus) dominate. The Chrysomelidae, including *P. tigrina*, cause rapid defoliation and have the greatest obvious impact on oil recovery. All growers considered *P. tigrina* to be the most significant pest.

INDUSTRY PERSPECTIVES ON PESTS

A survey of 30 producers, accounting for 60% of the estimated NSW plantation area, was conducted in 1996. The plantations ranged from 0.5–1,200 ha in size, with a median of 12 ha. Small producers dominate the industry. Ten plantations were less than 5 ha in size while an additional 5 were from 5–10 ha. Only 10 plantations were greater than 20 ha in size, of which 4 claimed organic status. For holdings of less than 10 ha, 14 claimed to be organic. Smaller producers have the poorest understanding of the pests present and the available control options. All growers, irrespective of size and production philosophy, had experienced losses due to insects at some time.

Most growers, whilst not stating exactly how they control insect pests, were generally happy with their current pest management strategies. Nevertheless, identification of the smaller sap-sucking groups (mites and psyllids) was a recurring problem. All producers were acutely aware of the difficulties in presenting a product with a "clean green" image in the market place. For the ease of marketing, growers are now seeking organic production status. This shift towards organic production, if closely adhered to, should insure residue-free oil in the market place.

Insecticidal residues were detected in oil in 1995. However, the industry as a whole does not consider insects to be a major issue. Only 12% of producers now regarded them as a major problem when they occurred. Producers consider Pyrgo beetle, psyllids and mites as the worst pests (Table 1). More than 82% of the growers believed marketing of oil was the largest single issue facing the industry.

INSECTS DAMAGING NEW SEEDLINGS

African Black Beetles, Mole Crickets and Cut Worms

The introduced African Black Beetles (*Heteronychus arator*) periodically damages newly set transplants in the field. Transplants are ringbarked at or below ground level, which causes desiccation and death. Adult beetles attracted into plantations from the surrounding pastures cause the damage. It is unlikely the beetle or its larvae could survive

Table 1 Industry survey—insect pests causing problems for growers and warranting control

Pest	*Grower response* (%)
Pyrgo	53.3
Psyllids	50.0
Mites	30.0
Tree hoppers	16.7
Saw flies	6.7
Scarabs	23.3
Weevils	6.7
Stem borers (moths)	3.3
Mealy bugs	3.3
Monolepta beetles	3.3
Scale	3.3

Source: Campbell and Maddox (1996).

the soil preparation work associated with plantation establishment. Damage is most severe in spring and late summer-autumn on light, well-drained soils.

No evidence exists to indicate Black Beetles are a pest in established plantations. The main effect of damage is the loss of plants within an area that may warrant replanting if severe. Since the adult beetles seldom fly (Goodyer 1995), maintenance of a clear uncultivated buffer may reduce the potential for damage. The worst damage is likely to occur in areas previously cropped with, or near, sugarcane.

Mole Crickets

Gryllotalpa spp. are easily recognised by their powerful forelegs used for digging. Mole crickets can cause significant damage to newly set tea tree transplants by severing their roots. Crickets usually forage in horizontal tunnels in the top 5 cm of soil but may feed on the soil surface during periods of high humidity following rain or irrigation. Damage relates to soil type and is worst in sandy loams near rivers or creeks.

Cutworms

Agrotis caterpillars occasionally attack newly set transplants, cutting through their stems near ground level and feeding on the felled plants. Both *Agrotis ipsilon* (black cutworm) and *A. munda* (brown or pink cutworm) occur in spring through autumn along the coast. Moths breed locally in pasture or weedy areas, or are blown in (often over considerable distances) on the wind associated with storm fronts. Cool, wet conditions in spring through autumn favour outbreaks of black cutworms. The brown cutworm prefers a mild dry winter to give good survival of over-wintering pupae. Early emergence of moths and rapid population development occurs with good growing conditions in spring through autumn.

Cutworms are more likely to infest new plantations next to pasture or weed areas. Transplants held near lights at night, that attract *Agrotis* moths, may become infested before field planting. Feeding damage if severe can kill or stunt plant growth and replanting may be necessary. Cutworms are only a problem after the transplanting stage and before the onset of rapid plant growth.

INSECTS DAMAGING FOLIAGE OF ESTABLISHED PLANTS

Chrysomelids

Paropsisterna tigrina has emerged as the most significant pest of plantation tea tree. The beetles over-winter in sheltered crevices within poorly managed plantations or in the surrounding wooded areas. Adults begin feeding and egg production in spring on expanding flush growth. Soil temperatures above 15°C (Curtis 1993) and adequate soil moisture triggers plant growth. The larvae are initially gregarious. On consumption of the young flush growth near the oviposition site the larvae spread out over the plant. Both larval and adult stages feed on young foliage produced in spring and autumn.

Like most paropsines on their respective myrtaceous hosts, *P. tigrina* is a specialist feeder and has adapted to its host's chemical defences. It is strictly a flush growth feeder although it will excise mature leaves when high population densities (>50 per plant) occur.

The rate at which new foliage matures on the plant is the key factor when assessing the resistance of a particular plant to *P. tigrina* attack. Many changes occur in the leaves as they age. The oil constituents change from precursor to end product forms (Southwell and Stiff 1989) and volatile emissions decrease (Murtagh 1994). Leaves become harder and their colour changes dramatically (Maddox 1996). Leaf colour changes from yellow to dark green. This occurs anywhere from near the bud to below the 10th pair of leaves, depending on their growth rate. Young regrowth is more vivid with a higher leaf colour chroma and value (based on Munsell notation), and this is the basis for the use of glossy yellow plates as a monitoring tool.

Changes in leaf oil are more rapid and occur in the first green leaf below the terminal bud (Southwell and Stiff 1989). In spring and autumn when flush growth occurs, *P. tigrina* females detect (either physically or chemically) these oil changes in the foliage and this stimulates oviposition after feeding. The preferred oviposition site is on the first "mature" leaves back along a twig from the terminal bud. On hatching the first instar larvae move to the terminal foliage to feed. Their survival, like other paropsine beetles, is dependent on the presence of young soft foliage (Ohmart *et al.* 1987; Larsson and Ohmart 1988; Ohmart 1991; Maddox 1996; Patterson *et al.* 1996). Trees must be actively growing to support a larval population. Once a tree has had the flush growth removed it is no longer suitable for *P. tigrina* oviposition. The probability of forecasting an outbreak is high if colonising adults can be monitored because oviposition is locally synchronised.

The number of *P. tigrina* generations within a plantation depends on temperature and soil moisture levels. The maximum daily rate of egg production is 13 eggs/female/day, at 29°C and 85% humidity. Exposure of eggs for 2–6 hours at 38–40°C reduces hatching to less than 15%; while exposure of 1st and 2nd instar larva to 35°C for 4–6 hours killed more than 75%. Exposure of small larvae at 40°C for 4 hours causes 100% mortality. Field temperatures were corrected for solar radiation affects, otherwise body temperatures are underestimated by 8°C for larvae and 3–4°C for eggs (Maddox 1996).

Temperature influences beetle activity and egg production. Outside the temperature range of 15–35°C egg production is negligible. Females resume oviposition after feeding on new flush growth for 5–7 days at 25°C and 80% relative humidity. Field temperatures outside the threshold range of 11.5 and 35°C, in spring and autumn, cause the bimodal distribution pattern of *P. tigrina*. On the north coast of NSW 5 generations of beetles are theoretically possible, but a maximum of 3 generations is more likely (Figure 1).

Yellow sticky traps located above the plants and throughout a plantation to survey adults maybe of more benefit than counting beetles on plants (Campbell and Maddox 1996). The traps intercepted beetles up to 2 weeks before finding beetles on the plants. Trap counts give managers a warning of the location and likely time of damage by the beetles and their larvae.

Trap colour was determined by comparing counts of *P. tigrina* on different coloured plates in the field. A high gloss yellow consistently gave the highest counts. Given the variability within plantations and the size of *P. tigrina* aggregations, traps on a 50 m grid allow accurate detection and plotting of activity. Placing traps in the field before the flights of the over-wintering adults can significantly reduce the use of insecticide. Traps allow site specific spraying to minimise foliar damage, reduce follow up insecticide use and restrict the expansion of the pest within the plantation.

Detection of the invading over-wintering population of *P. tigrina* is the crux of successful monitoring and control strategy in managed plantations.

In managed plantations, colour traps located around the perimeter indicate migratory activity. Within a plantation, *P. tigrina* populations establish in spring then leave and return to the same areas in autumn. The mechanism responsible for this behaviour is unclear and understanding this process could allow refinements to the monitoring technique. Evaluation and field testing of potential tree marking compounds (namely thujanes or cineole metabolites) identified by Southwell *et al.* (1995) are required.

Significant *P. tigrina* outbreaks have occurred at most plantations in NSW. The outbreaks monitored were at Wyrallah in February and November 1991, Grafton in February 1992 and Cudgen in February 1996. These outbreaks were linked to the occurrence of major rainfall events (150–250 mm in 48 hours) followed by rapid plant growth. At these times because of wet conditions, growers were unable to adequately protect the foliage during the build up phase of *P. tigrina*. Favourable weather patterns with no hot dry periods and maximum temperatures less than 32°C for 5–6 weeks lead to the development of 2 or 3 generations of *P. tigrina*. Leaf cutting occurred when the adults swarm, and little or no flush growth remained on the trees. Management of *P. tigrina* under these conditions depends on access to the plantations, otherwise losses will remain high. Such weather patterns are not uncommon during autumn on the north coast of NSW.

Scarabaeids (Pasture Scarabs)

Diphucephala lineata, the small green pasture scarab causes extensive localised defoliation of tea tree in early to mid summer on some plantations. The beetles feed and mate on

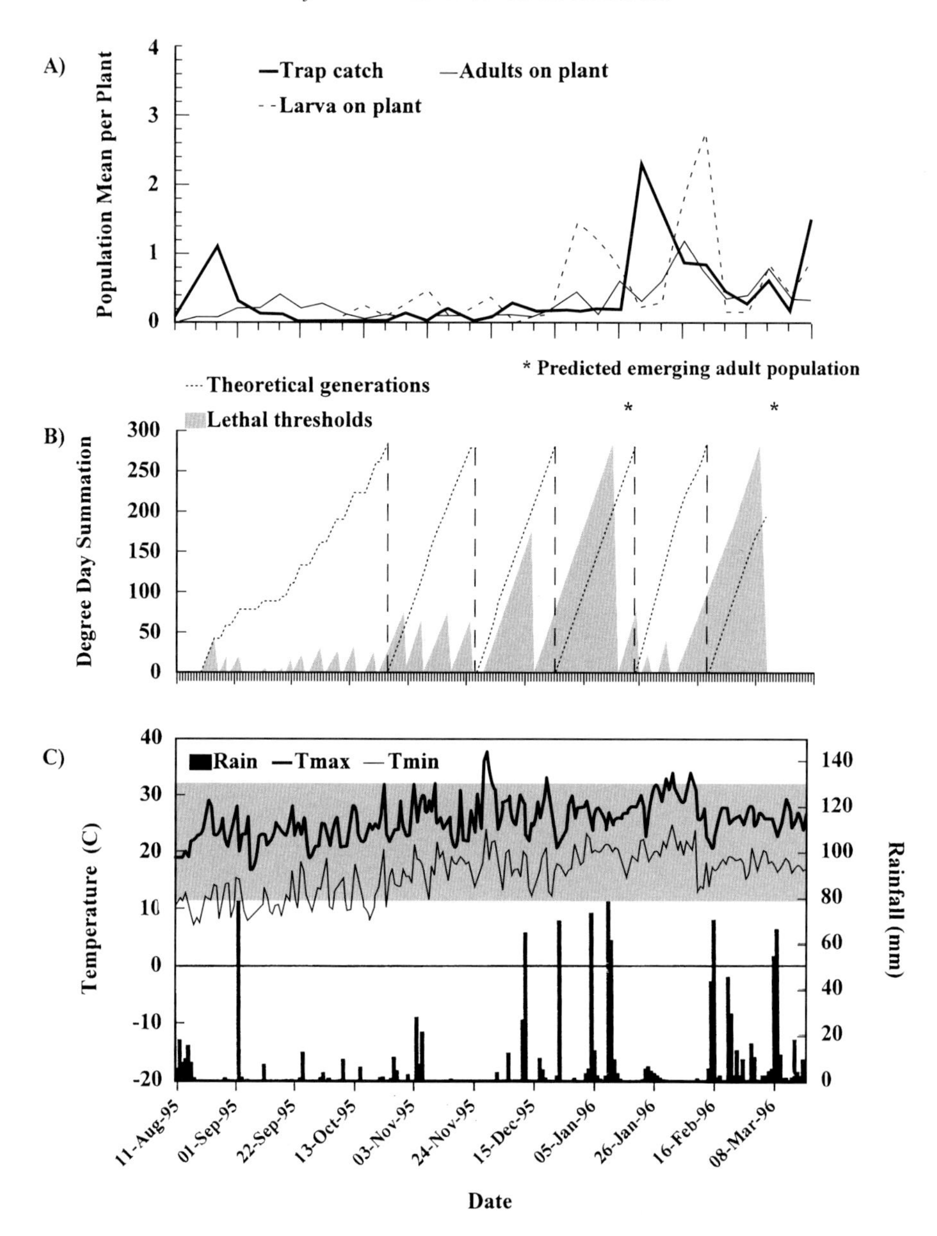

Figure 1 The use of temperature and insect development data to describe the field activity of *Paropsisterna tigrina* on *M. alternifolia* at Ballina during the 1995–96 growing season. Chart A plots the incidence of adults on yellow flight traps (n = 44) within the plantation, and the larval and adult activity on the plants (10 plants neighbouring each trap) over 31 weeks. Chart B shows the theoretical insect development possible over the same time scale based on available heat (291 Degree Days for each adult). If the threshold temperatures (11.5 and 32°C) are considered lethal then predicted adult emergence (*) matches field trap catch patterns in chart A. Chart C contains the daily weather information gathered from an automatic station scanning hourly.

the trees then return to the soil to lay their eggs. Trees at the edge of plantations bordering remnant forest or pasture are at the greatest risk and the worst damage occurs in plantations near lighter sandy soils. The larvae are subterranean pests of pasture and grass species.

Nectar Scarabs (*Phyllotocus* spp.)

Nectar scarabs are small light brown beetles with a dark brown tip to each wing cover and long hind legs. These beetles can swarm, particularly on plants in flower but cause no damage. Oviposition occurs in the soil, and on hatching the larvae feed on decaying organic matter and pasture roots. When plantation managers see large numbers of beetles they occasionally panic and want to spray.

Psyllids

The taxonomy of tea tree psyllids is poorly understood. The commonly occurring *Trioza* sp. prefers young flush growth and occur within plantations all year. However, population explosions occur in late summer through spring, when plants may be under water stress. Feeding by the mature and immature psyllids causes pitting and some distortion of the leaves and shoots. Psyllids deposited their characteristic yellow eggs that are easily visible at low magnification along the leaf margins. During dry periods in summer drops of sugary exudate can leak from the feeding sites, crystallise and fall to the ground. As with the other sap-suckers, ants often indicate the presence of psyllids.

Psyllids can be introduced into clean areas on the planting material. Poor nursery hygiene and failing to insure pest-free transplants is the source of the problem. The impact of psyllids on oil production needs quantifying and their control measures investigated.

Eriophyid Mites

Like the psyllids, the eriophyid mite complex associated with *M. alternifolia* is unknown and no species are described. However, eriophyids occur all year but the damage is more apparent in the winter and spring.

The eriophyids distort new foliage and cause the leaf margins on the ventral surface to bend; cells on the leaf surface within the distorted area bubble and become hairy. The mites live within the distorted areas. Eriophyids prefer new growth but also occur on basal growth near the ground. Removal of foliage from the residual stems after harvesting, for example with a flame weeder, may delay the colonisation of new growth with mites. The value of flame weeders in this role needs confirming. Growers can check for the presence of mites by gently heating the distorted foliage and examining it with a hand lens. Heat causes the mites to become active and move about within the distorted foliage.

Work by the authors indicates varietal differences in susceptibility of monoclonal plants to the eriophyids. Current work by D. Knihinicki (NSW Agriculture, Orange) should resolve the taxonomy and ecology of mites associated with *M. alternifolia.*

Leaf Galls

Distortion and stunting of buds on new growth due to the gall fly *Dasineura* sp. are common from November to March. Individual buds become brown, fail to expand then die and a shoot below the damage then assumes apical dominance. Multiple larvae can develop within a shoot after female flies place their eggs at the base of the bracts. Because of the ability of the adults to fly, edge effects are not obvious and superficial damage occurs across plantations. Despite the obvious damage in the field, gall fly is not a major pest. Variations in the susceptibility of individual plants to the gall fly are apparent in the field.

Aphids

As with most crops, aphids occur in tea tree. *Aphis gossypii* Glover (cotton or melon aphid) occurs particularly from October to May. Aphids prefer the terminal shoots of new flush and are often attended to by ants that collect the honeydew. Black sooty mould grows on the honeydew and indicates aphid or other sap-sucker activity.

Scale Insects

A number of scale species, including lac and nigra scales, which are major pests on other tree crops has been identified. Plants growing on marginal sites, or already weakened by defoliating or sap-sucking insects, appear the most prone to scale infestation. Scales are not a major problem within plantations as the cutting cycle does not favour the development or spread of this pest. Predators including coccinellids (generalist predators) and wasp parasites exert some control.

Leaf Hopper

Both the adult and immature stages suck sap from the expanding shoots on flush growth. Feeding causes the wilting of the shoots, that can then develop a distinct purple colouration. As with other sap-sucking groups leaf hoppers are tended by ants. Black sooty mould commonly grows on the exudate produced. Leaf hoppers occur all year, but appear most active when the host plant has new growth in spring and autumn.

Characteristically, adult leaf hoppers are less than 10 mm long, however the dominant species on tea tree (including *Erythroneura* sp.) are less than 5 mm long. Adults hold their wings tent-like when at rest. The wing colour varies from being transparent to pale green, grey or yellow depending on the species. Immature stages hop when disturbed, while the adults either hop or fly, making them difficult to collect.

Sawflies

The tea tree sawfly larvae (*Pterygophorus* sp.) causes major defoliation in native stands resulting in significant economic losses for bush cutters. In late summer wasps insert eggs in rows into the leaf tissue of young foliage. Hatching occurs within 10–15 days and once emerged the larvae begin feeding on the leaf. The larvae have an orange-brown head capsule, a dull body with cream and green longitudinal stripes and a pointed tail that they raise in defence when disturbed. Larval pupation occurs within the thick soft bark of mature trees, in residual stumps or in fallen debris.

In plantations with short cutting cycles and complete removal of the above ground biomass, sawflies are unlikely to become established. Plantations next to native stands should be checked in late summer and autumn when adults could fly in and lay eggs that could lead to larval defoliation. Neglected plantations could become a breeding site for sawflies in time.

Moth Larvae

The larvae of loopers, leafrollers, leaftiers, etc. cause some defoliation. However, moth larvae are often controlled when spraying for other pests such as *P. tigrina*, psyllids etc. Loopers feed on exposed leaves and may be 30–40 mm when fully grown. They arch up the middle of the body into an inverted U-shape when moving. When disturbed they remain attached at the rear and wave the front part of the body around.

The leaf-roller and tiers shelter during the day and emerge in the evening to feed. Disturbance of the leaf shelters often causes the larvae to drop down on a silk thread.

Adult moths can have a wing span of up to 50 mm and depending on species vary from dark brown to creamy white, often with patterned fore and hind wings. They fly near dusk and deposit eggs singly on leaves or stems.

Predatory bugs, e.g. *Oechalia schellenbergi* (Guérin-Méneville), exert significant control on these moths, particularly the looper caterpillars. These bugs also prey upon the larvae of *P. tigrina*, particularly late in the season. The potential of this bug is being investigated on a large plantation producing oil to organic certification. Insectary produced bugs will augment natural field populations.

INSECTS FEEDING ON WOOD AND BARK

Cerambycidae (Longicorn Beetles)

Too few beetles have been collected from plantations to form an opinion of the pest status of this group. Longicorns feed on wood. In plantations the harvest schedule and the size of the woody components are not conducive for these pests. Never-the-less plantations near remnant native stands may find occasional beetles on plant stems. In the future the old cut stems near ground level could act as an entry point for some cerambycids.

Weevils

A range of adult weevils, including *Aades* sp. and *Amnemus quadrituberculatus* (Boheman), have been collected feeding on the bark of stems within plantations. Depending on the density of adults and the extent of their feeding, stems can be ringbarked and die. The incidence of weevil damage is increasing and these feeding sites may allow entry of spores of *Dothionella ribis*, which can kill whole plants causing significant gaps within a plantation.

Successful control of weevils will depend on understanding the biology and ecology of the pests. The significance of weevil damage is currently an unknown quantity. Since the weevils feed on the plant stems, application of chemicals using current plantation technology will be inadequate.

Termites

Termites occur along the east coast, but they are not a problem in tea tree plantations. However, plantations not protected by a high water table are at risk. At West Wyalong, inland New South Wales, termites are a significant pest causing the death of trees (Cumming 1997). This probably indicates the trees are growing on a poor-quality site outside their natural geographical range. The termites enter the plant through the root system and then continue to feed within the stems. Control of termites will be both difficult and costly.

Stem-boring Lepidoptera

An unidentified stem borer is causing concern for some growers in northern New South Wales. The moth lays its egg within 30 cm of the apex of the dominant leader and just above a leaf base. On hatching the new larva enters the center of the shoot and chews its way either up or down before pupation. Before pupation the larva chews the wood in the shoot down to the cambial layer. Over time the shoot desiccates before snapping off in the wind and releasing the moth. The extent of damage only becomes obvious once the shoots desiccate and become a straw colour. Despite its apparent increase in some plantations, for reasons unknown, the damage caused by the moth is probably not significant.

VALUE OF PARASITES AND PREDATORS

With the push for producers to achieve organic status, the use of predators and parasites to control pests is an attractive management option. However, despite the existence of many beneficial insects attacking all stages of the pests in tea tree they are a relatively ineffective form of pest control and cannot be relied upon. The main problem is the lag in the development of the pests and the beneficial species. Beneficial insects usually become effective towards the end of a cycle of pest activity. They are at their greatest abundance after most of the damage has occurred. Current management practises on plantations do not allow for refugia in which beneficial insects can shelter or over-winter. The development of beneficial insect populations recommences each season within the plantation and depends on migration. The parasites and predators found are generalists and attack most pests found within tea tree.

PROBLEMS ASSOCIATED WITH CHEMICAL USAGE FOR PEST CONTROL

When *P. tigrina* or other pests invade a plantation, and the foliage practically disappears over night, the incentive to use an insecticide is high. This practice is safe if only registered chemicals are used in accordance to their label. Withholding periods before harvesting must be followed. Registered chemicals applied at inappropriate rates or frequencies could cause residues and their presence will influence marketing opportunities.

The unique nature of the crop, i.e. oil in oil sacks within the leaf, the solubility of most insecticides in the oil itself and the extraction process increases the chances of residues. Contract distillers must be aware of the possibility of residue carry over from one job to the next and take steps to clean the still between runs. Given the potential for tea tree oil, the industry must take responsibility for registering new chemicals after their proper screening for efficacy and residues. The industry should not rely on chemicals registered for use in other crops. Softer control options like *Bacillus thuringiensis* (Elliot *et al.* 1992) and organic methods are available, but need evaluating for use on tea tree.

CONCLUSIONS

On the evidence available *P. tigrina*, mites and psyllids cause the greatest damage within plantations and *P. tigrina* is the best recognised pest. Understanding conditions necessary for the development of pest outbreaks, i.e. threshold temperatures for the survival of eggs and larvae, allows the generation of a model that accounts for field behaviour. In conjunction with a reliable trapping method, e.g. colour trapping for *P. tigrina*, viable management strategies and action thresholds can be developed. These strategies if applied will be cost effective, minimise the use of insecticides and reduce the risk of residues in the oil.

The lack of biological data prevents the development of models for pests other than *P. tigrina* and the collection of such data should be an industry priority. Without such data the misuse of chemicals will continue to occur, increasing the possibility of oil contamination in the marketplace.

Finally, not all insects found in plantations are pests. The full pest complex of tea tree remains unknown, action thresholds are non-existent and the significance of any pest depends on the weather patterns.

REFERENCES

Campbell, A.J. and Maddox, C.D.A. (1996) Insect Pest Management in Tea Tree. *RIRDC Final Report DAN-9LA*, NSW Agriculture Tropical Fruit Research Station Alstonville, December 1996.

Clarke, B. (1996) NSW Agriculture, Casino. Personal communication.

Colton, R.T. and Murtagh, G.J. (1991) Tea Tree Oil Plantation Production. Agfact P6.4.6 NSW Agriculture & Fisheries.

Cumming, A. (1997) Mount Mulga Pastoral Co., West Wyalong. Personal communication.

Curtis, A. (1993) Growth and oil production of Australian *M. alternifolia*. M.Ag.Sc. Thesis, University of Queensland, Agriculture Faculty.

Elliot, H.J., Bashford, R., Greener, A. and Candy, S.G. (1992) Integrated pest management of the Tasmanian *Eucalyptus* leaf beetle, *Chrysophtharta bimaculata* (Olivier) [Col.: Chrysomelidae]. *Forest Ecology and Management*, **53**, 29–38.

Goodyer G. (1995) African Black Beetle. Agfact AE.54 NSW Agriculture.

Larsson, S. and Ohmart, C.P. (1988) Leaf age and larval performance of the leaf beetle *Paropsis atomaria*. *Ecological Entomology*, **13**, 19–24.

Maddox, C.D.A. (1996) Aspects of the biology of *Paropsisterna tigrina* (Chapuis) the major pest of *Melaleuca alternifolia* (Cheel). M.Sc. thesis, Department of Entomology University of Queensland, Brisbane, Australia.

Murtagh, G.J. (1994) Oil gland research techniques. *Rural Industries Research and Development Corporation Final Report*, NSW Agriculture, Wollongbar, January 1994.

Ohmart, C.P., Thomas, J.R. and Stewart, L.G. (1987) Nitrogen, leaf toughness and the population dynamics of *Paropsis atomaria* (Olivier) [Col.: Chrysomelidae]—a hypothesis. *Journal of the Australian Entomological Society*, **26**, 203–207.

Ohmart, C.P. (1991) Role of food quality in the population dynamics of chrysomelid beetles feeding on *Eucalyptus*. *Forest Ecology and Management*, **39**, 35–46.

Patterson, K.C., Clarke, A.R., Raymond, C.A. and Zalucki, M.P. (1996) Performance of first instar *Chrysophtharta bimaculata* larvae (Coleoptera: Chrysomelidae) on nine families of *Eucalyptus regnans* (Myrtacae). *Chemoecology*, **7**, 1–13.

Southwell, I.A., Maddox, C.D.A. and Zalucki, M.P. (1995) Metabolism of 1,8-cineole in tea tree (*Melaleuca alternifolia* and *M. linariifolia*) by pyrgo beetle (*Paropsisterna tigrina*). *Journal of Chemical Ecology*, **21**, 439–453.

Southwell, I.A. and Stiff, I.A. (1989) Ontogenetical changes in monoterpenoids of *Melaleuca alternifolia* leaf. *Phytochemistry*, **28**, 1047–1051.

Treverrow, N.L. (1992) The insect fauna of *Melaleuca alternifolia* with emphasis on three known pest species. *Rural Industries Research and Development Corporation Final Report*. NSW Agriculture, Wollongbar, November 1992.

6. BIOMASS AND OIL PRODUCTION OF TEA TREE

G. JOHN MURTAGH

Agricultural Water Management, Goonellabah, NSW, Australia

INTRODUCTION

The economic viability of the production of tea tree oil is heavily dependent on the oil yield from a plantation (Reilly 1991). Whilst other factors such as operating costs and the price of oil also affect profitability, the oil yield can vary considerably making it a key variable in any analysis of plantation profitability.

Tea tree oil is an essential oil that consists of a complex mixture of secondary plant products. The synthesis and accumulation of such products is typically complex and can be endogenously controlled, dependent on development processes that are related to cell differentiation, and sometimes regulated by exogenous factors including light, temperature and wounding (Wiermann 1981). This chapter explores the range of factors that appear to affect the production of tea tree oil, and indicates similarities and differences to other essential oil crops. With tea tree, the major components of oil yield are the oil concentration in leaves and the leaf yield. Both are affected, but in different ways, by a number of factors including the environment, plantation management and genetics (Murtagh 1991).

Most tea tree oil that is used in commerce is sourced from selected chemotypes of *Melaleuca alternifolia* (Murtagh 1998). Suitable oil can also be obtained from chemotypes of *M. linariifolia* (Williams 1995), *M. dissitiflora* (Brophy and Lassak 1983) and *M. uncinata* (Brophy and Lassak 1992), but as most experimental and commercial experience is with *M. alternifolia* the discussion will refer to this species unless indicated otherwise.

Terminology and Plant Parts

In this chapter, the term oil concentration is used to describe the amount of oil in a unit weight of the plant, or part of the plant such as the leaf. The standard unit is milligrams of oil per gram of dried leaf (mg/g). When the amount of oil was given as a volume in the referenced publications, it was converted to a weight by multiplying by the average oil density of 0.9 (Penfold and Morrison 1950). Another common unit of concentration expresses the weight of oil as a percentage of the plant weight. This can be converted to mg/g by multiplying by ten. To provide consistency, all published results were converted to the standard unit in this chapter.

When the weight of the plant or oil is referenced to a unit area where it grew, it is referred to as a yield with units of kg/ha or g/m^2. Some authors use the term oil yield to describe the concentration as defined above, but in this chapter such use was altered to maintain consistency.

Much agronomic work refers to the yield of above-ground parts and does not include the weight of roots. The above-ground growth is referred to as biomass. As tea tree oil is found only in leaves, it is useful to subdivide the total biomass into three components; leaves, fine stems and main stems. Fine stems are defined as stems of less than 2.5 mm diameter (Murtagh 1988), and main stems are the remainder. Fine stems carry virtually all the leaves. The total of leaf plus fine stems are called twigs. Such measurements are expressed on a dry-weight basis using separate estimates of the moisture content in each fraction to make the conversion. When twigs were distilled and the oil concentration was expressed per unit weight of twig, it was converted to a concentration per unit leaf weight by dividing by the measured ratio of leaf in twig. If this was unavailable and the measurements being compared were taken over a short interval, a typical ratio of 0.68 (Murtagh 1996) was used.

OIL CONCENTRATION

Tea tree oil is stored in subepidermal glands that are adjacent to the epidermis and equally distributed on both sides of a leaf (List *et al.* 1995). Oil glands are first apparent in immature leaves, with the number per leaf increasing as the leaf expands, to reach a maximum just before the leaf is fully expanded. The oil gland density appears to be under some degree of genetic control (List *et al.* 1995).

Analytical Methods

Two principal methods are available to determine the oil concentration: steam distillation or solvent extraction. A specific case of steam distillation that has the biomass immersed in the boiling water is sometimes termed hydrodistillation. More specialised methods such as head space analysis are usually confined to detailed experiments. Each of the two methods has its own advantages. Steam distillation can accommodate the larger samples that arise when all positions on a single plant or a number of plants in an agronomically sized plot are sampled. The method mimics the commercial process and gives oil of a similar chemical composition. It is important that the system incorporates a cohobation or reflux return to obtain complete recovery of the alcohol components of the oil (Kawakami *et al.* 1990; Murtagh 1991a).

On the other hand, solvent extraction followed by quantitative gas chromatographic analysis can use much smaller samples that can vary from a single leaf up to at least 5 g. Solvent extraction also avoids the conversion of precursor compounds in young leaves to the major constituent, terpinen-4-ol, as occurs with steam distillation (Southwell and Stiff 1989; Cornwell *et al.* 1995), and thereby suits biochemical investigations. It is the system of choice when there are a large number of samples, as in plant breeding programs, and in such sampling the issue of extracts having a different oil composition to the commercial product, which is distilled, can be avoided by not using young growth (Baker 1995). Tea tree has a sufficiently high oil concentration to enable a direct chromatographic analysis of the solvent extract without the need to concentrate the extract by evaporation, thus avoiding the accompanying losses that can be a drawback of the solvent extraction method (Charles and Simon 1990).

In a comparative study with tea tree (Baker *et al.* 1995), the two methods gave a similar recovery of monoterpenes, but more sesquiterpenes with solvent extraction. A major problem in doing the comparison was to obtain equivalent samples of very different size.

Either fresh or air dried samples can be distilled without affecting the result (Murtagh and Curtis 1991). Although dried samples distill more slowly, distillation is virtually complete within two hours (Murtagh and Smith 1993). Similar delays occur with solvent extraction (Doran *et al.* 1996). Because of oil losses, samples should not be oven dried before distilling. Losses increase with the drying temperature, and also vary between samples (Curtis and Murtagh 1989). The widest variation they measured between samples ranged from no oil loss at 45°C and 50% loss at 125°C in samples from one tree, to 33% loss at 45°C and 93% loss at 125°C from another tree. Points of note in these results were the consistent pattern of relatively high or low losses across all temperatures within a batch, and the lack of a relation between the relative batch loss and the moisture content of the samples before drying.

Variation in Oil Concentration

From first experience, the oil concentration was observed to vary over time (Penfold *et al.* 1948) and this gave rise to a number of studies that attempted to document and explain the variation (Table 1). Of the 12 studies that measured variation over time, five recorded a variation of more than 100% above the lowest value in the study, six had a variation between 15–57%, and only one recorded no variation. Similar variation is often found with other essential oils or secondary metabolites, as instanced in reviews by Flück (1963), Wiermann (1981), Harborne and Turner (1984), Lawrence (1986), Gershenzon and Croteau (1991).

Seasonal Variation

The oil concentration is generally highest in summer and lowest in late winter/early spring. Figure 1A shows the average seasonal variation in repeated tests on two plantations (Murtagh 1992; Murtagh and Smith 1996) in the humid subtropical environment of northern New South Wales (NSW). Although insufficient samples were taken during June–August to complete part of the trend line on the plantation that experienced winter frosts, the seasonal range was greater where winters were cooler. The trends shown in Figure 1A are means over a number of years, and while all years have a seasonal trend it can vary both in absolute magnitude and extent of variation (Murtagh and Smith 1996).

Tea tree is also grown in the dry tropics of north Queensland. Here there was almost no seasonal variation in a stand of a low concentration type, but more than 50% variation in a high concentration type (Figure 1B) (Drinnan 1997). Seasonal variation in oil concentration of more than 50% has also been observed in young leaves of *Eucalyptus camaldulensis*, another myrtaceous species with subepidermal oil glands (Doran *et al.* 1995).

The upper leaves on a tea tree plant often have a higher oil concentration than lower leaves (Curtis 1996), leading to the suggestion that the seasonal trend in concentration reflects oil losses during autumn–winter, followed by the production of new leaves with

Table 1 A literature survey of the variation in oil concentration in tea tree

Situation	*Source of variation*	*Oil conc.*		*Reference*	*Adj.**
		mg/g	%†		
Natural stand	Months	44–57	30	Penfold *et al.* (1948)	a
	Population	31–64	106	Butcher *et al.* (1994)	
Plantation stand	Years	39–45	15	Small (1981)	a,c
	Months, high conc.	63–99	57	Drinnan (1997)	a
	Months, low conc.	45–59	31	Drinnan (1997)	a
	Months	18–98	444	Murtagh and Smith (1996)	
	Months	24–57	138	Williams and Home (1988)	d
	Weeks	34–81	138	Murtagh (1992)	a
	Weeks	25–58	132	Murtagh and Baker (1994)	a
	Days (Nov.)	23–48	109	Murtagh and Etherington (1990)	a,b
	Days (Jan.)	64–95	48	Murtagh and Etherington (1990)	a,b
	Hours	37–47	27	Murtagh and Baker (1994)	a,e
	Trees	12–67	458	Williams and Home (1988)	d
Polyhouse	Hours	40	0	List *et al.* (1995)	
Glasshouse	Populations	22–36	64	Butcher *et al.* (1994)	

* Factors used to convert units used in published data to mg/g DW; [a]Oil density of 0.9 g/ml, [b]Leaf: twig ratio of 0.68, [c]Leaf dry weight = 36% wet weight, [d]Assumed published values were %w/w, [e]Range taken from greatest variation measured within a day.

† Range expressed as a percentage of the lowest value.

a high concentration during the following spring. In other words, the whole-plant oil concentration increases during spring because of the increasing proportion of young leaves rather than an increase in oil concentration in older leaves. List *et al.* (1995) obtained two pieces of evidence that support this view, that they termed the one-way development path. Their anatomical study suggested that immature glands were lined with metabolically active cells, whereas mature glands were lined with highly vacuolate cells that are unlikely to be involved in oil synthesis. Secondly, they found no variation in oil concentration over 48 hours.

However, not all data supports the one-way development hypothesis. Penfold *et al.* (1948) found that the oil concentration increased rapidly from the lowest to near the highest value for the year between October and November. This rate of increase was too rapid to be explained by the production of new leaves. A similar result was

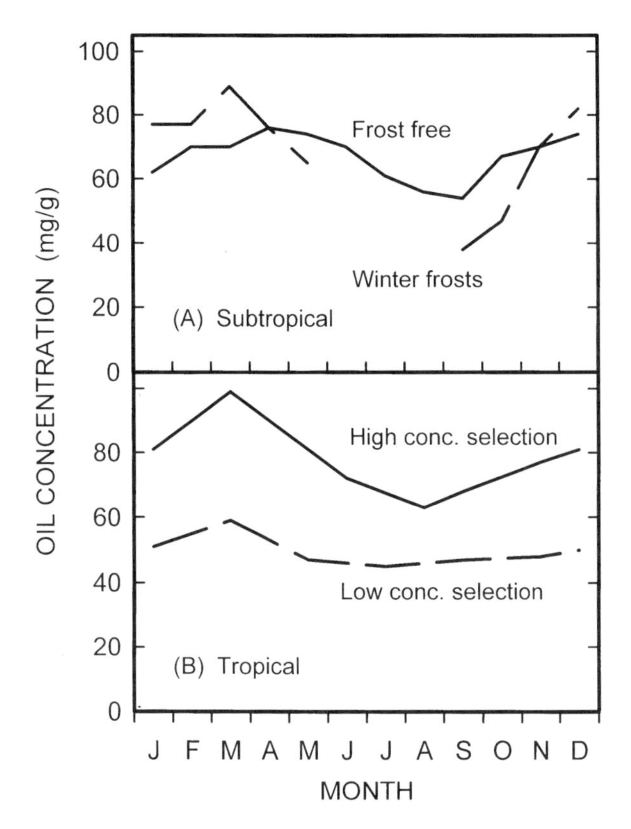

Figure 1 Seasonal changes in the oil concentration in (A) the humid subtropical environment of northern NSW, and (B) the dry tropical environment of northern Queensland

reported by Murtagh (1988) who, in one year, measured a 62% decline between August and September followed by a complete recovery in October.

Another experiment reported by Murtagh and Baker (1994) sampled plants of different regrowth ages on the same day. The regrowth was aged between 145 and 354 days at one site, and 82 and 292 days at another. The longest regrowth period included autumn–winter when concentrations usually decline but the concentration per unit weight varied more with the specific leaf area rather than the regrowth age. The specific leaf area is a measure of the area per unit leaf weight. Young leaves tend to have a high specific leaf area because they are thin and have less secondary thickening of cell walls. Thus the area, and perhaps the number of oil glands, is greater per unit weight with young leaves leading to a higher oil concentration per unit weight. When this effect was removed by expressing the oil concentration per unit leaf area, there was little difference in oil concentration between regrowth ages at one site, but it was more than double in the older regrowth at the other site. These results illustrate the difficulty of finding a suitable expression of oil concentration when contrasting leaves with very different specific leaf areas, as occurs between young and medium to old leaves.

Daily Variation

Effects due to the leaf age distribution and specific leaf area can be removed from a comparison by sampling over a short period when they would not change. When this was done by sampling at about the same time each day over a sequence, the oil concentration has been shown to vary. In one sequence over 8 days in November, the concentration halved over 2 days and completely recovered by the next (Murtagh and Etherington 1990). There was also a smaller 21% decline, followed by recovery, later in the sequence. These changes were statistically significant and occurred on days that followed the warmest nights (minimum temperatures of 17.7 and 15.9°C) during the sequence (Murtagh 1988). Subsequent sampling during summer at daily intervals on 3 sites also showed significant changes in oil concentration (Murtagh and Etherington 1990), but the changes were not related to temperature (Etherington 1989).

The inverse correlation between night temperature and oil concentration prompted the notion that the concentration declined because warmer nights increased the respiratory load and the oil supplied at least some of the substrate for the process. Monoterpenes are the major constituents of tea tree oil and are thought to be sometimes available for catabolism (Croteau 1988). Warm nights reduced the oil concentration in peppermint (*Mentha piperita*), especially during the flower initiation stage (Loomis and Croteau 1973). Curtis (1996) also obtained evidence that a warm night can decrease the oil concentration in tea tree. That result is discussed under the *Environmental Effects* heading.

Diurnal Variation

Studies that examined the diurnal pattern in oil concentration of tea tree have generally found no significant changes. Etherington (1989) in three sequences found no variation. Murtagh and Baker (1994) investigated 14 sequences of which three showed significant variation. List *et al.* (1995) and Curtis (1996) examined one sequence each and found no variation. The study by List *et al.* (1995) was done with potted plants in a polyhouse over 48 hours. All other studies were done in the field and a sequence occupied a day.

Another set of diurnal measurements, not included in the above, suggested that diurnal fluctuations were related to the water vapour pressure deficit (VPD) of the atmosphere (Murtagh 1991b). The samples were taken over three consecutive days from two watering treatments; irrigated and rain watered. The soil on the second treatment was dry when sampled. The tea trees were about 1.5 m tall, and separate samples were taken from the upper 40 cm of the canopy (upper strata), and below 40 cm (lower strata). The oil concentration in the upper strata was significantly higher than the lower strata, and it declined significantly within each of the first two days (Figure 2). There was a trend, not always significant, for the concentration in the upper strata to be higher on the rain-watered than the irrigated treatment, but the reverse applied in the lower strata. When data from both strata were pooled, there was virtually no difference between the two watering treatments. There was no change in the percent composition of the major constituents in the oil over the three days.

The weather conditions differed between the three days. Day one was hot and dry and the VPD continued to increase until sampling ceased. Day two was similar in the

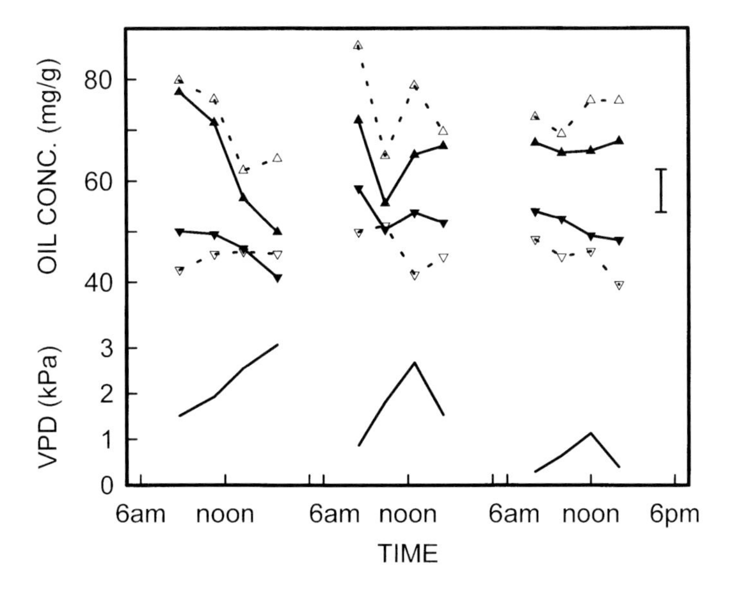

Figure 2 Diurnal variation in oil concentration over three consecutive days, in the upper (▲) and lower (▼) strata of the canopy, on irrigated (solid symbol) and rain-watered (open symbols) treatments. The water vapour pressure deficit (VPD) is also shown

morning, but a sea breeze arrived at noon and lowered the VPD. On this day, the decline in the oil concentration was arrested even before the instruments detected the change in the weather. Day three was overcast with a low VPD throughout. The relation between the oil concentration in the upper strata and VPD was summarised by two intersecting straight lines, that showed a significant decline in the concentration once a threshold VPD was exceeded (Figure 3). Separate relations were fitted to the irrigated and rain-watered treatments.

The threshold VPD were 1.5 kPa on the irrigated treatment, and 1.0 kPa if rain-watered. Whereas the threshold occurred at a lower VPD on the rain-watered treatment, the decline was greater on the irrigated treatment and equalled 22% at the highest VPD that was measured. The slower rate of decline on the rain-watered treatment suggested that these plants had acclimated to dry conditions.

However subsequent data obtained by Murtagh and Baker (1994) indicated the VPD effect was not the sole factor involved in diurnal fluctuations in oil concentration. In nine of the 14 sequences mentioned above, the VPD exceeded 2 kPa at sometime during the day, but only one of the nine sequences showed a significant decline in concentration. It is relevant that the sequence with the decline had the highest oil concentration of the nine sequences, starting the day at 55 mg/g. The declines discussed earlier occurred at higher concentrations, suggesting that diurnal fluctuation is more likely at high concentrations. One sequence in the 14 showed a significant increase of 26% in the oil concentration during a day that was warm and particularly humid, with the VPD being less than 0.9 kPa for most of the day.

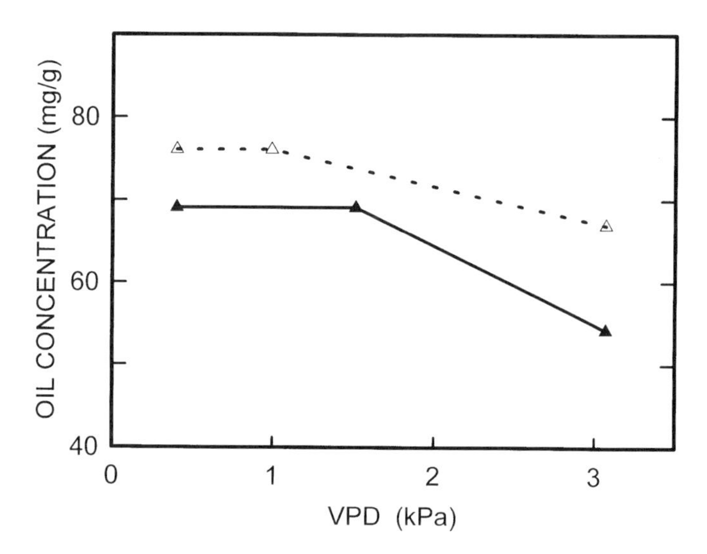

Figure 3 The pattern of change in oil concentration in the upper strata shown in Figure 2 with increasing vapour pressure deficit (VPD) on irrigated (solid symbol) and rain-watered (open symbols) treatments

Post-harvest Oil Concentration

Two studies have shown that the oil concentration remains constant for a considerable period after harvest in treatments that are handled the same whilst fresh. Murtagh and Curtis (1991) exposed samples taken from a common batch of twig material to a range of drying treatments that were designed to provide varying rates of drying, respiration, and opportunities for volatilisation. They found no effect on the oil concentration over 13 days in one experiment, and seven days in another. Twig samples were distilled, and although not presented in their paper, repeated sampling showed that the proportion of leaf in the twig sample was constant throughout each experiment, thus eliminating variation in the leaf to twig ratio as a possible source of error.

Whish and Williams (1996) also tested for post-harvest losses of oil. They distilled leaf samples, and when the leaf was stripped from the fine stems whilst fresh, there was no change in the oil concentration between distillations done immediately or 20 days later. However, when the twigs were air dried before the leaf was removed, the oil concentration was 28% greater. They suggested that oil movement from the stem gave the higher concentration in the latter result, but an alternative explanation is that the striping of green leaves from stems caused some oil loss. As discussed later, detaching branches or rough handling can increase volatilisation losses in other species. Zrira and Benjilali (1991) observed a 62% increase in oil concentration of *Eucalyptus camaldulensis* after shade drying, but it is not certain if the results were corrected for the varying water contents in the leaves. This issue must be clarified because if oil is lost when leaves are stripped while fresh, there is a strong argument to either distill twigs and adjust the concentration to a leaf basis by separately determining the leaf to twig ratio, or air dry the twigs before stripping leaves.

Oil losses after harvest vary between species, and those with a high water content tend to lose more (Guenther 1948). Fresh tea tree leaves are relatively dry with 26–47% dry matter (Murtagh and Smith 1996), and this combined with the subepidermal position of the oil glands would tend to reduce the loss of tea tree oil after harvest (Flück 1963; Murtagh and Curtis 1991). However, the relatively thin epidermal cap cell above oil glands could offset the positional protection in both *Eucalyptus* (Welch 1920) and *Melaleuca* species (List *et al.* 1995).

Factors Related to Oil Concentration

Environmental Effects and Water Content

Curtis (1996) studied the effect of temperature on the oil concentration in plants. When tea trees were grown in controlled environment chambers, the oil concentration increased from 13 mg/g at 15°/10°C (day/night temperatures) to 32 mg/g at 30°/25°C; a rate of increase of 1.27 mg/g/°C. The concentration was less at 35°/30°C. In a field study, Murtagh and Smith (1996) estimated that an increase in the mean temperature over the 3 months preceding harvest increased the oil concentration at a rate of 1.02 mg/g/°C. Although the trees carried a mixture of leaves of 0–12 months age, the temperature over the final 3 months was a better predictor of the temperature effect than the temperature over the complete regrowth period.

When Curtis (1996) measured the oil concentration in leaves under a range of environmental conditions in the field, the concentration was principally related to leaf age and the minimum temperature during the morning preceding harvest. In 100 day old leaves, the predicted oil concentration was 50 mg/g with a 10°C minimum, 46 mg/g with 15°C, and 42 mg/g with 20°C. The temperature effect was absent at 300 days when the predicted concentration was 35 mg/g. Curtis (1996) also found that the proportion of terpinen-4-ol in oil in leaves aged more than 100 days increased from 32% when the daily mean air temperature was 10°C, to 45% at 25°C.

Several studies have noted a positive correlation between oil concentration and water content in fresh tea tree leaves, and this was interpreted as an indication of the effect of water supply on oil concentration (Murtagh 1988). However, subsequent tests showed that irrigation increased the water content without a corresponding change in oil concentration (Murtagh 1992). Consequently, the effect of water stress on oil concentration is an open issue at this time. Gershenzon *et al.* (1978) noted a negative relation between the monoterpene concentration in field populations of *Satureja douglasii* and moisture stress. Many species of herbs and shrubs, but not trees, respond to water stress by slowing growth while continuing to produce secondary metabolites, leading to higher concentrations (Gershenzon 1984).

The relation between the concentration of oil (dependent variable) in growing tissue and water (independent variable) is difficult to interpret for two reasons. The water concentration is included in the dependent variable as it is used to adjust the oil concentration to a dry weight basis, and also as the independent variable in the relation. This commonality to both variables introduces autocorrelation and the possibility of creating a false correlation. It can be avoided by using the direct measurement of concentration per unit wet weight, but if so the results will be confounded by changes in the water concentration between samples. Also the positive correlation between oil and water content of leaves will dampen the true variation in oil concentration if expressed on a wet weight basis (Murtagh and Smith 1996). This is a problem for commercial growers who typically obtain a wet weight of their biomass and use this to calculate the oil concentration.

The second problem with interpretation relates to the difference between water content and water availability in leaf tissue (Waring and Schlesinger 1985). Factors such as changes in cell wall elasticity, membrane permeability, and the concentration of solutes in cells can all counter the effect of changing water concentration. Another factor could be the ratio of apoplastic to symplastic water in leaves. *E. globulus* has a relatively high proportion of cell wall water that is associated with the relatively high proportion (41%) of dry matter in fresh leaves (Gaff and Carr 1961). Tea tree has a similar proportion of dry matter, and it could be that the amount of apoplastic or symplastic water, rather than total water content, is required to interpret the effect of water content on the synthesis and loss of tea tree oil.

The effect of humidity is also not clear cut. In a controlled environment experiment (Lowe and Murtagh 1995), the oil concentration was 38% lower in plants grown at a continuous daytime VPD of 1.9 kPa (40% relative humidity at 25°C) than at 0.6 kPa (81%). Also, as described earlier, a high VPD (drier air) was implicated in the diurnal decline in concentration when the concentration was at a high level. Both experiments had a large difference or change in the VPD, but from a regional perspective where the differences in the mean VPD are less, there is no obvious effect of humidity on oil concentration. For instance, the mean daytime VPD is almost always higher at Mareeba in northern Queensland than at Lismore in northern NSW, but there is no obvious difference in the general level of oil concentrations that can be attributed to humidity effects (Figure 1). The mean daytime VPD in January is 1.43 kPa at Mareeba, and 1.39 kPa at Lismore. Corresponding values in July are 0.98 kPa and 0.80 kPa. Both districts also have the same seasonal trend in oil concentration but because of the high correlation between monthly temperature and VPD, the possible effect of temperature or VPD cannot be distinguished. Overall, it appears that large changes in humidity are required to affect the oil concentration, and while such changes can occur over short periods they are not present in long term means.

Oil Gland Density

As detailed in Chapter 7 the oil concentration is under strong genetic control, and this control could be expressed through the density of oil glands in a leaf. List *et al.* (1995) examined the number of oil glands in recently fully expanded leaves, and found slight variation within a plant, but a large variation between plants from a common seed source. In addition, the oil concentration was not correlated with oil gland density.

Hojmark-Andersen (1995) studied the frequency distribution of differently sized oil glands throughout several tea trees. He found the distribution varied between trees, between the north/east and south/west aspects, and between the two surfaces of a leaf but not within a surface. The oil concentration was correlated with the number of large oil glands, and the total oil gland area on the abaxial, but not the adaxial surface of leaves. These results highlight the need for thorough sampling to obtain a representative sample to measure either the gland density or oil concentration in a complete tree. Dawkins (1915) also commented on the difficulty of obtaining a representative sample when relating the oil content of several tree species to the number and size of oil glands.

Plant Vigour and Juvenile Effect

One experiment showed a positive relation between the biomass yield at one harvest and the oil concentration 12 months later at the next (Murtagh and Smith 1996). The effect was most marked when the regrowth age at the first harvest was extended beyond 12 months giving a corresponding increase in biomass. The following harvest had some of the highest oil concentrations, but also some of the lowest biomass yields in the 4 years of the experiment (Murtagh 1996). The low biomass yields were thought to reflect environmental conditions during the growing season, and not the increased yield at the first harvest. These results, but not all, fit Gershenzon's (1984) observation of a negative correlation between growth and production of secondary compounds in many species. They could reflect an adaptive response or a preferential allocation of resources when stress reduced growth. However the result does not fit the average seasonal trend when both growth rate and oil concentration are highest during summer. A likely explanation is that the low growth/high oil concentration years are an occasional event that is masked in average trends.

The same experiment differed from most field experiments in that there was a significant variation in oil composition. In 31% of samples, there was a significant decline in the proportion of γ-terpinene, α-terpinene and terpinolene, with the proportion of each being strongly, negatively correlated with the proportion of *p*-cymene in the oil. This is consistent with the oxidation of the former compounds to *p*-cymene (Guenther and Althausen 1949). There was no relation between the formation of *p*-cymene and the oil concentration, and it was only observed when 400 g samples were distilled in laboratory flasks and not when 5–10 kg samples were distilled in a larger vessel. Thus the formation of *p*-cymene was viewed as a distillation artifact, as did Koedam *et al.* (1979) with a number of other species.

The remaining 69% of samples had a normal proportion (0.3–1.7%) of *p*-cymene, but compositional changes affected the other four major monoterpene olefins, γ-terpinene, α-terpinene, terpinolene and α-pinene. The changes differed from those discussed above in that they were related to the oil concentration. When the concentration exceeded a threshold of 63–67 mg/g, there was virtually no further increase in the weight per leaf of the compounds (Figure 4). Whilst the effect was small for the minor compounds, it was statistically significant for all four. At the lower, unaffected oil concentrations, these compounds represented 39% of the total oil, and 55% of the samples in this analysis exceeded the threshold of 63–67 mg/g oil concentration.

The other constituents in the oil were not affected. While the decreasing proportion of olefins must have caused a corresponding increase in the proportion of the other compounds, the change was not statistically significant and the weight of the other compounds

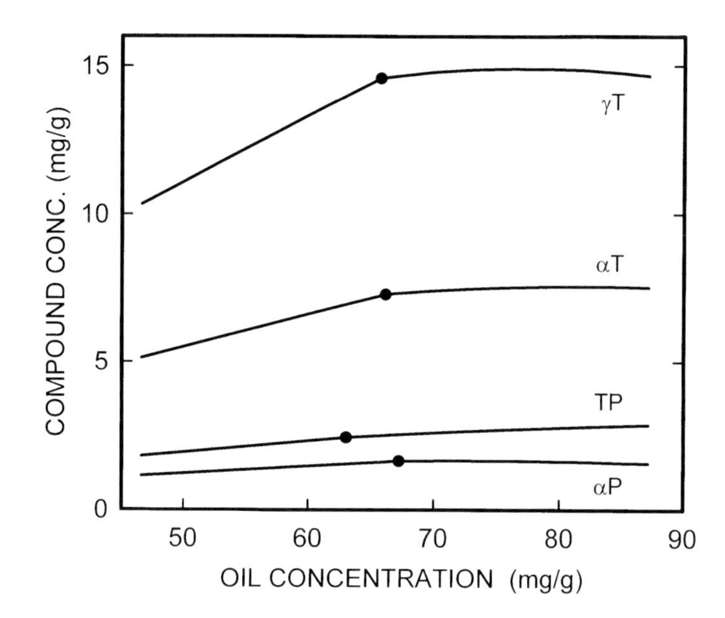

Figure 4 Changes in the weight per leaf of four olefins with an increasing oil concentration per leaf; γT, γ-terpinene; αT, α-terpinene; TP, terpinolene; αP, α-pinene. The symbol marks the threshold concentration at which the normal rate of accumulation changed. Taken from Murtagh and Smith (1996)

per leaf changed at a constant rate over the full range of oil concentrations. Murtagh and Smith (1996) concluded that selective volatilisation, perhaps promoted by the pressure of oil in cells or glands at high concentration (Dussourd and Denno 1991), was the most likely explanation of observed changes in oil composition. Olefins have higher octanol/water partition coefficients, and hence a higher permeance through cuticular membranes (Schönherr and Riederer 1989), than the other major compounds in the oil.

In a series of experiments that were conducted in a controlled environment glasshouse, the oil concentration was low across all treatments even when some treatments provided very favourable growing conditions (Lowe *et al.* 1996). Most treatments had oil concentrations of less than 35 mg/g, and all were less than 50 mg/g. By growing plants under similar conditions in the open, it was shown that the low concentration was not related to the environment in the glasshouse. Most experiments were done in 5 L pots, and while there was little response to increases in pot volume, the concentration increased from 36 mg/g in the first, to 47 mg/g in the third growth cycle, with each cycle growing over 5 months (Lowe and Murtagh 1997). In other words, the generally low oil concentrations reflected a juvenile plant effect. This result, together with the observation that fluctuating oil concentrations are generally restricted to the higher concentrations, suggests that mature plants should be used in experiments.

Controlling Processes

The oil concentration is a state variable that represents the balanced outcome from a number of processes that can include biosynthesis, catabolism, transport, interconversion and accumulation. Wiermann (1981) emphasised three aspects relating to the production of secondary compounds of which tea tree oil is representative:

1. Changes in the oil concentration represent a balance between gains from anabolism and losses from catabolism. During periods with a constant concentration, turnover might or might not be present.
2. An increase in the rate of accumulation could result from direct synthesis, or trans location from another organ.
3. Changes in the accumulation rate may reflect interconversion rather than *de novo* synthesis.

Using the definitions of Barz and Köster (1981), turnover occurs when secondary products are further metabolised and even degraded. Catabolism is a form of turnover where there is partial or complete degradation of the compound. When terpene turnover was measured in intact plants, the rate of turnover was either slow (half-lives of 5–170 days) or undetectable (Gershenzon *et al.* 1993, and references cited therein). This result contrasts with the rapid turnover measured in many earlier studies that used detached plant parts (Mihaliak *et al.* 1991). The Gershenzon *et al.* (1993) study included tea tree and found no turnover over 14 days in a cohort of young leaves that were unfolding or just unfolded at the start of the experiment.

Young leaves were sampled because their rapid rate of oil synthesis gave a measurable incorporation of radiolabelled $^{14}CO_2$ in the oil. The technique is not suited to older leaves, and given the uncertainty regarding results obtained on detached plant parts, there is no definitive picture regarding turnover in other than very young leaves.

Many essential oil plants synthesise the oil near, but not in, the storage organ. Symplastic transport, probably through the many plasmodesmata in the walls of synthesising cells, moves the oil to the storage organ (Cutter 1978). Under these conditions, factors that interfere with short-distance transport could affect the rate of accumulation in secretory cells and perhaps control synthesis by a feedback mechanism. Virtually all the relevant experiments with tea tree have centered on the oil concentration, and only Gershenzon *et al.* (1993) have provided a direct measurement of synthesis. They found a rapid rate of synthesis during early leaf expansion, a result that agreed with the general situation with terpenoid producing plants (Gershenzon and Croteau 1991).

When an increase in oil concentration is too rapid to be explained by an increasing proportion of young leaves, it reflects either *de novo* synthesis or conversion from bound forms (Wiermann 1981). A third outside possibility is translocation from other organs, but the storage glands for tea tree oil are virtually confined to leaves. Small numbers of glands can be observed on fine stems, but produce no measurable quantities of oil when stems alone are distilled.

Essential oil constituents are lipophilic compounds, and interconversion to more water-soluble forms is essential if the oil is to move from the normally well sequestered storage organ. Glycosides are common in plants and are thought to be important in interconversion and transport (Stahl-Biskup 1987). Croteau (1988) provided evidence that monoterpenyl glycosides are transport derivatives that can be found well away from the site of synthesis.

A possible explanation for the rapid decline followed by a recovery to the original oil concentration is a process of conversion to a water soluble form, then a reversion to the original form. The water soluble form need not move within the plant for the distillation process to record a drop in oil concentration, as water soluble compounds such as glucosides do not distill (Banthorpe *et al.* 1972). Even if the water soluble forms did distill they would not be recovered during the oil/water separation phase unless breakdown occurred during distillation.

Loss Pathways

Of the three potential loss pathways, volatilisation, interconversion and catabolism, information with tea tree is only available for volatilisation. Using intact tea trees with 16 week old regrowth placed in a darkened chamber at 25°C, Murtagh *et al.* (1993), measured a volatilisation rate of 1.3 μg/g DW leaf/h, a rate that is towards the lower end of the general range of 0.1 to 10 μg/g/h for monoterpene emissions (Tingey *et al.* 1991). If the measured rate held constant it would take 3.6 days to volatilise 1% of the oil, which is too slow to explain the negative movement in oil concentration discussed earlier.

Kawakami *et al.* (1990) used a stream of nitrogen to collect the emissions from detached tea tree twigs that were cut into 10 cm lengths and held at 15°C. They measured an emission rate of 0.5–2.5 mg/100 g fresh leaf/h, that translates to approximately 5.6 μg/g DW leaf/h; a much higher rate than the first study. The difference could arise from the use of twig sections in the second study. Removing a branch from *Salvia mellifera* more than doubled the short-term volatilisation (Dement *et al.* 1975), and rough handling was observed to increase the volatilisation from *Pinus radiata* (Juuti *et al.* 1990).

Murtagh *et al.* (1993) measured the volatilisation from tea tree at 15, 25 and 35°C. A Q_{10} value was used to quantify the temperature effect and equalled 3.1 with the olefins and 5.6 with the oxygenated compounds. Both values exceeded the effect of temperature on the saturated vapour pressures of the major compounds in emissions. Tingey *et al.* (1991) also noted the same difference across a number of species, and attributed it to either changes in pathway conductance or differences in pool size. The Q_{10} of 3.1 with the olefins is close to the expected value of 2–3 for enzymatically controlled processes (Salisbury and Ross 1992), and values in this range have been presented as evidence that pool size regulates the emission flux (Tingey 1981). Pool size had the greatest effect on the emission rate in the Murtagh *et al.* (1993) study. The even higher Q_{10} with the oxygenated compounds is well above the expected value for metabolic processes, and suggests that diffusion is controlled by the cuticular membrane (Schönherr and Bukovac 1978).

The volatilisation of individual compounds was also measured in a study by Murtagh *et al.* (1993) (Table 2). This was done with the 35°C treatment because the discrimination was best at a high temperature. The leaf oil concentration, given in Table 2, was measured on a solvent extract to avoid converting precursor compounds to terpinen-4-ol (Southwell and Stiff 1989). In agreement with studies with other species (Tingey *et al.* 1991), the tea tree emissions had a different chemical composition to the leaf oil.

Table 2 The composition of leaf emissions at 35°C, and leaf oil, from regrowth of *M. alternifolia*. Also listed are the standard emission rates ($E/C^{0.9}$) after correction for pool size effects, and saturated vapour pressures (SVP) at 35°C

Compound		*E'miss.* µg/m²/h	*Conc.* mg/m²	*E/C^0.9* µg/mg/h	*SVP* Pa
Type	*Name*				
Olefin	α-pinene	19.2	16.2	1.9	1008
	β-pinene	19.0	7.4	2.8	717
	p-cymene	5.4	3.2	2.2	463
	Limonene	1.7	12.0	0.5	390
	α-terpinene	19.6	25.9	0.8	373
	γ-terpinene	30.9	72.0	0.6	312
	Terpinolene	8.6	11.8	0.9	285
Oxygenated	1,8-cineole	39.1	127.7	0.5	372
	trans-sabinene hydrate	10.0	4.4	1.8	55
	cis-sabinene hydrate	12.7	14.6	0.9	43
	Terpinen-4-ol	50.9	153.7	0.3	25
	α-terpineol	43.8	20.2	1.2	17

The rate of emission (e'miss.) of each compound is expressed as the weight per unit double-sided leaf area per hour. The concentration (conc.) is given as the weight per double-sided leaf area (mg/m^2).

At 35°C, the emissions contained 40% olefins whereas the leaf oil had 32% olefins. The mean cineole concentration shown in Table 2 is higher than the concentration in commercial tea tree oil because a tree with an exceptionally high cineole concentration was included in the study.

The leaf oil concentration, sometimes referred to as the pool size, had the greatest effect on the emission rates of the individual compounds (Table 2). The effect was non-linear, meaning that the compounds present at the larger concentrations did not volatilise at an equivalent rate to those with a smaller concentration. A standardised emission rate that accounted for much of the concentration effect was obtained by dividing the rate by the concentration raised to the power of 0.9. The method of estimation used a statistical procedure to balance out the uneven weighting of results when some compounds were not detected in all runs. The high standardised emission rate for α-pinene, β-pinene and *p*-cymene relative to the other olefins can be attributed to their relatively high SVP, but the reason for the high standardised rate with *trans*-sabinene hydrate is not clear. Both sabinene hydrate isomers are evident in flush growth but are converted to end products and little or none remains as leaves mature (Southwell and Stiff 1990). With ongoing synthesis, the recently formed juvenile oil with both sabinene hydrate isomers could be held at a location that is more susceptible to volatilisation loss.

Comparisons of the relative volatilisation rates of various compounds are invariably confounded by their different physical properties. As these differences are less with α- and β-pinene these compounds were used for comparison. In both the Murtagh *et al.* (1993) and Kawakami *et al.* (1990) studies, the ratio of α- to β-pinene was about 2:1 in the plant oil, and 1:1 in the emissions. In other words, β-pinene was favoured in emissions relative to the composition of the plant oil. This contrasts with the likely effect of the relative physical properties that would favour the volatilisation of α-pinene, as it has the higher SVP (727 Pa and 524 Pa at 25°C for the α and β isomers respectively) and a lower viscosity than β-pinene (Drew *et al.* 1971). As the ratio of α- to β-pinene is 1:1 in the juvenile oil found in flush growth (Southwell and Stiff 1990), a better explanation is that the emissions were sourced from juvenile oil. This would explain the 1:1 ratio in the absolute volatilisation rates of the pinene isomers. Furthermore, for those compounds that were sourced from the juvenile pools of unknown concentration, the absolute emission rate given in Table 2 is a better measure than the standardised rate.

The presence of *cis*- and *trans*-sabinene hydrate in the emissions provided further evidence that they were sourced from juvenile oil. Southwell and Stiff (1990) found that both isomers were well represented in flush leaves but declined to trace or zero levels as the leaves developed to leaf node 13 and beyond. However, the observed lack of turnover in young leaves aged to 15 days (Gershenzon *et al.* 1993) complicates the issue. With typical leaf emergence rates of 0.5–1.0/d (Curtis 1996), leaf node 13 corresponds to a leaf age of 13–26 days. The majority if not all of this period coincides with the measured period of no turnover and hence no volatilisation from the young leaves. This suggests that the emissions were sourced from juvenile oil in mature leaves. The presence of juvenile oil in nonflush growth is consistent with ongoing synthesis in mature leaves, a process that could give the positive changes in oil concentration described earlier. The quantity of juvenile oil could be quite small and difficult to detect in a leaf-oil analysis because of the diluting effect of the bulk oil. Also, its presence could be transient according to the extent of replacement synthesis of oil as exhibited in the changing oil concentrations.

Double Pool Conceptual Model

A number of authors including Barz and Köster (1981) have noted that ongoing synthesis or interconversion is required to provide positive changes in oil concentration. In contrast with this view is the concept that oil stored in glands is well sequestered and isolated from the normal functioning of the plant (McKey 1979). The isolation need not be absolute and solubilising reactions can promote translocation and catabolism (Croteau 1988).

With tea tree, the fluctuations in oil concentration are more marked in mature, but not aged leaves, and in leaves with a high oil concentration. Also, emissions appear to be sourced from pools with a more juvenile oil than the bulk leaf oil. These observations can be accommodated in a double pool model of oil accumulation and loss. One pool represents a stable storage, with some seasonal decline in oil concentration, especially during winter. The second pool has a more variable concentration, and is subject to both additions and loss even in the short term. This is not a new idea. Banthorpe *et al.* (1972) suggested that plants may have two distinct pools of terpenes; one of which is affected by outside influences and the other is more inert. Loomis and Croteau (1973) used the concept of metabolic and storage pools to explain some of the periodic changes in oil concentration. Janson (1993) and a number of other authors proposed the equivalent of the double pool model for emissions from conifers, with the standing pool in resin ducts providing a relatively stable source of emissions and synthesising cells providing a variable source.

The issue of quantitative limits within a pool is uncertain, but it is interesting that the composition of leaf oil changed when the concentration exceeded about 65 mg/g in the Murtagh and Smith (1996) study. As a first step for the genetic lines currently used by the industry, 65 mg/g could be taken as the upper limit for the stable pool. With the highest observed oil concentrations near 100 mg/g, the difference of 35 mg/g would be the capacity of the variable pool. While the more variable oil concentrations tend to occur at the higher concentrations, this is not an absolute effect. Consequently, the model would not require the stable pool to be full before some accumulation could start in the variable pool.

BIOMASS YIELD

The leaf yield is the second major component of the yield of tea tree oil. As with the oil concentration it varies, but for different reasons and in a different pattern to the oil concentration (Murtagh 1991a). In plantation production, the tea trees are grown as a row crop and are influenced by many of the agronomic factors that affect cropping in general. These are reviewed in Chapter 3 and the current chapter concentrates on the physiology of biomass production.

The leaf yield is strongly correlated ($r=0.94$) with the total yield of biomass, and the two vary only through changes in the proportion of leaf in twigs, and the proportion of twig in total biomass (Murtagh 1996). As explained earlier, twigs consist of fine stems and leaves, and contain virtually all the leaves on a tree. Working with twigs that were not subject to leaf shedding, the proportion of leaf in twig on a dry weight basis was usually between 0.60 and 0.73, and cool conditions appeared to restrict the growth of leaves more than fine stems, giving a lower ratio (Murtagh 1996). In the same experiment, the proportion of twig in total biomass decreased from 0.69 to 0.43 as trees increased in size and the main stems occupied an increasing proportion of the total biomass.

Environmental Effects

Tea trees grow best at high temperatures. In a controlled environment experiment, Curtis (1996) measured a near linear increase in biomass weight between day/night temperatures of 15/10°C and 35/30°C. The leaf emergence rate also increased with temperature, but the rate of increase began to slow at the higher temperatures. The emergence rate was 0.1/d at 15/10°C, 1.9/d at 30/25°C, and 2.1/d at 35/30°C. In a field calibration of the temperature response, Murtagh (1996) found that a one degree increase in temperature above a threshold for growth of 16°C, increased the biomass yield by 105 g/plant/yr. As discussed later however, the temperature response was greater at some stages of growth than others. When tea trees are grown at the higher temperatures quoted above, the young leaves are limp, or nearly so, depending on the evaporative demand for moisture. Richards, quoted by McKey (1979) has suggested that in such leaves expansion precedes differentiation. If so the early differentiation and filling of oil glands, discussed above, may be delayed in high temperature growth.

The effect of water stress on biomass production depends heavily on the availability of subsoil moisture. When tea tree was grown on a site with a permanently damp subsoil at about 600 mm depth, the response to irrigation was largely restricted to one growth stage (Murtagh 1992). The situation is different when the subsoil dries out. While severe water stress may not kill trees, it will cause extensive defoliation and yield loss. Hence irrigation is essential on such soils (Drinnan 1997).

Murtagh's (1992) experiment measured the response to irrigation during the typically dry spring-early summer period on the north coast of NSW. The period can be divided into a pre-flush stage that extends until early October, flushing until early December, and a post-flush stage. The pre-flush stage was characterised by low soil temperatures and a period when the dominant resistance to water flow through a plant was in the liquid rather than the vapour phase (Figure 5). This differed from the more usual situation where the dominant resistance is in the vapour phase and reflects stomatal movements. When the mean air temperature was less than about 19°C, the liquid-flow resistance was dominant and was not altered by irrigation. Consequently the resistance probably arose within the plant rather than in the soil. Many factors can increase the liquid-flow resistance within a plant including temperature effects on root hydraulic resistance (Jones *et al.* 1985), cavitation or plugging within the xylem pathway (Zimmermann and Milburn 1982), and the effectiveness of mycorrhizas in water absorption (Jones *et al.* 1985). Regardless of the mechanism, the effect was temperature driven but exhibited as a water stress. The results in Figure 5 were measured on mature

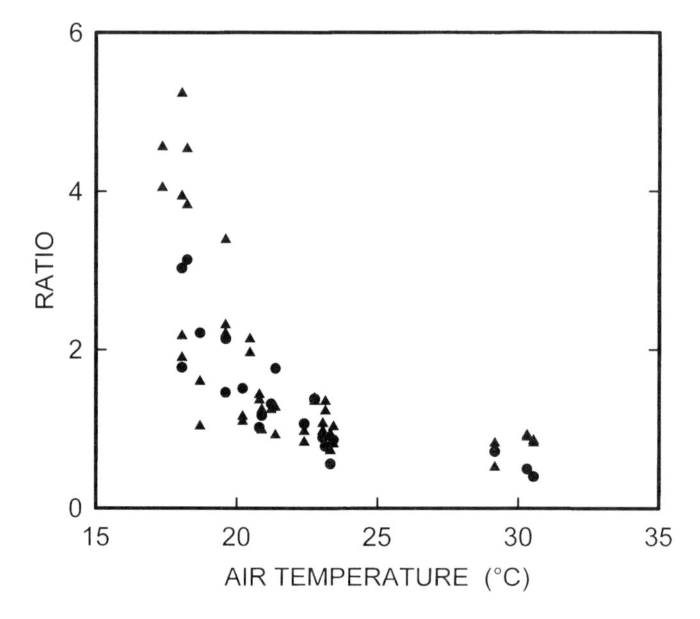

Figure 5 The ratio of the resistance to water flow in the liquid to the vapour phase at different temperatures on irrigated (▲) and rain-watered (●) treatments. All readings were taken at least 3.5 hours after sunrise

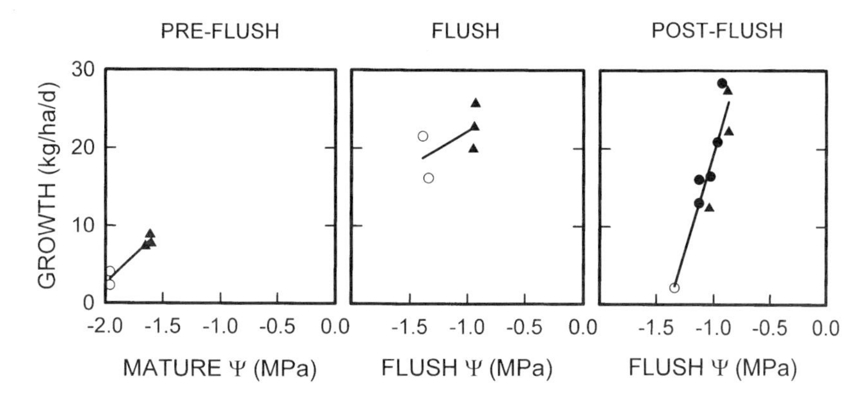

Figure 6 The relation between the leaf growth rate and shoot water potential (Ψ) during three growth stages. The results were collected over two years on irrigated (▲) and rain-watered (●,○) treatments, with closed and open symbols indicating predominantly wet or dry topsoil respectively

shoots, whereas all fresh shoots had a low ratio. No units are given for the ratio of the resistances in the liquid and vapour phases as two phases are physically dissimilar (Cowan and Milthorpe 1967), and the ratio was used as an index only.

The growth rates were related to the water potential on the various treatments (Figure 6). During the pre-flush stage, both shoot water potentials and growth rates were low, and the absolute response to irrigation was only moderate. Growth rates were much higher, even on dry soils, during the flushing stage but the response to irrigation remained moderate. In contrast there was a marked response to irrigation and increases in the water potential in the post-flush stage. Regardless of the stage, wetting the soil increased the shoot water potential by about 0.5 MPa.

Two factors are particularly important in the interpretation of these results. The experiment was done at a site with a permanently damp or wet subsoil, and the tea tree stand had a strong spring flush that typically follows winter frosts. Under these conditions, the only appreciable response to irrigation can be expected during the post-flush phase. In the Lismore district where the experiment was done, the post-flush period just precedes the summer rains and irrigation responses are small over a full regrowth cycle.

Murtagh (1996) used a growth model to describe the effect of temperature and water stress on the field growth of tea tree. The temperature effects are described above. Water stress began when the top soil had dried to less than 69% of the total available water content, and declined in a linear manner with further drying of the soil.

Other Effects

Tea trees are harvested by cutting the main stem near ground level, leaving a bare stump that will produce the coppice regrowth. However it takes time to produce a new canopy and during the first 3 months after harvest, the growth rate was 46% of that during months 4–6, the most efficient stage (Murtagh 1996). After 6 months, new shoots were less vigorous than before and the growth rate declined to 71% of the stage two rate. The relative rates were calculated with all other conditions remaining constant. The best yields of biomass were obtained by timing harvests so that the second and third regrowth stages coincided with the best environmental conditions for growth. The analysis of the effect of water stress on yield provided a good example of the gains from following this approach. Across all harvest times, water stress reduced biomass yields by an average of 24%. However, an October harvest that matched the first stage of regrowth to the driest period, had only a 7% reduction in yield due to water stress. This result reflected in part the assumption, supported by the functional analyses, that there is no water stress soon after harvesting because of the large root to shoot ratio (Blake and Tschaplinski 1986).

Melaleuca species are reputed to be very tolerant of water logging. Gomes and Kozlowski (1980) found no effect of 30 days flooding in stagnant water on the growth of *M. quinquenervia*, but longer periods did reduce growth. Bolton and Greenway (1996) obtained good growth from *M. alternifolia* growing in 100–150 mm deep, flowing sewage effluent over 20 months, but it is unlikely if the same tolerance would be present in stagnant water with a higher oxygen tension. Colton and Murtagh (1990) noted that growth was depressed on waterlogged soil.

OIL YIELD

Because the two are not closely related, the combined effect of oil concentration and leaf yield gives a wide range in potential oil yield. Colton and Murtagh (1990) indicated that the oil yield could range from a low of 43 kg/ha/yr to a high of 392 kg/ha/yr, with a yield of 150–200 kg/ha/yr representing a realistic target for most plantations in northern NSW.

Since these projections were recorded in 1990, there has been no confirmed advance in the potential oil concentration on a plantation scale, but the situation could change in the near future. Doran *et al.* (1996) recorded a 60% increase in oil concentration in plants grown from seed from a selected provenance, over the concentration in plants from selected lines used in commercial nurseries. Williams (1995) has developed clones that produced more than twice the oil yield of unselected trees. Efforts are proceeding to confirm that these gains can be achieved in commercial plantations, but at this stage it would be premature to use either set of work to adjust the projected oil yields. Apart from the use of genetic improvement to increase the oil concentration, it might be possible to use preharvest treatments to increase the concentration, but little work has been done in this regard.

Some progress has been made since 1990 towards producing higher leaf yields by fine-tuning the agronomic procedures in growing a crop (Murtagh 1998). The expected yields in Table 3 were obtained by increasing the previous estimates of leaf yield (Colton and Murtagh 1990) by 5%, and using the leaf : twig and twig : biomass ratios in Murtagh (1996), and a mean dry matter content of 40% to calculate the other plant yields. In addition to the above gain, more growers are producing crops in the higher yielding categories, so the industry average has increased by more than 5%. A realistic yield target would be 170–220 kg oil/ha.

Table 3 Typical yields of tea tree under three growing conditions

		Growing Conditions	
	Poor	Good	Excellent
Leaf yield (t DM/ha)	1.7	3.6	5.1
Twig yield (t DM/ha)	2.5	5.3	7.5
Biomass yield (t DM/ha)	3.6	8.9	13.6
Biomass yield (t GW/ha)	9.0	22.3	34.0
Oil yield (kg/ha) @ low conc.*	51	108	153
@ medium conc.	94	198	281
@ high conc.	136	288	408

*Representative oil concentrations are 30 mg/g (low), 55 mg/g (medium), and 80 mg/g (high). See text for explanation of plant fractions.

SUMMARY

Tea tree oil is produced from trees grown as a row crop. The cultural aim is to maximise the oil concentration in leaves and the yield of leaf at harvest time.

The oil concentration follows a seasonal trend, with the highest concentration in summer and the greatest amplitude between seasons in the cooler localities. Additional short-term variation is superimposed on the seasonal trend and is more marked in plants with a high oil concentration. The rapid recovery in concentration following a short-term loss indicates that new oil is obtained either from direct synthesis or interconversion from another chemical form. The oil concentration increases with increasing temperature and humidity, but was not altered by irrigation on a site with subsoil moisture. The observed changes in oil concentration are consistent with a double pool conceptual model where some oil is held in a stable storage, and the remainder in an organ that is subject to gains and losses of oil.

The yield of leaf is primarily determined by the total biomass yield of a tree. Trees grow best at high temperatures, and the effect of water stress on growth is most marked in the post-flush stage of growth. Growth is most efficient during 4–6 months after harvest when a new canopy has developed and the shoots are relatively young. Timing a harvest to synchronise this optimum growth stage with the best seasonal conditions for growth gave the highest biomass yields.

REFERENCES

Baker, G.R. (1995) NSW Agriculture, unpublished results.

Baker, G.R., Lowe, R.F. and Murtagh, G.J. (1995) NSW Agriculture. Unpublished results.

Banthorpe, D.V., Charlwood, B.V. and Francis, M.J.O. (1972) The biosynthesis of monoterpenes. *Chem. Rev.*, **72**, 115–163.

Barz, W. and Köster, J. (1981) Turnover and degradation of secondary (natural) products. In E.E. Conn, (ed.), *The Biochemistry of Plants. vol 7. Secondary Plant Products*, Academic, New York, pp. 35–84.

Blake, T.J. and Tschaplinski, T.J. (1986) Role of water relations and photosynthesis in the release of buds from apical dominance and the early reinvigoration of decapitated populars. *Physiol. Plant*, **68**, 287–293.

Bolton, K.G.E. and Greenway, M. (1996) A feasibility study of *Melaleuca* trees as candidates for constructed wetlands, SE Queensland, Australia. *Proc. 5th Int. Conf. Wetland Systems for Water Pollut. Control*, Vienna, Sept. 1996.

Brophy, J.J. and Lassak, E.V. (1983) The volatile leaf oils of *Melaleuca armillaris, M. dissitiflora and M. trichostachya. J. Proc. Roy. Soc. NSW*, **116**, 7–10.

Brophy, J.J. and Lassak, E.V. (1992) Steam volatile leaf oils of some *Melaleuca* species from Western Australia. *Flav. Frag. J.*, **7**, 27–31.

Butcher, P.A., Doran, J.C. and Slee, M.U. (1994) Intraspecific variation in leaf oils of *Melaleuca alternifolia* (Myrtaceae). *Biochem. System. Ecol.*, **22**, 419–430.

Charles, D.J. and Simon, J.E. (1990) Comparison of extraction methods for the rapid determination of essential oil content and composition of basil. *J. Amer. Soc. Hort. Sci.*, **115**, 458–462.

Colton, R.T. and Murtagh, G.J. (1990) *Tea-tree oil—plantation production.* Agfact P6.4.6, NSW Agriculture & Fisheries, 1990.

Cornwell, C.P., Leach, D.N. and Wyllie, S.G. (1995) Incorporation of oxygen-18 into terpinen-4-ol from the $H_2{}^{18}O$ steam distillates of *Melaleuca alternifolia* (tea tree). *J. Essent. Oil Res.*, **7**, 613–620.

Cowan, I.R. and Milthorpe, F.L. (1967) Resistance to water transport in plants—a misconception misconceived. *Nature*, **213**, 740–741.

Croteau, R. (1988) Catabolism of monoterpenes in essential oil plants. In B.M. Lawrence, B.D. Mookherjee and B.J. Willis, (eds.), *Flavors and Fragrances: A World Perspective*, Elsevier, Amsterdam, pp. 65–84.

Curtis, A. (1996) *Growth and Essential Oil Production of Australian Tea Tree (Melaleuca alternifolia* (Maiden and Betche) Cheel), Master of Agricultural Science Thesis, The University of Queensland.

Curtis, A. and Murtagh, G.J. (1989) NSW Agriculture. Unpublished results.

Cutter, E.G. (1978) *Plant Anatomy*, Pt 1, *Cells and Tissues*, 2nd edit, Edward Arnold, London, pp. 214–241.

Dawkins, A.E. (1915) The calculation of the oil content of foliage from measurements of the number and size of the oil glands. *Proc. Roy. Soc. Victoria*, **NS27**, 153–154.

Dement, W.A., Tyson, B.J. and Mooney, H.A. (1975) Mechanism of monoterpene volatilization in *Salvia mellifera. Phytochem.*, **14**, 2555–2557.

Doran, J.C., Baker, G.R., Murtagh, G.J. and Southwell, I.A. (1996) Breeding and selection of Australian Tea Tree for improved oil yield and quality. Final Report on Project No. DAN-87A, Rural Industries Research and Development Corporation, Canberra, October 1996.

Doran, J.C., Caruhapattana, B., Namsavat, S. and Brophy, J.J. (1995) Effect of harvest time on the leaf and essential oil yield of *Eucalyptus camaldulensis. J. Essent. Oil Res.*, **7**, 627–632.

Drew, J., Russell, J. and Bajak, H.W. (1971) *Sulfate Turpentine Recovery*, Pulp Chemicals Association, New York, pp. 30–44.

Drinnan, J.E. (1997) Development of the North Queensland Tea Tree Industry. Final Report on Project No. DAQ-184A, Rural Industries Research and Development Corporation, Canberra, 1997.

Dussourd, D.E. and Denno, R.F. (1991) Deactivation of plant defence: correspondence between insect behaviour and secretory canal architecture. *Ecology*, **72**, 1383–1396.

Etherington, R.J. (1989) *Essential Oils in Australia and Environmental Effects on Short Term Variation in Oil Yield of Tea Tree (Melaleuca alternifolia)*. Bachelor of Rural Science (Honours) Thesis, University of New England.

Flück, H. (1963) Intrinsic and extrinsic factors affecting the production of secondary plant products. In T. Swain, (ed.), *Chemical Plant Taxonomy*, Academic, London, pp. 167–186.

Gaff, D.F. and Carr, D.J. (1961) The quantity of water in the cell wall and its significance. *Aust. J. Biol. Sci.*, **14**, 299–311.

Gershenzon, J. (1984) Changes in the levels of plant secondary metabolites under water and nutrient stress. *Recent Adv. Phytochem.*, **18**, 273–320.

Gershenzon, J. and Croteau, R. (1991) Terpenoids. In G.A. Rosenthal and M.R. Berenbaum, (eds.), *Herbivores: Their Interactions with Secondary Plant Metabolites. vol. 1. The Chemical Participants*, Academic, San Diego, pp. 165–219.

Gershenzon, J., Lincoln, D.E. and Langenheim, J.H. (1978) The effect of moisture stress on monoterpenoid yield and composition in *Satureja douglasii. Biochem. Syst. Ecol.*, **6**, 33–43.

Gershenzon, J., Murtagh, G.J. and Croteau, R. (1993) Absence of rapid terpene turnover in several diverse species of terpene-accumulating plants. *Oecologia*, **96**, 583–592.

Gomes, A.R.S. and Kozlowski, T.T. (1980) Responses of *Melaleuca quinquenervia* seedlings to flooding. *Physiol. Plant*, **49**, 373–377.

Guenther, E. (1948) The production of essential oils. In E. Guenther, (ed.), *The Essential Oils*. vol. 1, van Nostrand, New York, pp. 85–226.

Guenther, E. and Althausen, D. (1949) *The Essential Oils*. vol. 2, van Nostrand Reinhold, New York, pp. 7–137.

Harborne, J.B. and Turner, B.L. (1984) *Plant Chemosystematics*, Academic, London, pp. 216–236.

Hojmark-Andersen, J. (1995) Investigations into the yield of essential oil from *Melaleuca alternifolia* (Maiden and Betche) Cheel. Fourth Year Report, Bachelor of Applied Science (Hort Tech), The University of Queensland, Gatton College, Lawes, October 1995.

Janson, R.W. (1993). Monoterpene emissions from Scots pine and Norwegian spruce. *J. Geophys. Res.*, **98 D2**, 2839–2850.

Jones, H.G., Lakso, A.N. and Syvertsen, J.P. (1985) Physiological control of water status in temperate and subtropical fruit trees. *Hort. Rev.*, **7**, 301–344.

Juuti, S., Arey, J. and Atkinson, R. (1990) Monoterpene emission rate measurements from a Monterey pine. *J. Geophys. Res.*, **95**, D6, 7515–7519.

Kawakami, M., Sachs, R.M. and Shibamoto, T. (1990) Volatile constituents of essential oils obtained from a new developed tea tree (*Melaleuca alternifolia*) clone. *J. Agric. Food Chem.*, **38**, 1657–1661.

Koedam, A., Scheffer, J.J.C. and Svendsen, A.B. (1979) Comparison of isolation procedures for essential oils. II. Ajowan, caraway, coriander and cumin. *Z. Lebensm. Unters. Forsch.*, **168**, 106–111.

Lawrence, B.M. (1986) Essential oil production. In T.H. Parliament, and R. Croteau, (eds.), *Biogeneration of Aromas*, American Chemical Society, Washington DC, pp. 363–369.

List, S., Brown, P.H. and Walsh, K.B. (1995) Functional anatomy of the oil glands of *Melaleuca alternifolia* (Myrtaceae). *Aust. J. Bot.*, **43**, 629–641.

Loomis, W.D. and Croteau, R. (1973) Biochemistry and physiology of lower terpenoids. *Recent Adv. Phytochem.*, **6**, 147–185.

Lowe, R.F. and Murtagh, G.J. (1995) NSW Agriculture. Unpublished results.

Lowe, R.F. and Murtagh, G.J. (1997) NSW Agriculture. Unpublished results.

Lowe, R.F., Murtagh, G.J. and Baker, G.R. (1996) NSW Agriculture. Unpublished results.

McKey, D. (1979) The distribution of secondary compounds within plants. In G.A. Rosenthal and D.H. Janzen, (eds.), *Herbivores*, Academic, New York, pp. 55–133.

Mihaliak, C.A., Gershenzon, J. and Croteau, R. (1991) Lack of rapid monoterpene turnover in rooted plants: implications for theories of plant chemical defense. *Oecologia*, **87**, 373–376.

Murtagh, G.J. (1988) Factors affecting the oil concentration in tea tree. *Proc. 4th Australasian Conf. Tree Nut Crops*, Lismore, August 1988, pp. 447–452.

Murtagh, G.J. (1991a) Tea tree oil. In R.S. Jessop and R.L. Wright, (eds.), *New Crops*, Inkata, Melbourne, pp. 166–174.

Murtagh, G.J. (1991b) NSW Agriculture, unpublished results.

Murtagh, G.J. (1992) Irrigation as a management tool for production of tea tree oil. Final Report on Project No. DAN-19A, Rural Industries Research and Development Corporation, Canberra, May 1992.

Murtagh, G.J. (1996) Month of harvest and yield components of tea tree. I. Biomass. *Aust. J. Agric. Res.*, **47**, 801–815.

Murtagh, G.J. (1998) Tea tree oil. In K.W. Hyde (ed.), *The New Rural Industries*, Rural Industries Research and Development Corporation, Canberra, pp. 272–278.

Murtagh, G.J. and Baker, G.R. (1994) Factors affecting Oil Yield in Tea Tree. Final Report on Project No. DAN-58A, Rural Industries Research and Development Corporation, Canberra, November 1994.

Murtagh, G.J. and Curtis, A. (1991) Post-harvest retention of oil in tea tree foliage. *J. Essent. Oil Res.*, **3**, 179–184.

Murtagh, G.J. and Etherington, R.J. (1990) Variation in oil concentration and economic return from tea-tree (*Melaleuca alternifolia* Cheel) oil. *Aust. J. Exp. Agric.*, **30**, 675–679.

Murtagh, G.J., Gershenzon, J. and Croteau, R. (1993) Washington State University, unpublished results.

Murtagh, G.J. and Smith, G.R. (1993) NSW Agriculture. Unpublished results.

Murtagh, G.J. and Smith, G.R. (1996) Month of harvest and yield components of tea tree. I. Oil concentration, composition, and yield. *Aust. J. Agric. Res.*, **47**, 817–827.

Penfold, A.R. and Morrison, F.R. (1950) "Tea Tree" oils. In E. Guenther, (ed.), *The Essential Oils.* vol. 4, van Nostrand Reinhold, New York, pp. 526–548.

Penfold, A.R., Morrison, F.R. and McKern, H.H.G. (1948) Studies in the *Myrtaceae* and their essential oils. Part 1. The seasonal variations in yield and cineole content of *Melaleuca alternifolia* Cheel. In *Researches on Essential Oils of the Australian Flora. vol. 1*, Museum of Applied Arts and Sciences, Sydney, pp. 5–7.

Reilly, T.L. (1991) The economics of tea tree. *Reports: Tea Tree Marketing & Planning Conference*, Ballina, NSW, 31 Oct.–2 Nov. 1991.

Salisbury, F.B. and Ross, C.W. (1992) *Plant Physiology*, Wadsworth Publishing, Belmont, Calif.

Schönherr, J. and Bukovac, M.J. (1978) Foliar penetration of succinic acid-2,2-dimethylhyrazide: Mechanism and rate limiting step. *Physiol. Plant*, **42**, 243–251.

Schönherr, J. and Riederer, M. (1989) Foliar penetration and accumulation of organic chemicals in plant cuticles. *Rev. Environ. Contam. Toxicol.*, **108**, 1–70.

Small, B.E.J. (1981) Effects of plant spacing and season on growth of *Melaleuca alternifolia* and yield of tea tree oil. *Aust. J. Exp. Agric. Anim. Husb.*, **21**, 439–442.

Southwell, I.A. and Stiff, I.A. (1989) Ontogenetical changes in monoterpenoids of *Melaleuca alternifolia* leaf. *Phytochem.*, **28**, 1047–1051.

Southwell, I.A. and Stiff, I.A. (1990) Differentiation between *Melaleuca alternifolia* and *M. linariifolia* by monoterpenoid comparison. *Phytochem.*, **29**, 3529–3533.

Stahl-Biskup, E. (1987) Monoterpene glycosides, state-of-the-art. *Flav. Frag. J.*, **2**, 75–82.

Tingey, D.T. (1981) The effect of environmental factors on the emission of biogenic hydrocarbons from live oak and slash pine. In J.J. Bufalini, and R.R. Arnts (eds.), *Atmospheric Biogenic Hydrocarbons*, vol. 1, Ann Arbor Scientific Publishing, Ann Arbor, Michigan, pp. 53–73.

Tingey, D.T., Turner, D.P. and Weber, J.A. (1991) Factors controlling the emissions of monoterpenes and other volatile organic compounds. In T.D. Sharkey, E.A. Holland and H.A. Mooney, (eds.), *Trace Gas Emissions by Plants*, Academic, San Diego, pp. 93–119.

Waring, R.H. and Schlesinger, W.H. (1985) *Forest Ecosystems*, Academic, Orlando, pp. 7–37.

Welch, M.B. (1920) Eucalyptus oil glands. *J. Roy. Soc. NSW*, **54**, 208–217.

Whish, J.P.M. and Williams, R.R. (1996) Effects of post harvest drying on the yield of tea tree oil (*Melaleuca alternifolia*). *J. Essent. Oil Res.*, **8**, 47–51.

Wiermann, R. (1981) Secondary plant products and cell and tissue differentiation. In E.E. Conn, (ed.), *The Biochemistry of Plants. vol. 7. Secondary Plant Products*, Academic, New York, pp. 85–115.

Williams, L.R. (1995) Selection and breeding of superior plants of *Melaleuca* to increase the production and antimicrobial activity of tea tree oil. In K.H.C. Baser, (ed.), *Flavours, Fragrances and Essential Oils, vol. 2*, AREP Publ., Istanbul, pp. 408–417.

Williams, L.R. and Home, V.N. (1988) Plantation production of oil of *Melaleuca* (tea tree oil)—A revitalised Australian essential oil industry. *Search*, **19**, 294–297.

Zimmermann, M.H. and Milburn, J.A. (1982) Transport and storage of water. In O.L. Lange, P.S. Nobel, C.B. Osmond and H. Ziegler, (eds.), *Physiological Plant Ecology, vol. 2*, Springer-Verlag, Berlin, pp. 135–151.

Zrira, S. and Benjilali, B. (1991) Effect of drying on leaf oil production of Moroccan *Eucalyptus camaldulensis*. *J. Essent. Oil Res.*, **3**, 117–118.

7. TEA TREE BREEDING

GARY BAKER

Wollongbar Agricultural Institute, Wollongbar, NSW, Australia

INTRODUCTION

Australian tea tree oil from the leaves of *Melaleuca alternifolia* (Maiden and Betche) Cheel is an important commercial product due to its antimicrobial properties. Oil production has expanded in recent years to meet increased demand for the oil from both Australia and overseas. The oil is distilled from leaf material harvested from natural stands and plantations. Production from natural stands is limited to about 10% (30 tonnes/year) while expansion of the industry is reliant on plantation production.

The economic viability and continued expansion of plantation (Plate 1) production in Australia is very sensitive to yield and oil price (Reilly 1991). Increasing oil production from both within and outside Australia has the potential to reduce oil prices. For producers to be competitive with potentially lower oil prices, productivity and profitability will need to be improved. Currently, productivity in the industry is considered below potential. A significant contributing factor is seed quality, as plantations are established largely from seed collected from a limited number of wild trees with only rudimentary selection for oil quality.

There is ample justification, therefore, for efficient breeding programmes with resources adequate to provide progressive, economic gains in oil yield and quality.

GENETIC RESOURCES

The Australian tea tree oil industry is based on the terpinen-4-ol rich oil distilled from the leaf of *M. alternifolia* (Plate 12). In addition to *M. alternifolia*, populations of *M. linariifolia*, *M. dissitiflora* and *M. uncinata* also produce terpinen-4-ol rich leaf oils which fall within the specification of the Australian Standard AS 2782-1985 (Standards Association of Australia, 1985) for oils of this type (Brophy *et al.* 1989; Brophy and Lassak 1992).

Tea tree oil is a complex mixture of mainly monoterpene and sesquiterpene hydrocarbons and alcohols. Brophy *et al.* (1989) identified 76 constituents in the oil, with the active constituent thought to be terpinen-4-ol (Southwell *et al.* 1993) and possibly also γ-terpinene (Lassak and McCarthy 1983). Williams *et al.* (1988) have shown that an increasing ratio of terpinen-4-ol to cineole increases the antimicrobial activity of the oil.

Natural Stands

Melaleuca alternifolia (family Myrtaceae), commonly known as Australian tea tree, is native to eastern Australia. Its natural distribution is mainly confined to the coastal watercourses of northern New South Wales (NSW) and the 'granite belt' of southern Queensland (Figure 1). Isolated populations extend south to Port Macquarie and northwest to Stanthorpe, 31°30'S and 28°30'S respectively (Butcher *et al.* 1994). The majority of natural stands grow in the flood plains of the Richmond and Clarence river systems. The range of altitude of *M. alternifolia* is 1–60 m in NSW for the two river systems and up to 800 m in southeast Queensland where the species is found near Ballandean and Stanthorpe.

The species grows in warm subhumid conditions. In this climatic zone, the mean daily maximum temperature during mid summer is 27–31°C and the mean daily minimum temperature during mid winter is 5–7°C. Frosts occur infrequently, on average

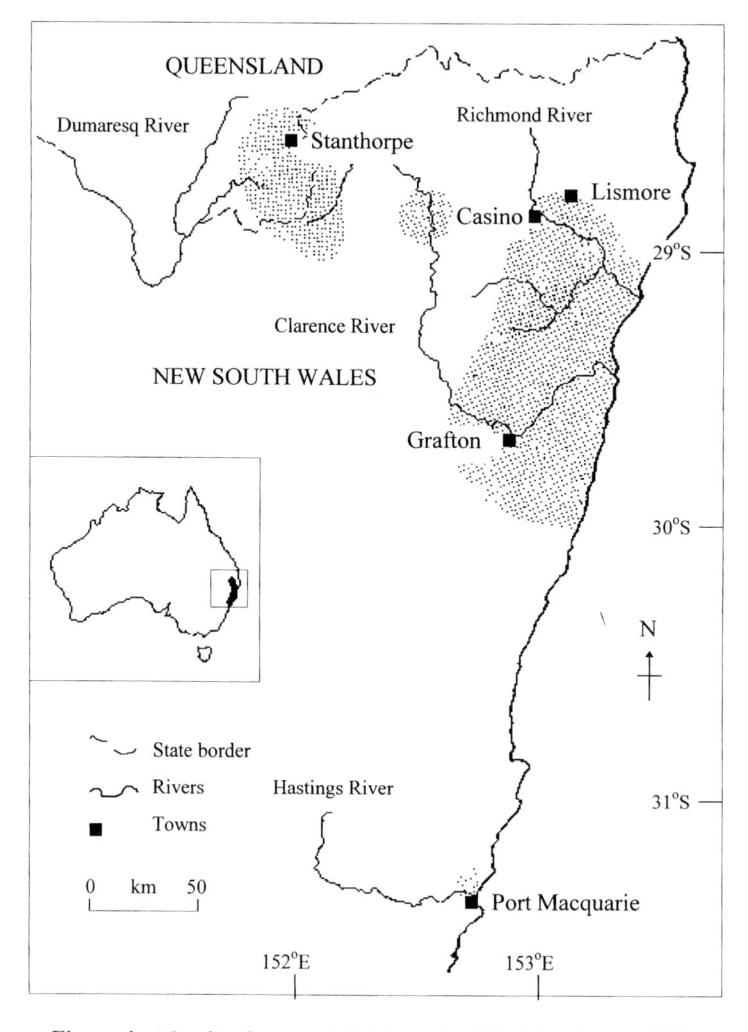

Figure 1 The distribution of *Melaleuca alternifolia* (after Byrnes 1986)

1–3 per year. The Ballandean-Stanthorpe area has, however, an average of 51 frosts per year. Rainfall for the Richmond and Clarence catchment areas ranges from 1000 to 1600 mm annually. Rainfall occurs predominantly in summer with a relatively dry spring.

Native stands are generally found on soils with a pH (measured in water) of 4.5–5.5 (Colton and Murtagh 1990) and a soil texture that ranges from sandy clay loam to heavy clay loam (Small 1981). These soil types in the watercourses and low lying areas of northern NSW are often swampy. *M. alternifolia* is tolerant of long periods (1–3 months) of inundation and even short periods (less than a week) completely immersed in flood waters (Colton and Murtagh 1990).

The natural form of *M. alternifolia* is a single stemmed paper bark tree with a clear bole for one to two thirds of the tree height (up to 15 m). Branch habit is mainly ascending and the crown broad (Plate 2).

Commercial quantities of oil from natural stands of *M. alternifolia* have been produced for over 50 years. Mature trees are harvested manually, leaving stumps about one metre high. Branches are trimmed and the foliage is taken for steam distillation. The cut stumps coppice vigorously (Plate 9) and the new regrowth is then harvested on a regular basis. The annual oil production of about 30 t from natural stands is limited by the restricted distribution of *M. alternifolia*, the small size of individual stands and poor accessibility of the resource particularly in the wet season. Oil production from stands in state forests is also dependant on licences issued by State Forests. Any restrictions to these licences will reduce future oil production from natural stands.

Plantations

The expansion of the Australian tea tree oil industry from natural stands to plantation production commenced in the 1980's to meet the increasing demand for tea tree oil from both overseas and domestic markets. By 1992/93, plantations of *M. alternifolia* were supplying approximately 70% of the market (Butcher 1994) and by 1996/97 the proportion is closer to 90%. Together with increasing the supply of oil, plantations were established to improve the productivity and profitability of oil production.

The selection of suitable areas to establish tea tree plantations was initially confined to the natural distribution of tea tree. It was assumed intuitively by growers that tea tree grown in plantations would require similar conditions to that required by the natural stands. More recently, however, plantations have been established both north and south of the North Coast of NSW. The warmer conditions north of NSW, has basically meant an extended growing season for tea tree while the use of irrigation enables the establishment of plantations in drier areas. High yields in areas outside the North Coast of NSW (Drinnan 1996), suggest that *M. alternifolia* should now be considered as a more widely adapted plant than first thought.

Most of the established plantations are however, of limited value as a genetic resource for a breeding programme because their genetic history is unknown or they were established from a very narrow genetic base. This occurs when the pedigree of the seed used to establish the plantation is unknown. Even today seed for plantations is often collected from a limited number of trees within a stand and often using only rudimentary selection to ensure that the quality of oil from those trees meets industry standards. Similarly, seedlings are often purchased from commercial nurseries with no certificate of origin. As oil yields and oil quality from plantations are variable, there is a need to improve both the amount and quality of the yield through an efficient breeding programme.

GENETIC VARIATION

A number of studies have assessed variation in natural stands of *M. alternifolia* for oil concentration (Bryant 1950; Butcher 1994) and oil composition (Penfold *et al.* 1948; Southwell *et al.* 1992; Butcher *et al.* 1994). Oil concentration can vary considerably between trees both within and between native stands. Bryant (1950) measured an oil concentration variation of 40–89 mg/g for a single stand. Butcher (1994) found a similar variation (25–88 mg/g) between 109 trees sampled from 11 populations throughout the species' natural distribution, with the total between-tree variation being evenly distributed within and between the different populations.

Melaleuca alternifolia produces an essential oil that can vary in composition. Three chemical forms, based on the proportion of 1,8-cineole in the oil were identified by Penfold *et al.* (1948). The three forms were classified as the terpinen-4-ol 'Type' containing 6–15% 1,8-cineole, 'Variety A' 31–40% 1,8-cineole and 'Variety B' 54–64% 1,8-cineole. A terpinolene-rich form of *M. alternifolia* has also been reported by Southwell *et al.* (1992). After the analysis of 109 trees from 11 different populations throughout NSW and Queensland, Butcher (1994) has proposed five chemical forms. Three forms, 'Type' (0.1–11% 1,8-cineole), 'Variety A' (36–48% 1,8-cineole) and 'Variety B' (65–71% 1,8-cineole), correspond approximately to those of Penfold *et al.* (1948). Two additional forms 'Variety C' and 'Variety D' are distinguished from within the terpinolene-rich form identified by Southwell *et al.* (1992). 'Variety C' (15–20% terpinen-4-ol; 30–36% 1,8-cineole; 10–18% terpinolene) and 'Variety D' (1–2% terpinen-4-ol; 17–34% 1,8-cineole; 28–57% terpinolene) are based on the relative proportion of terpinen-4-ol and terpinolene. The chemical variation of the oil produced by this species is of direct economic importance as commercial production of tea tree oil is limited to terpinen-4-ol rich or 'Type' oils.

Several progeny trials of *M. alternifolia* have recently been undertaken (Butcher *et al.* 1996; Doran *et al.* 1997) in northern NSW to examine the variation in oil yield and oil composition under plantation conditions. When assessing the potential to improve plantation productivity, variation in leaf yield and leaf oil concentration together with oil composition need to be considered. Progeny from 5 parent trees in each of 12 populations (Butcher *et al.* 1996) and from 200 parent trees representing 26 areas within 15 populations (Doran *et al.* 1997) were trialed. Significant variation was measured between populations for leaf yield, oil concentration and oil quality. Doran *et al.* (1997) concluded that seed collected from populations east of 153°E longitude out-performed seed collected from 'inland' trees. The partitioning of variation components between population, family, plot and tree for oil concentration (Butcher *et al.* 1996) revealed that the majority of variation occurred between populations while for growth traits the dominant source of variation was between individual trees.

BREEDING PROGRAMMES

The Need

In economic terms, a breeding programme is successful if revenues generated through the sale of genetically improved planting stock exceeds the cost of the programme. Additional benefits accrue to industry when the improved stock increases plantation productivity, profitability and thus competitiveness in the world market place. To be efficient, a breeding programme needs to match potential genetic gain with available genetic, physical, human and financial resources.

Australia is the custodian of the genetic resources of *M. alternifolia* and has the expertise and technology to utilise and improve this genetic resource (Doran 1992). The opportunity exists for the industry to undertake and maintain breeding programmes that will meet the needs of the industry and to maintain an advantage over competitors.

Breeding Strategy

The development of a tree improvement programme involves setting a breeding objective. From this a breeding strategy is developed to manage the genetic improvement and then a breeding plan evolves to achieve the set objectives. A breeding strategy works with a genetic base population and through the ongoing processes of selection and mating manages the mass propagation of the genetically improved population either as seed or cuttings.

Populations in a Breeding Strategy

An effective breeding strategy accumulates genetic gain over each successive generation in the cycle of progeny testing, selection and mating (Figure 2). This successive gain is

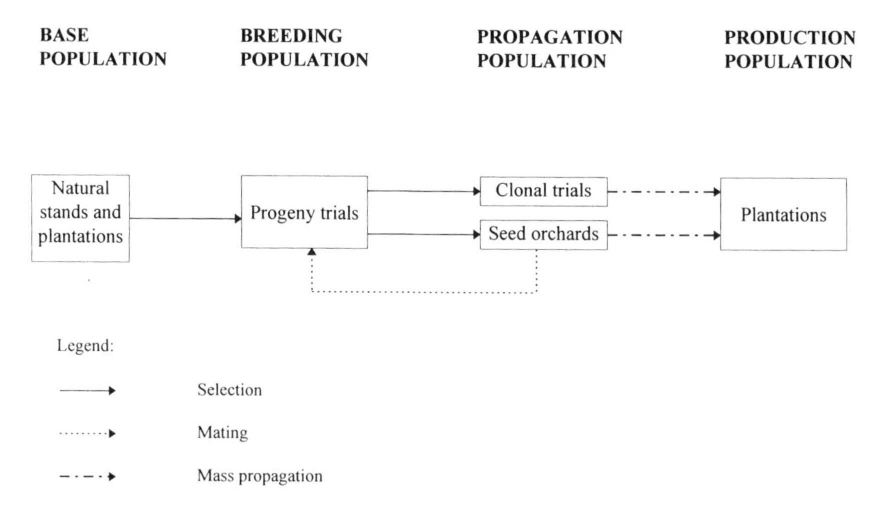

Figure 2 The populations and activities involved in tree breeding

achieved through maintenance of the three populations needed in the cycle to supply the genetically improved material for a fourth population (Eldridge *et al.* 1993), which for *M. alternifolia* is the oil producing plantation. These four populations are—

1. Base population—The gene resource, includes trees in natural stands and some plantations which are suitable for selection. This base resource of genetic variation will continue to be a source for selection to meet future breeding needs.
2. Breeding population—The trees and their progeny which are repeatedly tested, selected and mated over many generations to progressively improve genetic gain.
3. Propagation population—The trees selected from the breeding population to mass produce genetically improved planting material.
4. Production population—The trees in a plantation for the production of tea tree oil.

Selection and Mating

The identification of superior trees to mate is essential in the production of genetically improved seed. This process of selection and mating is needed for each generation to achieve the desired progressive genetic gain.

To enable selection in any trait there must be substantial variation of that trait in the population. To then achieve genetic gain for that trait in the next generation the trait must be readily passed on to the progeny. Genetic gain can also be captured asexually by cloning superior trees.

Efficient methods of selection include progeny testing on appropriate sites using suitable statistical designs, appropriate techniques to measure selectable traits and suitable selection technology such as index selection to take into account a number of selectable traits. Index selection can use genetic, economic and pedigree information to maximise the probability of selecting trees with the best genes (Eldridge *et al.* 1993).

Superior trees are selected for their greater number of favourable genes. Mating of these trees allows the favourable genes to recombine so that new and better trees arise, carrying even greater numbers of favourable genes (Eldridge *et al.* 1993). Mating or crossing of parents may be natural (open pollinated) or artificial (controlled pollinated).

A mating design is the pattern in which the female and male parents are crossed. Care is needed to minimise the potential of inbreeding from related parents, although the design may allow material from other sources to be incorporated. The adoption and then success of the mating design is therefore dependent on the availability of resources, particularly the knowledge and skills of the workers. Many mating designs exist. A general description of several designs is given by Zobel and Talbert (1984) who grouped them into incomplete pedigree designs (open pollinated and polycross) and complete pedigree designs (nested, factorial, bi-parent and diallel). Open pollinated progeny tests are inexpensive, simple and informative, while the value of known pedigree designs from controlled pollinated crosses lies in the additional information to be gained.

Breeding Plan

Once the breeding strategy has been developed to manage the genetic improvement of the species with the available resources, then the breeding plan is undertaken to detail the implementation of the breeding strategy. The typical breeding plan details goals and the methods needed to achieve objectives within the timeframe and resources of the programme. Once implemented, plans are subject to regular revision.

BREEDING STRATEGY FOR *MELALEUCA ALTERNIFOLIA*

Objectives

The objective of a tea tree breeding strategy would be to progressively improve oil yield and quality by increasing the yield of leaves per tree and the oil concentration in those leaves and ensuring that oil produced is of consistent quality to maximise desirability and price.

Selection Criteria

Plantations of *M. alternifolia* are managed as coppice crops to produce large quantities of terpinen-4-ol rich tea tree oil. To increase plantation production by improving the genetics of the plant, the selection criteria should deal with those plant characteristics that increase oil yield and quality. Those characteristics are—

(1) High leaf biomass as a seedling and when coppiced.
(2) High leaf oil concentration.
(3) High oil quality.
(4) Adaptable to different growing conditions.
(5) Resistant to diseases and pests.

Oil yield is a function of leaf biomass and the oil concentration of that leaf. To maximise gains from breeding, the selection should aim to achieve gains in both these traits. This is achieved when both traits are either independent of each other or if they are dependent, then they are positively correlated. This implies that gains in one trait will result in gains in the other. When a negative correlation occurs between selectable traits it is difficult to improve traits concurrently using recurrent selection (Eldridge *et al.* 1993).

A strategy is therefore needed to accommodate negative correlation in order to maximise genetic gains. Butcher (1994) has proposed three strategies for consideration. Firstly, the effect of the correlation can be minimised, by using a combined index selection (Cotterill and Dean 1990) with a restriction imposed on one trait at pre-selection levels (Cotterill and Jackson 1981). An alternative would be to select trees which are correlation breakers (Eldridge *et al.* 1993) and then to mass propagate those individuals. The third alternative would be to cross two separate breeding populations for plant dry weight and oil concentration. The hybrid progeny may express higher oil yields than could be achieved through simultaneous selection (Dean *et al.* 1983).

Oil quality should be determined on its antimicrobial activity against microorganisms. Currently, there is a market demand for oils with the lowest possible 1,8-cineole: terpinen-4-ol ratio (i.e. 1,8-cineole < 4% and terpinen-4-ol > 36%). The variation in oil composition between populations (see above) indicates the potential for selection to meet this demand. However, there are doubts about the need to lower the 1,8-cineole level in oils. Recent evidence has shown that 1,8-cineole is not detrimental to the oil's bioactivity or safety (Southwell *et al.* 1996, 1997) and that it may have beneficial effects (Southwell *et al.* 1993; Chapter 9). Other research indicates that relatively minor components in the oil (e.g. sabinene) may influence the efficacy of the oil against specific microorganisms (Williams *et al.* 1990). Selection for oil quality must meet current market demands, while allowing the flexibility to cater for the on-going changes in the market place.

Broad adaptability and resistance to diseases and pests are both selection criteria that influence leaf biomass more than oil concentration. Results from several progeny trials established in NSW show that, when seedlots were ranked for oil yield and again for oil quality, the order of ranking was similar for different sites (Doran *et al.* 1997). This indicates that there is a reasonable degree of stability for oil traits such that selected seedlots may retain their elite status over sites. From the 200 seedlots trialed by Doran *et al.* (1997), nine were selected to assess productivity in northern Queensland. Early results (Drinnan 1997) suggest that seedlot rankings for oil concentration are consistent to that when grown in NSW.

The major pest that defoliates tea tree plantations in NSW is *Paropsisterna tigrina* (Chapuis) or Pyrgo beetle (Campbell and Maddox 1996). Estimates of leaf loss to Pyrgo can vary with many reports of complete loss of crop (Maddox 1996a). The use of chemicals to control this pest is declining as the industry adopts a zero chemical use policy to ensure residue free oils. Thus the need for an alternative to manage this pest will increase. Selection for genetic resistance would offer the industry significant benefits. Pyrgo activity coincides with flush growth. Selection for traits to alter the flushing time, or rate at which the flush leaf matures may reduce leaf loss (Maddox 1996b). Maddox (1996a) reported also that feeding damage from Pyrgo can decline with an increasing proportion of α-terpinene and terpinolene in flush leaf oils.

Selection for disease resistance is currently not considered worthwhile as no serious diseases have been recorded on tea tree (Colton and Murtagh 1990).

To maximise genetic gain, the number of traits for selection should be kept to a minimum as the gains in individual traits decline as the number of traits for selection increases (Cotterill and Dean 1990).

Species Information

To develop a breeding strategy for *M. alternifolia* it is important to know certain aspects of the biology of the species. These include both sexual and asexual reproduction and the range of variation in economically important traits along with their relationships and heritabilities. These aspects are listed.

1. *Morphology.* The flowers of *M. alternifolia* are morphologically bisexual and are insect pollinated. Outcrossing predominates with less than 10% self-pollination (Butcher *et al.* 1992). This low rate of self-pollination is highly advantageous for a breeding strategy based on open pollination as it reduces the likelihood of inbreeding.
2. *Flowering.* The abundance and periodicity of flowering is variable both within and between populations (Doran *et al.* 1997). Flowering usually occurs during October

and November, for trees that are more than 2–3 years old. Flowering is heaviest during these months in years that have wet winters (Baker 1996, 1997). It appears that a wet winter is associated with both an increase in the number of trees flowering in a population and with the number of flowers per tree.

3. *Pollination.* Although the controlled pollination of *M. alternifolia* is possible (Moncur 1997), pollinating by hand is inherently slow and expensive. To undertake controlled crosses in a breeding strategy, the potential gains from such crosses have to be balanced against the costs, particularly when resources to a breeding program are limited. Potential gains from such crosses include the use of the progeny as a source of pedigree material for genetic studies as well as producing unrelated families for further selection.
4. *Propagation.* Mass propagation of this species is by seed. Cloning is possible by micropropagation (Hartney and Svensson 1992) or cuttings (Whish 1993) but the use of clones in yield trials over several years is needed to test if clones are suitable for plantation production. Cuttings from selected trees can be used in the establishment of clonal seed orchards to capture greater genetic gains.
5. *Distribution.* Information on the natural distribution of *M. alternifolia* (see above) is used when collecting seed for progeny trials. Seed is collected from the terpinen-4-ol rich oil trees located throughout the natural distribution of the species in NSW. Although seed normally matures 12–18 months after flowering (Colton and Murtagh 1990), mature seed can be collected at any time as some trees retain their seed crops for several years.
6. *Genetic parameters.* Genetic parameters express estimates of genetic and non-heritable variations of a population in respect to some characteristic (Allard 1960). In addition to being needed for the evaluation of different breeding strategies, these estimates together with their nature, magnitude and inter-relationships are necessary to assess improvement by selection. Heritability is a measure of how strongly a trait is influenced by genetics (Hanson 1963). When heritability is high, gain from selection will be high. Genetic parameters and expected gains from selection and breeding as estimated in several *M. alternifolia* progeny trials are—

   (a) Heritabilities of 0.51 (Doran *et al.* 1997) and 0.67 (Butcher 1994) for oil concentration, 0.21 for plant height and 0.14 for stem diameter (Doran *et al.* 1997), 0.25 for plant dry weight and 0.27 for coppicing ability (Butcher 1994). These estimates indicate that improvement, particularly in oil concentration would follow selection for single traits.

   (b) The absence of genetic correlation between oil concentration and growth traits (e.g. basal diameter, which is highly correlated with leaf yield (Doran *et al.* 1997)). This suggests that oil concentration and leaf biomass should be able to be improved simultaneously in a breeding programme. Butcher (1994), however, reported a negative genetic correlation of 0.42 between oil concentration and plant dry weight implying that genetic gain would have to be balanced between these two traits.

   (c) Calculated gains of 17% for oil concentration and 14% for coppicing (Butcher 1994). These estimated gains were derived from one generation of breeding at a selection intensity of one tree in ten when a combined index selection was used, restricting plant dry weight to pre-selection levels.

Resources to Implement the Strategy

Tree breeding is a long-term activity that requires both a considerable commitment and balanced use of genetic, physical, human and financial resources (Doran 1992). All breeding programmes need to efficiently use limited resources. In the case of a *M. alternifolia* programme, analytical requirements for the determination of leaf oil traits are a major criteria in the selection of a breeding strategy.

The major genetic resource available for use in a *M. alternifolia* breeding programme is seed collected from natural stands. Plantation trees do not represent a desirable genetic resource in terms of seed, however, they could be used as a source of clones for inclusion in the programme. Collection of seed from plantations is generally not feasible or desirable, as plantations are usually harvested annually and so do not produce mature seed. Additionally any seed that did result from a plantation, is likely to be highly inbred as plantations are usually established using seed from only a limited number of parent trees.

The physical resources include suitable sites to establish and maintain progeny trials and orchards. Sites for open pollinated orchards need to be located in isolation from stray pollen. Physical resources also include the facilities used to implement the breeding strategy, such as a nursery to grow seedlings and clones, an equipped laboratory to determine leaf oil characteristics and appropriate room and equipment to store, process and analyse samples and results.

For a strategy to be successful, funding has to be both adequate and long-term. An appropriate budget allows for the initial capital and then on-going costs of tree breeding, while providing funds for the employment of personnel, their transport and communication.

Breeding for tree improvement is becoming a highly specialised science, particularly in the field of mathematical statistics for efficient selection and quantitative genetics when predicting the consequences of selection (Eldridge *et al.* 1993). To develop and manage a strategy efficiently, a team of specialists and technical assistants is needed.

Monitoring Progress

Strategies should have the capacity for review. Progress needs to be documented and reviewed on a regular basis. Communication with research partners and industry facilitates feedback leading to greater efficiency in the chosen breeding strategy.

Determinants of a Breeding Strategy

Breeding strategies can range from simple and cheap to complex and expensive (Eldridge *et al.* 1993). Limited resources will often determine the complexity of the strategy and hence the potential gain. The right strategy for a breeding programme will maximise the genetic gain from available resources. When choosing the appropriate breeding strategy, there are four basic components to consider.

1. *Method of selection* (*mass or recurrent*). Recurrent selection maintains family identity for the base, breeding and propagation populations while mass selection does not. With fewer measurements, records or statistics, mass selection provides a cheaper source

of genetic improvement, however genetic gains are usually less than for other strategies. Also, mass selection has a greater potential for inbreeding as the identity of the propagation population is unknown.
2. *Method of mating* (*open or controlled*). The full pedigree of both parents is maintained with controlled pollination to maximise genetic gain. The cost, time and skill required however can be prohibitive.
3. *Type of breeding population* (*single*, *multiple*, *nucleus or interspecific*). The main breeding population for each strategy is usually established from open pollinated seed that has been collected from a large number of unrelated trees in natural stands or plantations. This is to maximise the likelihood of genetic variation from which to make selections. A single breeding population is the cheapest to establish. This population may be thinned to become a seed orchard to supply improved seed for the next generation or is used as a source of selected material that is propagated vegetatively or as seed from controlled crossing to a separate seed orchard. Multiple populations subdivide a breeding population to meet possible future selection needs while avoiding the possibility of inbreeding. Nucleus breeding concentrates on selecting a small nucleus population from the main breeding population. From this nucleus, improved seed or clones are then produced. Interspecific breeding maintains a breeding population of each parent species which is to be crossed. Outstanding individuals of each species are then crossed to produce hybrids which are usually propagated as clones.
4. *Method of mass propagation* (*seed or clones*). Seed production is relatively inexpensive while cloning, although expensive, offers some increase in genetic gain.

CASE STUDY: IMPROVEMENT OF *MELALEUCA ALTERNIFOLIA*

An outline of the breeding strategy and breeding plan undertaken by CSIRO Forestry and Forest Products, Canberra and NSW Agriculture, Wollongbar for improving both the amount and quality of oil yield in *M. alternifolia* is given below. The main breeding objective is to increase oil yields in new plantations by 30% in five years (by 1998) with a 60% increase at the conclusion of the first generation of breeding, about year 2001.

Breeding Strategy

In addition to maximising genetic gain, a breeding strategy for *M. alternifolia* must reflect aspects of the tea tree industry. Although, the industry is expanding rapidly, the commercial use of this Australian species is still relatively recent and limited. This limits the funding available to the industry for research.

A simple breeding strategy was therefore adopted for this species where the use of limited resources was matched to achievable goals. The strategy consists of two parts, interim activities and the main breeding component (Figure 3).

The first incorporates interim measures to provide some short-term modest improvements by provision of seed from a natural seed stand (SS) and best provenances (Natural stands). The quality and quantity of oil produced from the progeny trials are to be used to identify the best provenances.

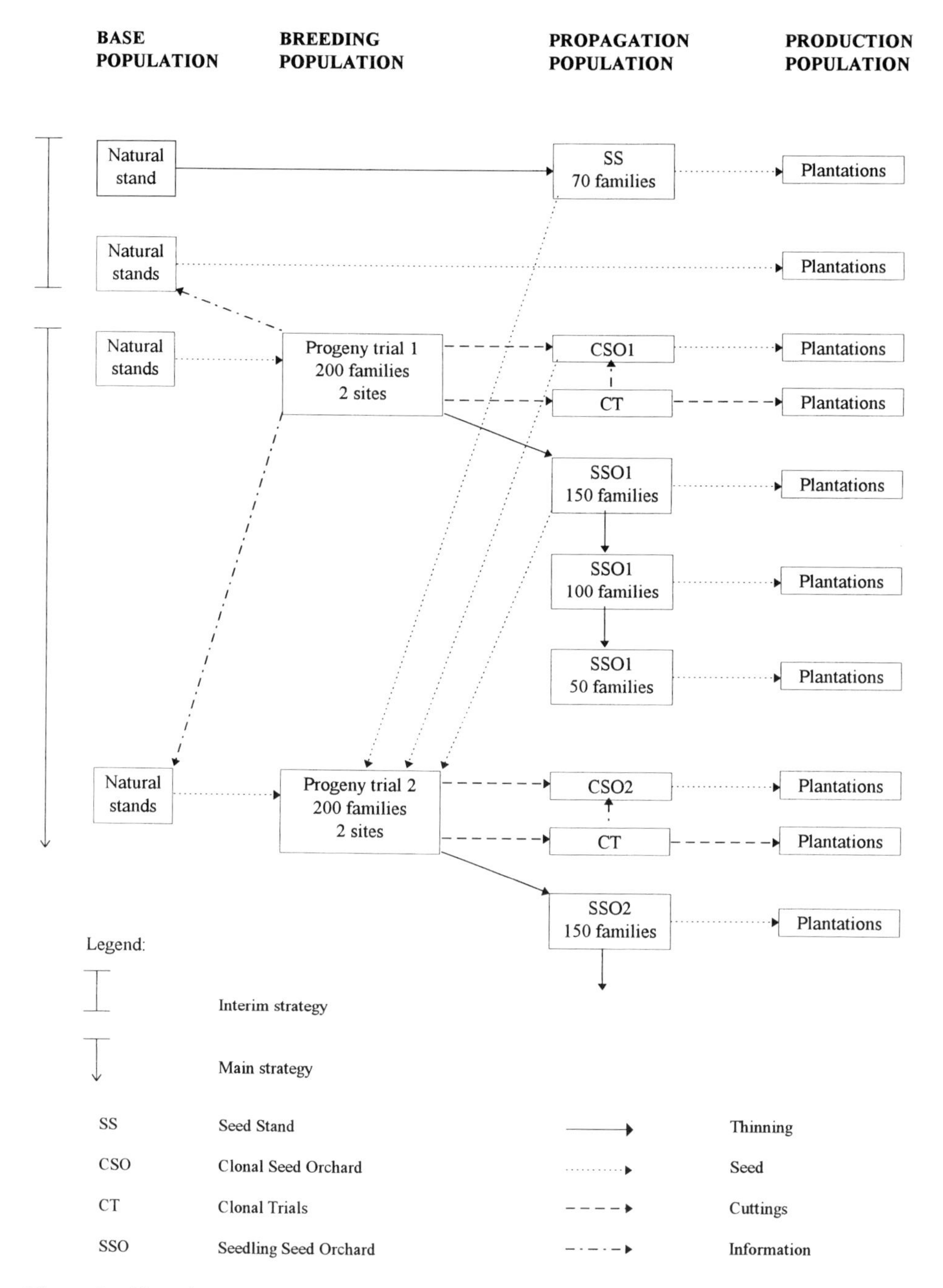

Figure 3 Flow chart representing the breeding strategy to supply progressively improved seed and cuttings of *M. alternifolia* for plantation production

The main strategy, which is designed to give substantial genetic gain, uses recurrent selection with open pollination in a single population. This population serves as a progeny trial and finally as a seedling seed orchard (SSO). This simple strategy has been used very successfully in improvement programmes for *Eucalyptus* (Eldridge *et al.* 1993), as it has achieved large genetic gain without excessive cost. The strategy also has the added flexibility to allow some limited controlled pollination in the seed orchards (CSO and SSO) and vegetative propagation in the progeny trials and clonal trials (CT) to meet any specific needs of the industry.

The first activity in the main strategy is to collect open pollinated seed (Plates 13, 14) from about 200 unrelated trees in natural stands. These families are used to establish three progeny trials on two contrasting sites. Two of the trials are harvested annually to provide estimates of family coppicing ability across sites. The remaining trial is used as a widely based breeding population which is to be progressively thinned (based on family and individual tree performance) to form the first seedling seed orchard (SSO1). Improved seed from this first seed orchard is then used to help establish the second generation of breeding. Each new generation is established from the best trees of the previous generation with an infusion of new unrelated material from other promising sources. Family thinning from 150 to 100 to 50 is based on growth, oil characteristics and information from the other progeny trials. As family numbers reduce, the increase in selection pressure results in progressively improved seed to be used in the establishment of plantations.

The two progeny trials that complement the main breeding trial, also serve as a reserve of genetic material should the programme need other seedling seed orchards.

Breeding Plan

Tree breeding is an expensive and long term activity. Industry associations and government funding bodies must be prepared for a substantial and on-going financial commitment for the breeding plan to be implemented.

Developing and then implementing the plan requires a wide-range of appropriate expertise and cooperation between the growers, the breeders and the funding consortium. Implementation on a cooperative basis encourages growers and breeders to share work and capital resources. With such cooperation significant savings to the programme can be made, particularly, at times of peak resource use, such as the growing of seedlings and establishing field trials on grower's properties. To assist growers to contribute to the programme, a form of in-kind payment to the value of their contribution should be considered. The payment may involve access to improved seed or cuttings.

Once the funding has been organised and a cooperative attitude adopted, a management team is required to implement the breeding strategy. The team requires or needs to access the personnel and facilities necessary to run the programme. Needs would include statistical expertise for the design and analysis of field trials, agronomic skills and facilities for plant propagation and field trial research and laboratory skills and equipment to determine leaf oil quality and quantity.

The breeding plan is developed to implement the breeding strategy within the timeframe and resources of the programme. A breeding plan for *M. alternifolia* would therefore

Table 1 The breeding plan activities for the first generation of improved *M. alternifolia*

Trial	*Major tasks*	*Activities*
Seed stand	Site selected and developed as a natural seed stand	Consult with growers to select site Peg site and label trees Sample trees and analyse oil Select best trees Harvest non-selected trees Monitor flowering and seed maturity
	Collect seed	Collect capsules and store seed
Natural stands	Best natural stands selected	Collect seed for progeny trials Sample seedlot trees and analyse oil Use progeny trials to select best stands Monitor flowering and seed maturity
	Collect seed	Collect capsules and store seed
Progeny trials	Sites selected and trials established	Consult with growers to select sites Grow seedlings Design and establish trials Assess growth and oil
Seedling seed orchard 1	Convert a progeny trial into a seedling seed orchard	Select best trees using index selection Progressively thin non-selected trees Monitor flowering and seed maturity
	Collect seed	Collect capsules and store seed
Clonal trials and clonal seed orchard 1	Sites selected and trials and orchard established	Consult with growers to select sites Select best available trees Take and propagate cuttings Design and establish trials and orchard Assess growth and oil
	Collect cuttings from clonal trials	Harvest trees in clonal trials Use regrowth for cuttings
	Collect seed from clonal seed orchard	Monitor flowering and seed in orchard Collect capsules and store seed
Coppicing trials	Convert remaining progeny trials into coppicing trials	Harvest progeny trials Assess regrowth and oil
Yield trials	Evaluate oil productivity of selected and improved seed	Consult with growers to select sites Grow seedlings Design and establish trials Assess growth and oil
Flowering studies	Evaluate the use of controlled pollination	Monitor flowering and select trees to cross Undertake controlled pollinations
	Collect seed from controlled crosses	Monitor capsule production and maturation Collect capsules and store seed

detail the activities and resources used to achieve the major tasks of the breeding strategy. A summary of the adopted breeding plan activities is presented in Table 1.

Crucial activities of the breeding plan involve progeny testing. Progeny trials evaluate the performance of offspring from parents with full or partially known identity (Eldridge *et al.* 1993). This evaluation achieves three important objectives: firstly, an estimation of the breeding value of parent trees used to identify best provenances for further seed collection; secondly, the provision of a source of selected trees for cuttings or improved seed if the trial is thinned to a seed producing orchard and finally, an estimation of the genotype by environment interaction and other genetic parameters for predicting genetic gain.

The justification for a breeding programme is the genetic gain realised in improved plantations and the financial returns on the programme investment (Eldridge *et al.* 1993). Well designed yield trials are an important aspect of the breeding plan as they demonstrate how much gain has been realised when progeny from the breeding programme are compared to other sources.

SUMMARY

The range of variation in oil traits for *M. alternifolia*, together with the high heritability for oil concentration and moderate heritability for growth, indicate that considerable gains can be made from breeding programmes.

A breeding programme can achieve initial genetic gain in *M. alternifolia* plantations through choosing seed from the best provenances. Then on-going gains are made by selecting trees in the best provenances, mass producing their seed in seed orchards and selecting a few outstanding trees for vegetative propagation.

The breeding programme detailed in the case study capitalises on the large initial gain achievable from provenance choice and later improvement by recurrent selection and mating in each future generation.

ACKNOWLEDGEMENTS

This chapter has been written as part of my activities as project officer for the research project "The Improvement of Australian Tea Tree Through Selection and Breeding". This project is jointly funded by the Rural Industries Research and Development Corporation (RIRDC) and the Australian Tea Tree Industry Association (ATTIA). Institutional support is provided by NSW Agriculture and CSIRO. Their funding and support is gratefully acknowledged. I also thank my colleagues John Doran, Ian Southwell and Robert Lowe for their commitment to the project and for their comments on this chapter.

REFERENCES

Allard, R.W. (1960) *Principles of Plant Breeding*, Wiley, New York.

Baker, G.R. (1996) The improvement of Australian tea tree through selection and breeding. Unpublished Milestone Report for Project DAN-151A—30/11/96. For the Rural Industries Research and Development Corporation, Canberra and NSW Agriculture, Wollongbar, NSW.

Baker, G.R. (1997) The improvement of Australian tea tree through selection and breeding. Unpublished Milestone Report for Project DAN-151A—30/11/97. For the Rural Industries Research and Development Corporation, Canberra and NSW Agriculture, Wollongbar, NSW.

Brophy, J.J., Davies, N.W., Southwell, I.A., Stiff, I.A. and Williams, L.R. (1989) Gas chromatographic quality control for oil of *Melaleuca*, terpinen-4-ol type (Australian tea tree). *Journal of Agriculture and Food Chemistry*, **37**, 1330–1335.

Brophy, J.J. and Lassak, E.V. (1992) Steam volatile leaf oils of some *Melaleuca* species from Western Australia. *Flavour and Fragrance Journal*, **7**, 27–31.

Bryant, L.H. (1950) Variations in oil yield and oil composition in some species of eucalypts and tea trees. *Technical Notes, Forestry Commission (Division of Wood Technology) NSW*, **4**, 6–10.

Byrnes, N.B. (1986) A revision of *Melaleuca* L. (Myrtaceae) in northern and eastern Australia, 3. *Austrobaileya*, **2**, 254–273.

Butcher, P.A. (1994) Genetic diversity in *Melaleuca alternifolia*: Implications for breeding to improve production of Australian tea tree oil. PhD. Thesis, Australian National University.

Butcher, P.A., Bell, J.C. and Moran, G.F. (1992) Patterns of genetic diversity and nature of the breeding system in *Melaleuca alternifolia*. *Australian Journal of Botany*, **40**, 365–375.

Butcher, P.A., Doran, J.C. and Slee, M.U. (1994) Intraspecific variation in leaf oils of *Melaleuca alternifolia* (Myrtaceae). *Biochemical Systematics and Ecology*, **42**, 419–430.

Butcher, P.A., Matheson, A.C. and Slee, M.U. (1996) Potential for genetic improvement of oil production in *Melaleuca alternifolia* and *M. linariifolia*. *New Forests*, **42**, 419–430.

Campbell, A.J. and Maddox, C.D.A. (1996) Insect pest management in tea tree. Final Report for the Rural Industries Research and Development Corporation, Canberra.

Colton, R.T. and Murtagh, G.J. (1990) Tea-tree oil—plantation production. Agfact P6.4.6, NSW Agriculture and Fisheries, Sydney.

Cotterill, P.P. and Dean, C.A. (1990) *Successful Tree Breeding with Index Selection*. CSIRO, Melbourne.

Cotterill, P.P. and Jackson, N. (1981) Index selection with restrictions in tree breeding. *Silvae Genetica*, **30**, 2–3.

Dean, C.A., Cotterill, P.P. and Cameron, J.N. (1983) Genetic parameters and gains expected from multiple trait selection of radiata pine in eastern Victoria. *Australian Forest Research*, **13**, 271–278.

Doran, J.C. (1992) Breeding strategy for genetic improvement of *Melalecua alternifolia*. Report to Australian Tea Tree Industry Association and Rural Industries Research and Development Corporation. CSIRO, Canberra.

Doran, J.C., Baker, G.R., Murtagh, G.J. and Southwell, I.A. (1997) Improving tea tree yield and quality through breeding and selection. Rural Industries Research and Development Corporation Research Paper Series No. 97/53, RIRDC, Canberra.

Drinnan, J. (1996) Tea tree research in North Queensland. *Tea Tree Oil Symposium*. August 1996 Abstracts. Wollongbar Agricultural Institute, Wollongbar, NSW, pp. 3–4.

Drinnan, J. (1997) Tea tree research in North Queensland. *Tea Tree Oil Symposium*. August 1997 Abstracts. Wollongbar Agricultural Institute, Wollongbar, NSW, p. 5.

Eldridge, K., Davidson, J., Harwood, C. and van Wyk, G. (1993) *Eucalypt Domestication and Breeding*, Oxford University Press, Oxford.

Hanson, W.D. (1963) Heritability. In W.D. Hanson and H.F. Robinson (eds.), *Statistical Genetics and Plant Breeding*, No. 982 (National Academy of Sciences Publication), (National Research Council, Washington), pp. 125–140.

Hartney, V.J. and Svensson, J.G.P. (1992) Micropropagation of superior clones of *Melaleuca alternifolia*. In *Breeding Strategy for Genetic Improvement of Melaleuca alternifolia*, (Report to Australian Tea Tree Industry Association and Rural Industries Research and Development Corporation), Appendix A, Canberra.

Lassak, E.V. and McCarthy, T. (1983) *Australian Medicinal Plants*, Methuen, Australia.

Maddox, C.D.A. (1996a) Aspects of the biology of *Paropsisterna tigrina* (Chapuis) the major pest species of *Melalecua alternifolia* (Cheel). MSc. Thesis, University of Queensland.

Maddox, C.D.A. (1996b) NSW Agriculture, unpublished results.

Moncur, M.W. (1997) CSIRO Canberra, unpublished results.

Penfold, A.R., Morrison, F.R. and McKern, H.H.G. (1948) Studies in the Myrtaceae and their essential oils. Part 1. The seasonal variations in yield and cineole content of *Melaleuca alternifolia* Cheel. In *Researches on Essential Oils of the Australian Flora*, Part 1 (Museum of Technology and Applied Science, Sydney), pp. 5–7.

Reilly, T.L. (1991) The economics of tea tree. *Tea Tree Marketing and Planning Conference*, 31 October–2 November, Ballina, NSW, pp. 30–38.

Small, B.E.J. (1981) Effects of plant spacing and season on growth of *Melaleuca alternifolia* and yield of tea tree oil. *Australian Journal of Experimental Agriculture and Animal Husbandry*, **21**, 439–442.

Southwell, I.A., Freeman, S. and Rubel, D. (1997) Skin irritancy of tea tree oil. *Journal of Essential Oil Research*, **9**, 47–52.

Southwell, I.A., Hayes, A.J., Markham, J. and Leach, D.N. (1993) The search for optimally bioactive Australian tea tree oil. *Acta Horticulturae*, **334**, 256–265.

Southwell, I.A., Markham, J. and Mann, C. (1996) Is cineole detrimental to tea tree oil? *Perfumer and Flavorist*, **21**, 7–10.

Southwell, I.A., Stiff, I.A. and Brophy, J.J. (1992) Terpinolene varieties of *Melaleuca*. *Journal of Essential Oil Research*, **4**, 363–367.

Standards Association of Australia, Essential oils—oil of *Melaleuca,* terpinen-4-ol type, 2782, Sydney. Standards Australia (1985).

Williams, L.R., Home, V.N. and Asre, S. (1990) Selection and breeding of superior strains of *Melaleuca* species to produce low cost, high quality tea tree oil. In L.R. Williams and V.N. Homes (eds.), *Modern Phytotherapy—The Clincal Significance of Tea Tree Oil and Other Essential Oils*, Vol. 2, Macquaire University, Sydney, pp. 73–91.

Williams, L.R., Home, V.N., Zhang, X. and Stevenson, I. (1988) The composition and bactericidal activity of oil of *Melaleuca alternifolia* (tea tree oil). *International Journal of Aromatherapy*, **1**, 15–17.

Whish, J.P.M. (1993) The selection and propagation of high oil-yield tea trees. Unpublished Report. Department of Agronomy, University of New England, Armidale.

Zobel, B.J. and Talbert, J. (1984) *Applied Forest Tree Improvement*, Wiley, New York.

Plate 10 Eight months of regrowth in a highly productive plantation (A. Manciagli)

Plate 11 Forage-harvested trees are taken to the distillery in trailer-bins for oil extraction (R. Colton)

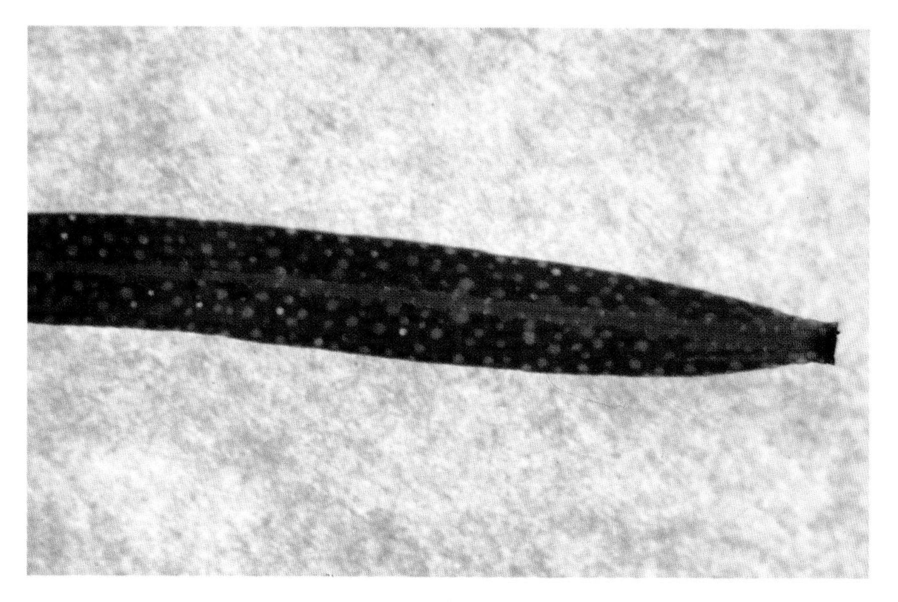

Plate 12 *Melaleuca alternifolia* leaf oil glands (A. Curtis)

Plate 13 *M. alternifolia* seed capsules (J. Murtagh)

Plate 14 *M. alternifolia* seed (R. Colton)

Plate 15 Tipping mobile still (G. Davis)

Plate 16 Simple direct-fired tea tree still (L. Davis)

Plate 17 Two-pot distillery with boiler and separator in the background, sealed mobile bin (left) and suspended bin lid (right) (R. Colton)

8. TEA TREE OIL DISTILLATION

GEOFFREY R. DAVIS

Hornsby, NSW, Australia

INTRODUCTION

This chapter considers mainly the practical aspects of steam distillation of tea tree oil, the essential oil of *Melaleuca,* terpinen-4-ol type.

The definition of an essential oil which was adopted by the Standards Association of Australia in 1968 and also by the International Standards Organisation (ISO) at the Ninth Plenary Meeting of the Technical Committee ISO/TC54 Essential Oils, held in Lisbon, 5th–9th March, 1968, was:

> Essential oils are volatile oils, generally odorous, which occur in certain plants or specified parts of plants, recovered therefrom by accepted procedures, such that the nature and composition of the product is, as nearly as practicable, unchanged by such procedures.

This is an important definition. It specifies clearly that the nature and composition of the oil must be unchanged, as nearly as practicable, by the process of extracting it, and is therefore one reason why steam distillation is an appropriate method of extraction. Furthermore, because steam distillation has been the extraction method since the oil was first produced, the market accepts steam distilled oil as normal oil. Oil derived by another technique might be of slightly different chemical composition and therefore might not be accepted by the market as normal tea tree oil.

The standards for this oil, both Australian (Standards Australia 1997) and International (International Standards Organisation 1996), define the oil as "Essential oil obtained *by steam distillation* of the foliage and terminal branchlets of *Melaleuca alternifolia* (Maiden et Betche) Cheel, *M. linariifolia* Smith and *M. dissitiflora* F. Mueller as well as other species of *Melaleuca* provided that the oil obtained conforms to the requirements given in this International Standard".

There are other methods of extracting essential oils from plants, including:

Solvent extraction followed by steam distillation of the extract.
Solvent extraction followed by vacuum distillation of the extract.
Liquid carbon dioxide extraction.
Enfleurage.
Expression from the pericarp of citrus fruits.
Use of microwaves has also been suggested.

THE DISTILLATION PROCESS

Steam distillation is the appropriate method of extraction, not only because it causes minimum change to the composition of the oil during extraction, but also because steam is readily available, cheap, not hazardous (chemically), can be used at low pressure and can be recycled. In particular, it allows the oil to be extracted at a temperature which is constant and low enough not to damage the oil. By using steam at atmospheric pressure, the oil will be distilled from the plant material at a temperature slightly below that of boiling water. This is because the vapour pressure of a two or more phase liquid mixture in which the phases are not miscible, is equal to the sum of the partial pressures of the phases. Thus if water boils at 100°C, a mixture of water and some other volatile liquid which is not soluble in water, i.e. the oil, will have a higher vapour pressure than water alone, and will therefore boil at a lower temperature (Guenther 1965).

This is a fundamental principle of steam distillation. Steam distillation is therefore the best method of extracting the oil, both in relation to cost of extraction and quality of oil.

In this chapter, the term "still" refers to the still-pot only and not the whole unit comprising the boiler, the pot and condenser and separator.

Pressure

Certain essential oils may be more advantageously steam-distilled at still pressures greater than atmospheric pressure, others at lesser still pressures. Pressure distillation is achieved by enhancing the still pressure which is then maintained independently of the boiler pressure. Fortunately, there is no advantage in either reduced or elevated pressure in the distillation of tea tree oil from the foliage. Thus elaborate pressure equipment or vacuum pumps are not called for. In fact, working at other than atmospheric, particularly at raised pressure, is likely to produce an oil of different composition from normal tea tree oil.

Solubility in Water

Some essential oils are partially soluble in water. Furthermore, some compounds in an oil might be more soluble than others. In the case of tea tree oil, while its solubility in water is low, the most valuable compound, terpinen-4-ol, is slightly soluble. Therefore, if tea tree oil is extracted by steam distillation there will be some removal of terpinen-4-ol as long as fresh steam is passed through the charge during distillation or comes into contact with oil in the receiver.

Cohobation

The traditional technique for preventing loss of oil in the distillate water is to return the water to the still during distillation. By so doing, the water in the system becomes saturated with the oil, or soluble parts of it, and, being saturated, no longer absorbs oil during the distilling process. This process is known as cohobation. In the case of some oils with soluble components, cohobation is necessary. With tea tree oil, the loss of oil through its solubility in water is small. Whilst cohobation is desirable, it is not essential. Cohobation, however, is certainly desirable where the water supply is limited. It is also useful in maintaining the water level in the still where steam is raised in the still. Where steam is raised externally, cohobation is more difficult. It is necessary to take the distillate water to a tank where it can be adjusted for pH and possibly have boiler water treatment compounds added before being reintroduced to the boiler.

It is worth noting that extraction of oil from various *Melaleuca* species is not difficult. It is, in fact, a very simple operation.

In order to extract the oil we need to pack the plant material into a container, pass steam through the container, collect the steam and oil vapour that distils off, condense the vapours to water and oil, and separate the oil from the water.

Tea tree oil is a very amenable oil to distil. It comes from the leaf readily. It is not adversely affected by the water with which it is distilled, or by atmospheric oxygen in the brief periods between removing it from the receiver and packing in an air tight container. Therefore no special equipment or action is required to preserve the oil at, or soon after distillation. Separation of the oil and water is easily effected by using the immiscibility and difference in density and of the two liquids of the distillate. The amount of oil dissolved in the distillate water is small (negligible, if cohobation is used) and therefore recovery of oil dissolved in the water unnecessary.

Until recently, tea tree oil was produced from natural stands and small, sometimes movable, apparatus was required. In recent years plantations have supplied most of the oil where larger, static and more sophisticated apparatus is used. In small plantations, and even in the early stages of larger plantations, the small, low capital cost equipment is appropriate and effective. A description of suitable equipment for both applications follows.

Plant Material

To obtain oil by steam distillation it is necessary to pass steam through the oil-bearing parts of the plant. In the case of tea tree oil this means the leaves and terminal branchlets. The main woody part of the plant does not contain oil. As the *Melaleucas* coppice well after being cut, (except for the initial harvest) it is the coppice regrowth that is collected. Consequently, the proportion of non oil-bearing material harvested is low and the whole of the coppice regrowth can be put into the still. Where the trees are harvested by hand, some stripping of the leaves from the larger woody stems is feasible, resulting in a higher proportion of oil-bearing material in the still. However, for mechanical harvesting it is uneconomical to separate the non oil-bearing part of the plant. Furthermore, there is nothing distilled out of the woody part that is detrimental to the oil. The wood simply occupies space in the still.

Fuel

For steam distillation of tea tree very simple apparatus can be effective. At present tea tree oil is at a price where use of petroleum oil or gas for raising steam is economically feasible. If the scale of operation is large, a steam boiler or steam generator can be used, as these devices are efficient steam producers and can be automated. If the operation is small, the high capital cost of establishing special steam raising equipment cannot be justified. Therefore raising steam within the still is just as effective and more economical for the small operator, particularly if wood or extracted leaf is the fuel used.

DISTILLATION EQUIPMENT

For a simple but effective direct fired still (Plate 16) the equipment consists of:

The firebox or boiler.
The still.
The condenser.
The receiver or separator.
The constant level tank.

Fire Box

The fire box is required if a boiler is not used and steam is generated in the still. It can be made of metal, brick/concrete or, in some cases, earth. Its functions are to provide a base for the still and to contain the fire. The dimensions will depend on the size of the still and the fuel used. In all cases the box should allow the maximum area of the base of the still to be exposed to heat from the fire, but ensure that no part of the base of the still that does not have water above it is exposed to direct heat of the fire.

For tea tree, the cheapest fuel is dried leaf after the oil has been extracted. This requires a fire box about 1 metre high. While the fuel is cheap and available, the necessity for constant stoking makes use of this fuel unattractive in most circumstances. For wood fuel the box needs to be about 500 mm high, while for an oil burner 300 mm is sufficient. The "box", which has neither top nor bottom, requires one side to be open to allow stoking of solid fuel, or to have an aperture to take an oil or gas burner. At the opposite end an aperture to take a flue is necessary. In the case of using an oil burner a baffle is required to distribute the flame from the burner so that it impinges on the base of the still instead of going directly up the flue.

The flue of at least 150 mm diameter needs to be carried horizontally 3 or 4 metres from the still to reduce smoke and heat near the still and then vertically about the height of the still to cause a draught, or a straight flue can be set at an angle to achieve the same effect. A tall flue, while more effective in getting smoke away from the still, causes too much draught, so dragging too much heat from the fire box. It is possible in some sites to set up a still by excavating a fire-box in the earth and having no made-up fire-box at all. However a flue is essential for comfortable operation of the equipment.

The Still

There has been a great deal of argument about shape and size of stills. My experience is that shape is of very little consequence. The essential factors in building an effective still are that steam passes uniformly through an evenly packed charge of leaf and can flow smoothly from the still without much pressure developing in the still.

It is worth noting that the simplest of stills, consisting of no more than a 200 litre standard drum, direct fired, and with a 50 mm galvanised iron pipe as a condenser running down through a creek or pond can produce an oil equivalent in both quantity and quality to an oil extracted with the most sophisticated equipment.

Therefore in designing a still the main consideration is ease of handling the plant material and economy of operation. The size of the still needs to be related to the amount of leaf to be processed. It is obviously uneconomical to establish a still with a capacity to handle much less leaf than is available. It is also undesirable to have a still which cannot be filled completely before commencing distillation. The yield of oil per unit weight of leaf tends to be lower in a partially filled still. Although tea tree leaf can be left for some time after harvest, some deterioration, particularly in hot humid weather, is likely if harvested leaf is kept too long before being distilled. In a properly charged still, a cubic metre of still space will take 250–300 kg of chopped plant material. If the still is to be moved from time to time, the size needs to be related to the equipment available for dismantling, transporting and setting up again.

Other factors to be considered in still design are:

1. The still can be made of mild steel or stainless steel. The former is cheaper and therefore suitable for bush stills and small scale plantations. Stainless steel stills, while more expensive to establish, last longer. The quality of the oil produced in mild steel stills is in no way inferior to oil produced in stainless stills.
2. There must be no obstructions to the loading and unloading of leaf, i.e. whether box-shaped or cylindrical, the top must open fully.
3. Steam distribution must be adequate. This can be achieved by means of tubes on the floor distributing steam evenly under the charge where steam is injected, or by placing the leaf on a mesh or perforated plate above the water in a direct fired still where steam is generated in the still. This method can also be used for injected steam. In all cases even and firm packing of the leaf is necessary.
4. The steam and oil vapour outlet must be near the top of the still and be large enough to facilitate rapid passage to the condenser, thus preventing pressure building in the still.
5. A drain, at least 50 mm diameter should be placed at floor level to facilitate cleaning the still, and in the case of the steam-injected still, to allow drainage of excess water.
6. A water inlet, usually a 50 mm socket, is best welded in the wall about 50 mm from the floor to allow for "make up" water during distillation or water from the receiver in the case of cohobation.
7. The mesh or perforated tray on which the leaf is stacked needs to be loose enough to be easily removed to allow cleaning the floor of the still. Where steam is injected, provision for cleaning the distribution tubes must be made, e.g. removable caps on the end of the tubes, or open ends where small diameter tubes are used.
8. The base of the still should extend beyond the walls by about 200 mm if the still is sitting on bricks so that almost the whole of the base is available to transfer heat from the fire to the water above, but the fire must not impinge on any part of the base that does not have water above it.
9. Adequate provision must be made for sealing the lid when closed.
10. The lid should be insulated, particularly if the still is in the open. Insulation of the walls is not so important.

Typical stills giving layout and dimensions are shown in Figures 1 and 2.

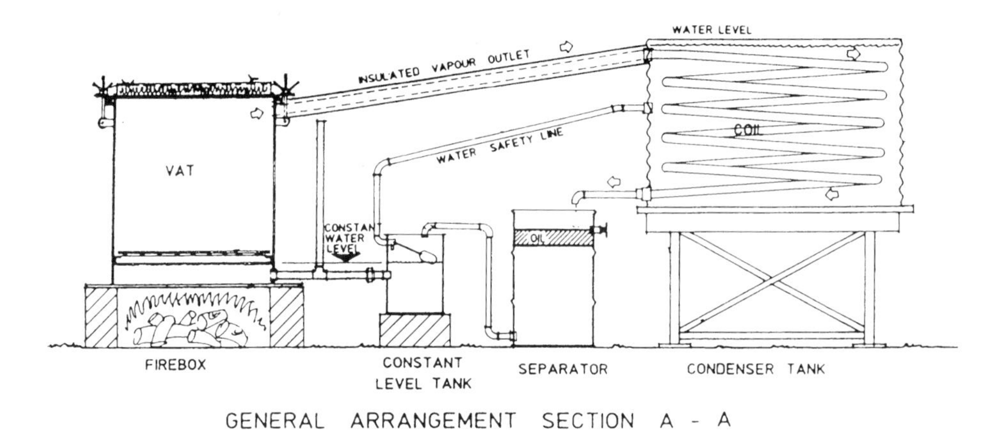

Figure 1 Layout of a simple still for the production of tea tree oil (Designed by G.R. Davis Pty Ltd for, and reproduced with, the permission of CSIRO Forestry and Forest Products)

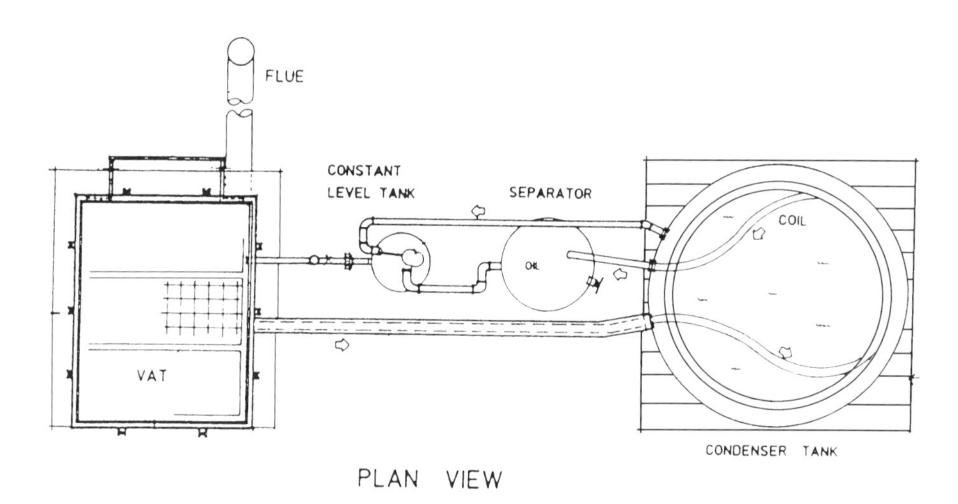

Figure 2 Plan view of the above tea tree still

The Condenser

The function of the condenser is to convert the steam and oil vapour that issues from the still to water and liquid oil at the desired temperature. The commonest types use water as the coolant although air-cooled condensers can be used in some circumstances, as can refrigeration.

If flowing water is available, very little equipment is required to construct an effective condenser. For field stills, or small plantation stills, metal pipes running from the still to and under a stream or pond are quite effective. The above-water section of the pipe should be short if possible, or if it has to be long, should be at least 75 mm diameter for a still of 3 m^3 or more capacity. Once under water a 50 mm diameter tube is sufficient. This needs to be at least 20 m long. With this system the fall in the stream has to be sufficient to allow the construction of a dam through which the condenser pipe runs so that the outlet is above water. As the condenser pipe must run downwards throughout its length, the receiver into which the condensate flows must be below the water level of the stream or pond.

This type of condenser, which is very simple and cheap is widely used in hilly country, but is not suitable for flat or swampy country. For flat country a condenser can be set up close to the still, and consist of a tank containing a volume of water with a pipe running through it, usually as a descending spiral. In this type the cooling water is static, requiring no pump. If the volume of water is sufficient, after cooling overnight or between distillations, it will cope with the next distillation. Occasional topping-up to replace water evaporated from the tank is all that is required. This is an effective type of condenser, easily made and cheap. About 20 m of tube beneath the water is sufficient; say 3 lengths of galvanised 50 mm tube.

A more compact and more portable condenser is the multitube type in which the hot distillate flows through a number of tubes while the coolant is pumped in a jacket around the tubes. In some cases the vapours flow in the space around the tubes while the coolant flows through the tubes. The advantages of this type of condenser are that it is much more compact, easily established and transported, easily cleaned and, because the hot vapours contact a large area of cold surface immediately on entering the condenser and collapse to liquid at that point, the resulting diminution in volume tends to pull the vapour from the still to the condenser.

The tubes, through which or around which the vapours flow are nowadays usually stainless steel. Other metals can be used provided they have reasonable conductivity and do not react with the oil.

The disadvantage of multi-tube condensers are the high cost and the necessity for a pump and motor to push the coolant through the condenser, unless it is possible to gravitate the water from a stream as is often done in hilly country. Where electric power is available, the necessity to pump is not much of a disadvantage, although it is a cost since the pump must run continuously during the distillation. Where no electric power is available from the grid, small petrol motors are the usual source of power. These are not always reliable and therefore the static tank type condenser is more attractive.

The capacity of the condenser has to be considered in relation to size of still and speed of distillation. With the static tank type it is not much more expensive to set up a 10,000 litre tank than a 5,000 litre. The cooling coil is the same. With the extra volume of water, the only time the condenser can be inadequate is when several distillations are done in rapid succession in hot weather. Even then, all that is necessary to get effective condensation is to push some cold water into the tank, displacing the hot water.

With a multi-tube condenser, if it is vapour-in-tube type, it may be mounted vertically or at any angle down to almost horizontal. There must always be enough slope to enable the condensed vapour to flow to the outlet. If the condenser is coolant-in-tube type, it is best mounted vertically. Unless it is very well designed a horizontally installed condenser of this type is likely to condense but not cool the distillate.

The effectiveness of a condenser depends upon the surface area available for heat transfer, the heat conductivity of the metal cooling surface, the difference in temperature of the distillate and the coolant, and the rate of vapour flow into the condenser. The third factor can be controlled to some extent by the rate at which the coolant (water) is pumped through or around the tubes. The first two factors are fixed for any condenser. It is therefore essential to make sure that the condenser is big enough before installation. If it is not, the rate of distillation will be limited.

Receiver or Separator

For tea tree the liquid oil and water flowing from the condenser during distillation are, for practical purposes, non-miscible and of different densities. Therefore separation can be achieved by allowing the oil to rise to the top of the mixture. There are essential oils with a higher density than water, and some with a density close to that of water. These require different apparatus or treatment, but for tea tree oil simple flotation is an effective means of separating oil and water.

The essential features of this type of separator are that it must be able to collect the distillate flow as it comes and be of sufficient volume to allow the liquids to settle so that there is no current flowing through. This can be done by introducing a second vessel within the main one, but if the volume of the primary vessel is large enough effective separation will occur. If cohobation is employed, any oil not separated will return to the still and come over later in the distillation.

An effective separator can be made by using a drum, e.g. 200 litre, into which the distillate flows, with an outlet for water from close to the bottom taken up to 3/4 of the height of the drum, and another outlet for the oil directly through the wall at the same height. Figure 1 illustrates this. For a lighter-than-water oil such as tea tree, introducing the distillate through a funnel with a turned up end has some merit as this directs the flow of oil globules upward.

If the condenser is of stainless steel, a separator of the same material tends to keep the oil free from colour, although stainless steel is not essential for either apparatus. Where the densities of the two liquids of the distillate are close, the temperature of the distillate is important, as, for oils lighter than water, difference in densities will increase as the temperature rises. However, for tea tree oil the difference at ambient temperature is sufficient for effective separation. There is no advantage in operating at other than ambient temperature.

Constant Level Tank

A constant-level tank in association with the still is desirable with a direct-fired still. This can be a 60 litre or so drum, with its base set no higher than the base of the still. From this drum about 50 mm from the bottom, a pipe, say 50 mm diameter, runs horizontally into the still. Thus, provided there is at least 50 mm depth of water, the water level in this drum and in the still will be the same. As distillation proceeds, the water level in the still tends to fall as steam is evaporated off. Water from the drum runs in to replace the evaporated water. This replacement water comes from the separator into which the condensed distillate flows. Although it is not necessary or in fact desirable to have much water in the still at the beginning of the distillation, it is essential that there is enough to ensure that the bottom of the still is covered at all times. At least 60 mm of water should remain in the still after water is evaporated up into the charge of leaf and before the vapours have condensed and run into the receiver, from which water will then flow to the constant level tank, and then back to the still. To guard against the water level falling too far before the distillate starts to flow, a pipe from the condenser tank, or the condenser water line where a multi-tube condenser is used, runs to the constant level tank where water enters through a float valve. Therefore, when the water level in the still falls to a pre-determined level coolant water from the condenser enters the constant level tank and flows through to the still. This safety device also comes into play if the distillate is not condensed for any reason, or if the still springs a leak.

EQUIPMENT ASSEMBLY

In putting the apparatus together it is essential to have the various components at the right levels. The vapour outlet duct may rise if necessary. If so, it is necessary to insulate the duct from the still to the condenser. If vapour condenses in an upward-sloping duct it will run back into the still and have to be re-vaporised. Because of the size of the drops running back, much more time is required to vaporise the oil the second time. The vapour duct needs to slope up only where the height of the condenser is greater than the vertical distance between the vapour outlet and the water return inlet to the still. There must be sufficient height to allow the distillate to flow from the condenser to the receiver and then to the constant level tank. See Figures 1 and 2.

When the apparatus is assembled, water is added to the still (if steam is to be generated in the still) to a level of approximately 100 mm. The leaf is carried on a mesh about 150 mm from the floor, i.e. 50 mm above water level. Water is also added to the condenser tank, or in the case of a multi-tube condenser, the pump that pushes water through the condenser is started. The receiver/separator is filled with water until it just overflows into the constant level tank which in turn overflows into the still to which it is connected. The still is packed with leaf, the lid of the still is closed and sealed and the fire is started or steam from a boiler injected.

DISTILLATION RATES AND TIMES

Packing the leaf into the still must be done with care. With machine harvesting the leaf material normally packs evenly, although some positioning of the chute of the harvesting machine might be necessary to achieve this. In the case of hand packing of the leaf, care must be taken to ensure there are no sections of loosely packed leaf through which most of the steam will travel. If there are loosely packed sections distillation will be prolonged, or possibly not completed.

The initial task in distilling is to raise the temperature of the leaf to a point where the oil and water vaporise and flow to the condenser. Therefore it is desirable to push the steam through the leaf mass rapidly in the early part of the distillation. The more steam that is pushed into the leaf, the more rapidly will the temperature rise to distilling temperature, i.e. just below 100°C. At this stage the condenser has not come into operation and there will be a slight increase in pressure in the still. As the distillate starts to flow, one must consider the capacity of the vapour duct and the condenser to handle the flow. If the vapour duct is large in diameter and short, and the condenser the type in which the vapours meet a large cooling area immediately on entering, e.g. multi-tube type, a vigorous rate of distillation can be maintained. If the vapour outlet from the still is restricted or the condenser is of the long, small diameter, single-tube type, where condensation occurs a long way down the condenser, the distillation rate, i.e. the amount of steam being generated in the still, or being injected from a boiler, will have to be reduced to avoid building pressure in the still or pushing the vapours through without complete condensation, thus losing oil to the atmosphere. Once condensation occurs, the collapse of vapours to liquid which occupy a very much lower volume, causes vapour to be pulled from the still, so reducing pressure in the still.

A steady distillation rate is desirable. If the rate slows too much, oil vapour will condense on the lid or the walls and fall back into the leaf, from which it has to be vaporised again.

The end of the distillation can be determined by collecting, in a narrow cylinder or bottle, a sample of the distillate as it flows from the condenser and visibly assessing the proportion of oil to water. Where fuel is not expensive, allowing the process to continue results in maximum yield of oil, and with tea tree, no serious decline in quality of the oil. Where fuel is costly, a point is reached where oil recovered in the late stages of the distillation is insufficient to cover the cost of the fuel for this stage. Demand for the still must also be considered. If distillation is not keeping up with harvesting, starting a new batch might be preferable to continuing the old distillation in order to gain a very small quantity of oil.

The time required to distil oil from plant material depends on a number of factors. In particular, the physical structure of the oil-bearing part of the plant, the chemical composition of the oil, the design of the distilling equipment and the packing of the still. For any oil, the first two factors are constant. The time required for distillation therefore is determined mainly by the amount of steam put through, which in turn depends on the design of the still, the capacity of the condenser, and the packing of the still.

SPENT LEAF

Leaf should be removed from the still within a few hours after distillation is complete. If the charge is left in the still for several days or more, removal is more difficult, and the life of a mild steel still is shortened due to rust when the metal is wet. The simplest method of removal is to lay chains on the platform on which the leaf is packed (one chain along the centre of the base and one more at right angles is usually sufficient) carry the chains over the sides until the still is filled, then lay the ends of the chains on top of the leaf. A lifting device is then used to pull out the leaf mass. In a still, other than a very small one, it is necessary to pull the leaf out in two or more layers, as the weight of the compacted wet distilled leaf plus the friction to be overcome calls for strong lifting equipment. In a tightly packed still, particularly if left for some time, endeavouring to pull the charge out in one lot might result in lifting the still off its base before the leaf moves from the still. The use of two or more slings overcomes this problem. A tractor with a lifting device is ideal for emptying the still. If this is not available, a stayed post with jib and winch is adequate. This allows the charge to be lifted and swung clear of the still.

THE MOBILE STILL

The still described so far is suitable for small operations. For large-scale plantation operations, while the principles of distillation are the same, a major consideration becomes the economic handling of the biomass including harvesting and distilling and disposal of the extracted material. In order to extract the oil from the leaf it is necessary to harvest the biomass, put it into a vessel and pass steam through it. There are several machines that will harvest effectively, without ill effect to the tree, and will elevate the leaf into a vessel. Using this vessel as the still eliminates the necessity to transfer the material to another vessel, and also allows the extracted leaf to be dumped as required, without being transferred. At this stage of the industry's development, mobile stills (Plates 11, 15, 17) appear to be the most economical way of producing tea tree oil. This technique is well established in the production of essential oils, including eucalyptus, lavender, mint and others.

Steam is injected into the mobile still at a suitable location where there is a good water supply and where a boiler, or steam generator, condenser and separator are located. The essential points, as in the small still, are that the leaf is uniformly and firmly packed in the still, that a steam-tight lid is clamped on, that the steam is uniformly distributed through the leaf, and that the condenser-duct and the condenser are of appropriate size. Cohobation is possible, but calls for the distillate water to be stored and adjusted before being returned to the boiler.

The tall cylindrical mobile still devised by E.F.K. Denny to distil lavender has the added merit of evenly packing the plant material by rotating the cylinder as the material is fed into the still by the harvester. Commonly used mobile stills in the tea tree industry are rectangular and some care is required to ensure that the leaf is evenly packed into the still by the harvesting machine.

A mobile still suitable for distillation of eucalyptus oils was developed by Mr. L.J. Davis of G.R. Davis Pty. Ltd. Improvements over 25 years have resulted in an efficient still for distillation of not only eucalyptus oils but also tea tree, peppermint, fennel and other oils.

The latest "Davis" type still, used widely in Australia, consists of a metal box approximately 2 m wide by 4 m long by 1.8 m deep mounted permanently on a trailer. The trailer to which the still is fixed consists of a rigid frame with the main axle about one metre from the rear and the front steering wheels being a standard tractor front wheel assembly mounted in front of the still. This machine is suitable for operating where the equipment needs to be turned sharply such as in natural stand harvesting or in small plantations with limited headland room. Where long distance hauling at higher speed is required the same still mounted on a single axle with tandem wheels on a rocking-bar suspension is more suitable (Plate 17).

The box, of 3 mm mild or 1.5 mm stainless steel, is supported by 100 by 50 mm rectangular hollow section vertical members with a similar RHS top frame on the front and sides. The sides slope 80 mm from the bottom outwards towards the top. This facilitates unloading and allows a narrower track for the main wheels so that the still can be towed on normal tracks. The front slopes outwards 30 cm from bottom to top. A still of these dimensions will carry about 3 tonnes of leaf material, and the apparatus will weigh about 2 tonnes. If a heavy forage harvester is used, a minimum 70 hp tractor is required.

A standard towbar at the front is required to connect the still to the tractor or harvester, and a towing device is also necessary at the back to allow two or more stills to be towed together.

A completely open top allows loading from the chute of the harvester (Plate 11). Hinged boards increasing the height of the rear half of the side walls and the back wall of the still minimise loss in windy conditions. The rear extension board should be of mesh so that air and dust can move through while leaves are retained. A ladder is built on the outside of the still, and footholds placed on the inside—these footholds should not have vertical members as steam will move rapidly up them, reducing distillation efficiency.

A rear-opening door on offset hinges must be able to swing completely clear to allow removal of spent leaf material. This door is secured by over-centre locking clamps.

The lid consists of braced 3 mm mild steel or 1.5 mm stainless steel plate with a 75 mm inverted channel around the edge. The lid is suspended so that, when the full still is towed into the still-house, the channel fits closely over the top edge of the still. Nitrile-bonded cork is placed in the channel to ensure a steam-tight seal between lid and still. The lid is secured by over-centre clamps attached to the lid itself.

From a 15 cm diameter outlet near the centre of the lid (Plate 17) a flexible vapour duct leads directly to a multi-tube condenser with circulating water. This duct needs to be flexible enough to allow the lid vertical movement of at least 20 cm. The lid is supported from a single central point to allow movement in any direction. The lid can be shaped to some extent for strength and rigidity.

The vapour duct needs to be at least 15 cm in diameter to allow rapid distillation. This is not compatible with a single tube coil-in-tank type condenser when steam from a boiler is used, as excess pressure will build up in the still during the early phases of distillation. Equipment with mobile stills and multi-tube condensers is not designed to work at pressure. The required capacity of the condenser will depend on the rate of steam application.

In the floor at the rear of the still, near one side, a 5 cm socket and cock is placed to act as a drain. This drain can be kept open so that excess condensed water can flow out. Air is also expelled, in the early stages of the process, thus reducing pressure build up. Oil is not lost through this drain.

Steam is introduced from a wood or spent leaf-fired boiler through a 50 mm flexible steam line with a female Kamlock fitting. It is desirable to have about 0.5 m of flexible pipe in the horizontal plane and another 0.5 m in the vertical plane for ease of connection.

It is also necessary to ensure that steam is evenly distributed beneath the leaf. For this purpose steam is injected through a 50 mm diameter tube running transversely across the floor at the front from which a series of smaller tubes about 300 mm apart run down the length of the floor. The smaller tubes have holes of about 6 mm diameter in the sides at about 300 mm intervals so that steam is released at points 300 mm apart under the whole charge. Dust and small pieces of leaf tend to clog these pipes, consequently it is necessary to be able to clean them periodically. This can be done by having caps on the end which can be removed to allow steam to be blown through, or in the case of 18 mm or less diameter tubes, leaving the ends open.

Stainless steel flexible tubing, most suitable for the vapour duct and steam lines, is expensive. White nitrile rubber tubes are adequate, particularly for the vapour duct. These are less expensive but have shorter lifetimes.

This is a modified version of a previously described mobile still (Davis and House 1991).

Oil is collected from the condenser in a receiver as in the simple still.

HARVESTING AND LOADING

Of the various machines tried in harvesting eucalypts, heavy forage harvesters have proven the most satisfactory. The type of machine where the flails create sufficient draft to elevate the harvested material is preferable to a machine which elevates by an auger. As most forage harvesters are designed to harvest soft herbage, only some will stand operating on woody material. Standard or modified standard machines are far more economical than specially designed machines, both in initial cost and running and repair costs.

While flail type harvesters are suitable for tea tree harvesting in dry conditions, in wet and soft soil there is a risk of beating the tree out of the ground instead of just removing the above-ground portion. Therefore, young trees and trees in soft soil are best harvested by a machine with horizontal cutters near ground level. Some standard types of corn harvesters are effective for this purpose.

UNLOADING

Unloading is achieved by attaching two chains, each 11 m long, to rear-facing hooks on the floor of the still near the door. The chains are laid along the floor to the front and carried up over the front wall, the remaining length (enough to reach the back wall on top of the leaf charge) being held on brackets mounted on the outside of the front wall. When the still is full, the chains are laid along the top of the charge.

To empty the charge, the door is opened, the chains secured to a fixed point about 3 m above the ground and the still towed forward. The leaf charge is thus rolled out.

An alternative to pulling the leaf out is to tip it out. This calls for more expensive equipment (Plate 15) but provides very rapid emptying. It also removes the necessity for a back door, thus allowing a more rigid still and less potential for steam leaks.

CONCLUSION

Successful distillation of many essential oils can be achieved with simple equipment. The technology is also simple, materials cheap, and the quality of oil produced high, provided the standards of construction and operation of the still are adequate. Mechanisation improves efficiency and makes the whole operation more cost effective, but at a greater initial financial outlay. For new oil ventures, the approach will depend largely on the scale of production envisaged.

REFERENCES

Davis, G.R. and House, A.P.N. (1991) Still design and distillation practice. In D.J. Boland, J.J. Brophy and A.P.N. House (eds.), *Eucalyptus Leaf Oils*, Inkata Press, Sydney.

Guenther, E. (1965) The production of essential oils. Methods of distillation, enfleurage, maceration, and extraction with volatile solvents. In Guenther, E. (ed.), *The Essential Oils*, Van Nostrand Co. Inc. (Reprint), Princeton, New Jersey, Vol. 1, p. 92.

International Standards Organisation (1996) Oil of *Melaleuca*, terpinen-4-ol type (Tea Tree Oil). *International Standard ISO 4730: 1996(E)*, International Standards Organisation, Geneva, 8pp.

Standards Australia (1997) Oil of *Melaleuca*, terpinen-4-ol type (Tea Tree Oil). *Australian Standard AS2782-1997*, Standards Australia, Homebush, 12pp.

9. BIOLOGICAL ACTIVITY OF TEA TREE OIL

JULIE L. MARKHAM

Centre for Biostructural and Biomolecular Research, University of Western Sydney, Hawkesbury, Richmond, NSW, Australia

INTRODUCTION

Captain Cook is reported to have used the leaves of the tea tree to brew a spicy tea and Sir Joseph Banks, the botanist with Cook's expedition, included samples of the plant in his collection. Exactly which plants were used is not known as the name 'tea tree' has been used for a number of similar plants in the genera *Melaleuca* and *Leptospermum.* Whilst essential oils from several species possess antimicrobial activity, the oil which is widely used today for its antiseptic properties is the oil of *Melaleuca alternifolia.*

The natural habitat of *Melaleuca alternifolia* is the Bungawalbyn region of north-eastern New South Wales, Australia and Drury (1991) reports that the Bundjalung Aborigines, who lived in this area, treated skin infections by crushing leaves of the tea tree over the injury and covering it with a warm mudpack. It is possible that early settlers also used this remedy.

Scientific studies on tea tree oil began in the 1920s at the Museum of Technology and Applied Sciences in Sydney. Penfold and Grant (1925) reported that the oil was a non-toxic and non-irritating antiseptic, which was more effective than phenol, the standard for comparison. Following this demonstration of the effectiveness of tea tree oil in laboratory tests, evidence of its therapeutic use began to accumulate. Humphrey (1930) tested a water-miscible preparation of tea tree oil and reported its value at concentrations of 2.5–10% for the treatment of a variety of conditions, such as wounds, peronychia (infection of the nails), coryza and sore throats, and suggested it would have potential applications in the treatment of vaginal infections and burns. In addition to its efficacy as an antiseptic, several other features of the oil were cited. These included the ability of the oil to dissolve pus in infected wounds, to cleanse dirty wounds and to resolve chronic infections, particularly of the nail, without any apparent damage to healthy tissue.

During the 1930's the reputation of tea tree oil as 'a medicine chest in a bottle' continued to develop. The major producers of the oil, Australian Essential Oils Ltd, published a report in 1936 on the medical and dental applications of Ti-trol, the name used for the neat oil and of Melasol, a water-miscible preparation (cited in Lassak and McCarthy 1983). The list of applications in external conditions continued to grow, and all reports stressed the non-toxic and non-irritating properties of the oil. The ability of the pure oil to penetrate unbroken skin was also reported (Drury 1991).

A reputation for tea tree oil as a safe, effective antiseptic had been established by the time of the outbreak of the Second World War, and its record impressive enough that cutters and producers were exempted from military service. In addition, the oil was incorporated in army and navy first-aid kits for use in tropical regions (Drury 1991) and added to machine cutting oils to reduce infection following injury to the hands by metal filings and turnings (Lassak and McCarthy 1983).

A number of factors led to a decline in the industry following the Second World War. As well as competition in the market-place from newly-developed synthetic agents, the product was inadequately promoted and the supply was inconsistent, both in quality and quantity. The natural stands of trees, located in swampy areas, were difficult to harvest and different chemotypes, with oils of variable composition, were difficult to distinguish in a simple manner.

During the 1970's the interest in 'natural' products was renewed, and over the last twenty years the production of oil has continued to increase and so have the applications. Tea tree oil is incorporated as the active antimicrobial or as a natural preservative in a wide range of pharmaceutical, cosmetic and 'cosmeceutical' products including antiseptic creams, face washes, pimple gels, vaginal creams, veterinary skin care products, tinea preparations, foot powders, shampoos and conditioners. It is claimed to provide rapid soothing relief when added to burns blankets and creams. Alongside this, there has been a recognition by the industry of the need to support the early anecdotal claims about the efficacy of the oil with scientific data which will enable registration of the oil as a pharmaceutical product. This chapter will review the published studies which have examined *in vitro* and *in vivo* activity of the oil.

IN VITRO TESTING OF ANTIMICROBIAL ACTIVITY

The major therapeutic claims for tea tree oil involve its antimicrobial activity and its effectiveness against a wide range of bacteria and fungi has been demonstrated. The methods used by various authors vary both in principle and detail and it is relevant here to review the methods and evaluate their usefulness in the testing of essential oils.

Test Methods

Methods commonly used for determining antimicrobial activity of a substance fall into two categories: those which measure whether the agent is microbiostatic (inhibits growth) and those which determine whether it is microbiocidal (kills the target organism). Inhibition of test organisms can be determined by diffusion tests or by measurement of the Minimum Inhibitory Concentration (MIC) in broth or agar dilution tests, while suspension tests, such as the Rideal-Walker test and the TGA disinfectant tests, measure a lethal effect.

Diffusion Tests (Disc or Well)

Diffusion tests involve placing a sample of the test compound on a disc or in a well in the centre of a plate containing a nutrient medium which has been seeded with the test organism. Water soluble compounds diffuse into the seeded medium producing a continuous concentration gradient of the agent. Following incubation, a clear zone, called the zone of inhibition, develops around the disc or well if the organism is sensitive to the agent. The size of the zone provides some indication of the relative activity of the substance, however, a number of factors including the volume and type of medium, the concentration and age of inoculum, the incubation conditions and the size, charge and conformation of the active ingredient(s) will all affect the result (Lorian 1986). This technique has been standardised for antibiotic testing to give reproducible results, which correlate with MIC values, but because of the different diffusion properties of various substances, it is only useful for comparing sensitivity of different organisms to a single antibiotic and not appropriate for the direct comparison of a range of antibiotics.

The method has been widely used to screen antimicrobial activity of essential oils, including tea tree oil (Kar and Jain 1971; Yousef and Tawil 1980; Janssen *et al.* 1985; Deans and Svoboda 1988; Biondi *et al.* 1993; Williams *et al.* 1993; Gundidza *et al.* 1994; Bagci and Digrak 1996). However, there are a number of disadvantages when the method is applied to the testing of water-insoluble essential oils and inconsistent results and a lack of correlation with MIC values have been reported (Morris *et al.* 1979; Janssen *et al.* 1986; Carson and Riley 1995). The lack of water solubility of the components of tea tree oil limit their diffusion through the agar. Only the more water soluble components, such as terpinen-4-ol, 1,8-cineole and α-terpineol, diffuse into the agar from the disc; hydrocarbon components remain on the disc or evaporate (Southwell *et al.* 1993). Consequently, the contribution of these components to the activity of the oil cannot be assessed and it is possible that the activity of the oil will be underestimated. Whilst the method is easy to perform and requires only small volumes of the agent, its use as a screening tool continues, but it is not appropriate where more quantitative measurements are required, or where the antimicrobial activity of oils of different composition are being compared.

Determination of Minimum Inhibitory Concentration

MIC measurements have been used extensively to quantify the antimicrobial activity of essential oils. This type of test is more useful than the diffusion test for comparing both the activity of oils of different chemical composition and the sensitivity of a variety of organisms, and a number of variations of the method have been published (Beylier 1979; Villar *et al.* 1986; Kubo *et al.* 1991; Patkar *et al.* 1993; Southwell *et al.* 1993; Chand *et al.* 1994; Dellar *et al.* 1994; Nguyen *et al.* 1994; Carson *et al.* 1995b; Griffin *et al.* 1998). These include broth (macro and micro) and agar methods, with or without the use of an indicator of cell viability to determine the endpoint.

The principle of the assay is that the test organism is added to a series of dilutions of the agent prepared in a nutrient medium, and presence or absence of growth is determined after a period of incubation. The Minimum Inhibitory Concentration is recorded as the lowest concentration of the agent which inhibits growth. This definition lacks precision as it does not clearly define the preparation and concentration of the inoculum, the test medium, the contact time or the method of determining the end-point. Due to different interpretations of these variables, it is difficult to directly compare the results of various authors, and there is a strong need for the tea tree oil industry to standardise procedures.

Broth Dilution Methods

Broth methods, especially those carried out in microtitre trays, have the advantages of lower workloads for a larger number of replicates and the use of small volumes of the test substance and growth medium. In broth methods, turbidity of the oil-water emulsion can interfere with the reading of the end-point, particularly in microtitre assays. For this reason, indicators such as fluorescein diacetate (Chand *et al.* 1994), *p*-iodonitrotetrazolium violet (Dellar *et al.* 1994) and triphenyl tetrazolium chloride (Carson *et al.* 1995b) have been used. However, autofluorescence is a problem with FDA (unpublished observations) and Carson and Riley (1994) report that colour changes of TTC do not correlate exactly with MIC. An alternative method using the redox dye, resazurin, has been developed (Mann and Markham 1998). The reported advantages of this method are the ease of end-point determination associated with a colour change from blue to pink and the reliability of prediction of MIC.

Broth methods can also be adapted to test Minimum Bactericidal or Fungicidal Concentrations by transferring aliquots from tubes or wells which show no growth after the incubation period, to fresh medium. Organisms are considered to be non-viable if there is no growth after this second period of incubation. A microtitre method modified for the testing of anaerobic organisms has also been reported (Shapiro *et al.* 1994).

Agar Dilution Methods

Agar dilution methods, in which various concentrations of the test substance are added to the agar medium prior to inoculation with the test organisms, overcome the turbidity problem associated with broth methods. Studies in our laboratory have shown excellent replication using Isosensitest agar and concentrations of Tween 20 of 0.25–0.5% (Griffin *et al.* 1998). This method has the advantage that with modifications to the growth medium and incubation conditions it can be adapted to suit any organism, but it is not readily adapted to the inclusion of MBC measurements. Some authors have reported that MIC values from agar dilution tests tend to be higher than those obtained by broth dilution tests (Atkinson and Brice 1955; Walsh and Longstaff 1987).

Measurement of Microbiocidal Activity

Although the methods above can determine Minimum Lethal Concentrations of agents they involve prolonged contact (generally 24–48 hours) between the agent and the test organism and give no evidence of the rate of kill. Because antiseptics and disinfectants are generally required to reduce the microbial load rapidly, kill rate or suspension tests are more relevant to their intended use. In principle, such tests involve the addition of an appropriate volume of inoculum to a clinically relevant concentration of the agent followed by the testing of the viability of the culture after set periods of incubation. This basic procedure can be adapted by changing variables such as inoculum density, concentration of the test agent, contact time and presence of potential inactivators, to satisfy the testing requirements. In a study of the death kinetics of *Staphylococcus aureus, Candida albicans* and *Aspergillus niger* treated with samples of thyme oil, it was reported that variations between oils with similar MIC values could be demonstrated by studies of death kinetics (Lattaoui and Tantaoui-Elaraki 1994).

Use of Solubilising and Emulsifying Agents

Broth and agar dilution methods have generally been developed for use with water-soluble preparations and require modification for use with essential oils of low water solubility. To ensure contact between the test organism and tea tree oil for the duration of the assay it is necessary to use a solubilising or emulsifying agent. The agents most commonly used are Tween 80 and Tween 20 at concentrations ranging from 0.001 to 20% (Beylier 1979; Walsh and Longstaff 1987; Chand *et al.* 1994; Carson *et al.* 1995b; Griffin *et al.* 1998). DMSO (Scortichini and Rossi 1991; Aboutabl *et al.* 1995), DMF (Kubo *et al.* 1991), ethanol (Morris *et al.* 1979; Deans and Svoboda 1988; Biondi *et al.* 1993) and 0.15–0.2% agar have also been used (Remmal *et al.* 1993; Mann and Markham 1998).

Use of emulsifying and solubilising agents may result in changes in the physicochemical properties of the test system, even though they have no antimicrobial activity when tested on their own. It has been reported that nonionic surfactants, such as the Tween compounds, form micelles above a concentration known as the critical micelle concentration. Lipophilic molecules, such as the components of tea tree oil, may become solubilised within the micelles and thus partitioned out of the aqueous phase of the suspension (Schmolka 1973). Kazmi and Mitchell (1978) have shown that antimicrobials solubilised within the surfactant do not contribute to the activity as they do not come into direct contact with the microorganisms. The amount of material solubilised increases at higher concentrations of surfactant (Van Doorne 1990). Results from our laboratory support this view: the antimicrobial activity of tea tree oil decreased as the concentration of Tween 20 in the test medium was increased from 0.1% to 5% (Mann and Markham 1997). Premixing of the oil with Tween before addition to the test medium would be expected to exacerbate this problem. The impact of such effects must also be considered in the formulation of pharmaceutical and cosmetic products, as the interaction between components may reduce the activity of the active ingredients.

Other studies suggest that low concentrations of surface-active agents in the mixture may actually enhance the activity of the antimicrobial through causing changes to the permeability of the cell membrane of the microorganism (Denyer and Baird 1990). This raises doubts as to the suitability of agents such as Tween 80 and 20 in the assay procedures. Van Doorne (1990) reports that ethanol, in concentrations as low as 5%, can have a marked potentiating effect on the activity of antimicrobial agents and these authors question its use as a solubilising agent.

To obtain consistent, reproducible results it is important that contact between the oil and the microorganism is maintained throughout the test period. Allegrini *et al.* (1973) reported that emulsions of essential oils in water containing 1% Tween 80 or Tween 20 disintegrated within one hour but that emulsions with 5% Tween were stable for 24 hours, the incubation period of many assays. However, as stated above, Tweens at this concentration exert an inhibitory effect. A more suitable dispersing agent for tests carried out in broth is bacteriological agar at concentrations of 0.15–0.2% (w/v)

(Mann and Markham 1998; Remmal *et al.* 1993). Emulsions containing 0.15% bacteriological agar were stable for 19 hours (Mann and Markham 1998). As well as the stability of the emulsion, the use of agar as a stabiliser has the advantages of its lack of chemical reactivity and, as reported in the study of Remmal *et al.* (1993), MIC values are lower than when Tween 20, Tween 80, Triton X100 or ethanol are included as emulsifiers or solubilisers.

To conclude this section, several aspects of the methods which have been described should be highlighted. Firstly, test conditions do not generally reflect the actual conditions of use of the preparation. Thus, an agent may be active when in direct contact with an organism in a liquid medium, but may be inactivated in the presence of blood or pus in a wound, or be unable to penetrate unbroken skin and hence not reach the contaminated site in sufficient concentration to be effective. Exposure time, particularly at concentrations near the MIC, may not be sufficient, particularly in external use, unless the agent has residual activity.

Secondly, some tests measure only inhibition of growth of the test organisms, not a lethal effect. Whilst the former are the tests most commonly reported in the literature, the link between these measurements, which generally involve prolonged contact (18 hours or more) between the organism and the agent, and *in vivo* use, are questionable. Even where such methods are adapted to also measure MBC no indication is given of the death kinetics (rate of kill).

Thirdly, the test methods which have been published in the literature have been developed for water-soluble compounds, and they do not always give reliable results with non-water-soluble oils and may underestimate the true antimicrobial activity. Adaptations are needed to ensure adequate and consistent contact between the oil and the test organism throughout the period of the test. The importance of standardisation of methods must be stressed.

Lastly, the method of formulation of the product can profoundly affect the physical and biological properties of the active agent. Thus, it cannot be assumed that a formulation will be effective simply because a specific amount of active ingredient has been included. This also highlights the need for microbiological testing of products to ensure that they are effective at inhibiting or eliminating microorganisms: chemical tests to measure concentrations of active ingredients are more precise, but do not, on their own, provide evidence of the efficacy of the product. Laboratory tests are important, but results of these tests can still only be regarded as a useful preliminary to clinical trials.

Antimicrobial Activity of Tea Tree Oil

Penfold and Grant (1925) first demonstrated the activity of tea tree oil in the Rideal-Walker test, a standard test of the period, which employed *Bacillus typhosus* (now known as *Salmonella typhi*) as the test organism. The Rideal-Walker coefficient was reported to be 11, indicating that tea tree oil is 11 times more effective than phenol. The results in Table 1 show that tea tree oil also compares very favourably with a number of other essential oils when tested by this method.

Very little further work was done until the 1970s when Low *et al.* (1974) reported MIC values of 1 : 16 against *S. aureus* and 1 : 32 against *Salmonella typhi,* and Beylier (1979)

Table 1 Antimicrobial acitivity of some essential oils (Rideal-Walker coefficient data adapted from Penfold and Grant (1925) and Schilcher (1985))

Source of oil	*R-W coefficient*
Melaleuca alternifolia	11
Melaleuca linariifolia	10
Eucalyptus dives 'Type'	8
Eucalyptus citriodora	8
Backhousia citriodora	16
Lavender	1.6
Clove	8
Thyme	13.2

included tea tree oil in a study of the antimicrobial activity of ten essential oils derived from Australian native plants. Tea tree oil was effective against the five test organisms, which included both Gram-positive and Gram-negative bacteria, a yeast and a mould. The MIC value against *S. aureus* (0.25–0.2%) was considerably lower than that reported by Low *et al.* (1974).

Both tea tree oil and formulations containing tea tree oil at concentrations of 5% have been reported to pass the Therapeutic Goods Act (TGA) test for antiseptics and disinfectants (Graham 1978), which includes type strains of the Gram-positive bacterium *Staphylococcus aureus* and the Gram-negative *Escherichia coli, Proteus vulgaris* and *Pseudomonas aeruginosa* as test organisms (Altman 1989). The oil has also been reported to pass both USP and BP preservative efficiency tests in semi-solid formulations at concentrations of approximately 0.5–1% (Altman 1991), although higher concentrations may be needed to satisfy the requirements of the test in relation to *Aspergillus niger* (this volume, Chapter 12).

The very broad spectrum of activity of tea tree oil, a highly desirable characteristic in an antiseptic or disinfectant, has been confirmed by a number of studies published during the last decade. Table 2 identifies the organisms which have been tested by various authors and presents MIC values for a large number of organisms from a single study (Griffin *et al.* 1998). These results suggest the enormous potential of tea tree oil in a variety of applications, including the treatment of external conditions such as acne, tinea, thrush and staphylococcal and streptococcal infections, in oral hygiene products, in the disinfection of cooling towers contaminated with *Legionella,* and in agricultural uses. Not only has the sensitivity of many species of bacteria and fungi to tea tree oil been demonstrated, but some studies have examined the susceptibility of large numbers of recent clinical isolates, as well as type strains, of particular species (see Table 2 for references). Such data provides valuable information about the variability in sensitivity of organisms likely to be encountered in therapeutic use of the oil.

The majority of studies report MIC values and a comparison of results for commonly tested organisms from different studies is presented in Table 3. Although there is variation in MIC values against different strains within a species and between studies, the majority of values are less than 1%. Lack of consistency of results may be accounted for by differences in oil composition, in test organisms and methods of determining

Table 2 Spectrum of antimicrobial activity of tea tree oil

Organism	*References*[1,2]	*MIC, %* (v/v)[k]
Gram-positive bacteria		
Staphylococcus aureus	a,b,c,d,e,h(105),j(69),k	0.2
MRSA	g(60),k	0.2–0.3
Staphylococcus epidermidis	c,e,j(15),k	0.5
Enterococcus faecalis	k	0.5
Propionibacterium acnes	b,e,f(32),k	0.4–0.5
Micrococcus luteus	j(4),k	0.2
Corynebacterium spp	j(10),k	0.2–0.3
Bacillus subtilis	k	0.3
Bacillus cereus	k	0.3–0.4
Streptococcus spp	l	
Gram-negative bacteria		
Escherichia coli	a,b,c,d,h(110),k	0.2
Enterobacter aerogenes	k	0.3
Klebsiella pneumoniae	j(14),k	0.3
Proteus vulgaris	k	0.2
Pseudomonas aeruginosa	a,c,j(10),k(3)	>2
Pseudomonas putida	k	0.5
Serratia marcescens	j(11),k	0.2–0.3
Legionella pneumophila	k	
Yeasts		
Candida spp	a,b,c,d,j(15),k,l(32)	0.2
Malassezia furfur	i(52),m(22)	
Pityosporum ovales	j	0.2
Moulds		
Aspergillus niger	a,d,k	0.3–0.4
Aspergillus flavus	b,k	0.4–0.5
Trichophyton mentagrophytes	b,k,m	0.3–0.4
Trichophyton rubrum	b,k,m	1.0

[1] The numbers in brackets refer to the number of isolates tested.

[2] a Beylier (1979); b Bassett *et al.* (1990); c Williams *et al.* (1993); d Southwell *et al.* (1993); e Raman *et al.* (1995); f Carson and Riley (1994); g Carson *et al.* (1995a); h Carson *et al.* (1995b); i Hammer (1996); j Hammer *et al.* (1996); k Griffin *et al.* (1998); l Carson *et al.* (1996); m Nenoff *et al.* (1996).

MIC values. For example, the relatively high MIC values reported by Raman *et al.* (1995) may result from leaving oil/agar combinations to stand overnight at 55°C to allow mixing before preparing dilutions and inoculating, a practice likely to result in the loss of volatile active ingredients.

Overall, there are no apparent differences in sensitivity between Gram-positive and Gram-negative bacteria in most studies. In studies where MBC values are also reported (for example, see Hammer *et al.* 1996) they are similar to the MIC value for some organisms, but several dilutions higher against other organisms. Generally, MIC and MBC values tend to be closer for Gram-negative than for Gram-positive bacteria,

Table 3 Comparison of tea tree oil minimum inhibitory concentration (MIC) data determined by an Agar[a] or Broth[b] dilution method and expressed as %, v/v

Organism	1[a]	2[b]	3[a]	4[b]	4[a]	5[a]
Staphylococcus aureus	0.2	0.12–0.5 (**n*=69)[2.1]	0.25–0.5	0.08	0.08	0.63–1.25
Staphylococcus epidermidis	0.5	0.12–0.5 (**n*=15)[2.2]	nd	nd	nd	0.63–1.25
Enterococcus faecalis	0.5–0.75	nd	nd	0.08	0.16	nd
Propionibacterium acnes	0.5	0.25–0.5 (**n*=32)[2.3]	nd	nd	nd	0.31–0.63
Escherichia coli	0.2	0.12–0.25 (**n*=110)[2.1]	0.12–0.25	0.08	0.16	nd
Pseudomonas aeruginosa	2, >2	2–5 (**n*=10)[2.2]	4	0.16	>0.16	nd
Candida albicans	0.2	0.12–0.25 (**n*=14)[2.4]	0.02–0.04	nd	nd	nd
Aspergillus niger	0.3–0.4	nd	0.008–0.015	nd	nd	nd

nd not determined. **n* refers to the number of isolates tested.
[1]Griffin *et al.* (1998); [2.1]Carson *et al.* (1995b); [2.2]Hammer *et al.* (1996); [2.3]Carson and Riley (1994); [2.4]Hammer (1996); [3]Beylier (1979); [4]Walsh and Longstaff (1987); [5]Raman *et al.* (1995).

suggesting that there may be differences in the mechanism of action between the two groups of organisms.

Pseudomonas aeruginosa has been consistently shown to be far more resistant to tea tree oil, as well as to a number of other disinfectants or antiseptics, than other organisms tested (Table 3; Janssen *et al.* 1986; Williams *et al.* 1993). Some variation does occur between strains, but MIC values are generally above 2%. It has been reported that oils high in sabinene-hydrate, the precursor of terpinen-4-ol in the plant, are more active against *Pseudomonas aeruginosa* than standard tea tree oil (Markham *et al.* 1995), as are some aged oils (Markham *et al.* 1996).

The study of Hammer *et al.* (1996) reports an interesting difference between the sensitivity of bacteria which are part of the normal skin flora, such as *Micrococcus* spp. and coagulase-negative staphylococci, and some pathogenic organisms, termed transients, which may be transmitted via the hands to people and surfaces. When a large number of strains were examined, MBC_{90} values for the first group were higher than for some of the transients organisms, suggesting that handwashes containing tea-tree oil may reduce transmission of pathogens via the hands, while causing minimal disturbance to the normal flora.

In order to assess the potential of tea tree oil in oral hygiene products, two studies of the sensitivity of anaerobic and facultatively anaerobic oral bacteria, including species of *Actinomyces*, *Bacteroides*, *Fusobacterium*, *Peptostreptococcus* and *Streptococcus* to tea tree oil have been reported. Walsh and Longstaff (1987) used both broth and agar dilution methods and recorded MIC values in the range 0.02 to 0.08%, indicating a greater sensitivity of these organisms than the aerobic organisms included in the study. Shapiro *et al.* (1994)

report higher MIC values (0.1 to >0.6%) for a similar group of organisms. There are a number of possible explanations of these differences. Firstly, different isolates of the same species and different species in the same genus do not always have the same sensitivity to an antimicrobial agent; secondly, as reviewed earlier in this chapter, results are affected by variations in test methodology and thirdly, the composition of the oils were not clearly defined.

The susceptibility of fungi and viruses to tea tree oil has been less extensively studied. With the fungi the focus has been on species included in standard tests such as *Aspergillus niger* and *Candida albicans*, on post-harvest fungal pathogens (Bishop and Thornton 1997) and on several dermatophytes implicated as causative agents of skin conditions such as tinea (Table 2). Published MIC data indicates that tea tree oil is inhibitory to the dermatophytes at concentrations up to 1%. Data of minimal fungicidal concentrations is lacking, but the results of preservative efficiency tests suggest that fungicidal concentrations will be at least several times higher than the MIC values. This is also suggested by the clinical study of Tong *et al.* (1992) in which a formulation containing 10% tea tree oil was evaluated in the treatment of tinea of the foot. Although subjects reported an improvement in symptoms, there was no mycological cure.

Inactivation of viruses has not been reported, however, a recent study has reported that plants treated with tea tree oil were more resistant to infection with tobacco mosaic virus than untreated controls (Bishop 1995). Testing of viruses, such as herpes simplex virus and human papilloma viruses, which cause infection of the skin and mucous membranes in humans, would be useful in determining whether there is a potential for products aimed at the treatment of cold sores, genital herpes or warts.

Relationship Between Chemical Composition and Antimicrobial Activity

Gas chromatography studies have shown that tea tree oil is a complex mixture of approximately 100 terpenes, the concentrations of which can vary widely between oils (Brophy *et al.* 1989). Although the composition which optimises antimicrobial activity has not been completely defined, there is strong evidence that terpinen-4-ol, a monoterpene alcohol, is the most significant antimicrobial component and the ISO Standard 4730 "Oil of *Melaleuca* Terpinen-4-ol Type" (International Standards Organisation 1996) stipulates a minimum concentration of 30% of this component. Antimicrobial activity increases rapidly as the concentration of terpinen-4-ol increases up to approximately 40%, with only a slight further increase at concentrations above this (Williams *et al.* 1993; Southwell *et al.* 1993; Griffin *et al.* 1998). MIC values begin to rise when the concentration of terpinen-4-ol drops below 30%. Terpinen-4-ol has a Rideal-Walker coefficient of 12–13.5 (Penfold and Grant 1925), and MIC values of 0.2–0.6% against *S. aureus*, 0.06–0.2% against *E. coli* and 0.2–0.25% against *C. albicans* have been reported (Carson and Riley 1995; Griffin 1995; Raman *et al.* 1995).

The significance of the other components is less clear. The antimicrobial activity of α-terpineol, which is also a monoterpene alcohol, and hence more water soluble than the hydrocarbon components of the oil, has been reported in a number of studies, with

MIC values in a similar range to those for terpinen-4-ol (Deans and Svoboda 1989; Nguyen *et al.* 1994; Carson and Riley 1995; Griffin 1995; Raman *et al.* 1995). This compound is only a minor component in tea tree oil (typically approximately 3%) and, as a result, its contribution to the oil's antimicrobial activity is probably limited.

Knobloch *et al.* (1989) reports a correlation between water solubility and antimicrobial activity of terpenoids, however, the diffusion method used in the study to determine antimicrobial activity does not adequately assess the activity of non-water-soluble components (Southwell *et al.* 1993). Other studies report that a number of the hydrocarbon components present in tea tree oil, including α- and γ-terpinene, terpinolene and *p*-cymene do possess varying degrees of antimicrobial activity, but the results are conflicting and difficult to compare because of differences in the purity of compounds, test methods and test organisms (Deans and Svoboda 1989; Nguyen *et al.* 1994; Williams and Lusunzi 1994; Carson and Riley 1995).

The contribution of the minor components to the properties of the oil is not well understood, but given the complexity of the composition of the oil, there is considerable potential for interactions between various components. There are reports in the literature that certain monoterpenes affect the functioning of cell membranes (Brown *et al.* 1987), but whether this is the primary effect of tea tree oil on microbial cells is yet to be determined. It is important to determine the mechanism of action before an understanding of the significance of the minor components can be achieved. In determining the optimal antimicrobial chemistry of tea tree oil, it will also be important to maintain other valuable properties of the oil, especially the reported antiinflammatory and mild anaesthetic effects, in the treatment of conditions of the skin and mucous membranes.

Two aspects of the chemistry of tea tree oil which deserve further consideration are the significance of 1,8-cineole and changes which occur in oils with aging. These are discussed below.

Significance of 1,8-cineole

In recent years there has been a great deal of controversy over the significance of 1,8-cineole in tea tree oil. The Standard stipulates a maximum concentration of 15% for 1,8-cineole. The purpose of this maximum limit for cineole is to exclude oils from the high cineole chemotypes of *Melaleuca alternifolia*, which have low levels of terpinen-4-ol, as such oils are less effective antimicrobials. In recent years, there has been a push towards minimising the level of cineole and as a result low cineole oils ($<5\%$) have become popular in the market place.

The most commonly stated reason for minimising cineole concentrations in tea tree oil is that it is claimed to be a skin irritant (Lassak and McCarthy 1983; Barnes 1990; Carson *et al.* 1995b; Raman *et al.* 1995). However, these claims are unsubstantiated. A recent study has reported no irritancy or allergenicity due to 1,8-cineole in any of the 25 participants, and, although there were three cases of allergy, none were due to cineole (Southwell *et al.* 1997). In another study, in which twenty-eight women were treated daily with vaginal pessaries impregnated with tea tree oil containing 9.1% 1,8-cineole, no irritation of the mucous membranes was reported (Bélaiche 1985a). The long-term use of eucalyptus oil, which contains approximately 75% 1,8-cineole, as a chest rub also supports this view. There is evidence that a small number of individuals do have an adverse skin reaction to tea tree oil, but the component responsible varies from person to person and there is only one documented case of a skin reaction to cineole (de Groot and Weyland 1992). There is also a report of irritancy due to terpinen-4-ol (Knight and Hausen 1994). The toxicity of tea tree oil is discussed in Chapter 10 of this monograph and readers are referred to it for further information.

Another possible reason for maintaining levels of cineole below 5% would be that such oils were more effective as antimicrobials. When tested as a pure compound, conflicting results for MIC values of 1,8-cineole are reported. A number of workers report low activity (Low *et al.* 1974; Cruz *et al.* 1989; Williams *et al.* 1993; Raman *et al.* 1995; Griffin 1995), while other studies report zones of inhibition comparable to those obtained with terpinen-4-ol and α-terpineol and MIC values between 0.16 and 0.5% against various test organisms (Deans and Svoboda 1989; Nguyen *et al.* 1994; Carson and Riley 1995). Williams *et al.* (1988, 1993) used the disc diffusion assay to compare several oils with different levels of 1,8-cineole and terpinen-4-ol and report that oils become less active against *C. albicans* as cineole concentrations increase. The difference was less marked when the same oils were tested against *E. coli* and *S. aureus.* However, the reported effects may be attributable to changes in concentration of terpinen-4-ol rather than 1,8-cineole. The lack of reproducibility of the disc diffusion method used also limits the significance of these results.

In a more comprehensive study of eight oils blended to contain various concentrations of 1,8-cineole between 1.5 and 28%, there were no significant differences found in MIC values (Table 4, Southwell *et al.* 1996). By comparing the results for Oil 6 with Oils 7 and 8, it can be seen that it is important that the level of terpinen-4-ol remains above 30%. Similar results were obtained with a further 12 Gram-positive and Gram-negative bacteria (Mann and Markham 1997), providing convincing evidence for the view that oils with ultra low levels of cineole are not superior in terms of antimicrobial activity. Thus there is no scientific basis for the market push for ultra low cineole oils.

Table 4 Effect of 1,8-cineole on MIC values determined by an agar dilution method using Isosensitest agar with 0.25% Tween 20

*Sample**	*Terpinen-4-ol* %	*1,8-cineole* %	*E. coli*	*S. aureus*	*C. albicans*
1	41.0	1.5	0.2	0.2	0.2
2	39.8	3.1	0.2	0.2	0.2
3	40.8	5.4	0.2	0.2	0.2
4	34.4	10.4	0.2	0.2	0.2
5	32.1	14.8	0.2	0.2	0.2
6	32.1	18.0	0.2	0.2	0.2
7	26.8	23.9	0.3	0.2	0.3
8	24.7	28.5	0.3	0.2	0.3

*Tea tree oil samples were mixed with 1,8-cineole to give final concentrations of 1,8-cineole ranging from 1.5–28.5%.

In fact, it has been reported that 1,8-cineole is very effective in enhancing the penetration of other compounds through the skin (Williams and Barry 1991), and thus its restriction to very low levels may reduce the clinical effectiveness of the oil.

Effect of Aging

p-Cymene, limited by the ISO standard to a maximum of 12% of the oil composition, generally represents less than 5% of freshly distilled tea tree oil, but has been reported to increase in oils as they age, due to oxidation of hydrocarbon components, mainly α- and γ-terpinenes. The rate and extent of these oxidation reactions varies among oils, but appears to relate more to the surface area of the oil exposed to the air, than to storage temperature or colour of glass of the storage bottle (Southwell 1993). Table 5 shows the change with aging of the levels of six of the major components of a tea tree oil sample in our laboratory. Oxidation of α- and γ-terpinenes has occurred and *p*-cymene content has increased from 2.78% to 15.06%, with no significant changes in the concentration of terpinen-4-ol or 1,8-cineole. Associated with these changes was a decrease in MIC values; that is, the oil became more active with aging, an effect also reported by Lassak and McCarthy (1983) in some oils.

Williams and Lusunzi (1994) have reported oils from *Melaleuca dissitiflora* with terpinen-4-ol concentrations above 40% and *p*-cymene at concentrations up to 16% and their data indicates that such oils possess greater antimicrobial activity than standard tea tree oil. When these authors tested individual components by the disc diffusion method, both terpinen-4-ol and *p*-cymene showed significant activity, with *p*-cymene giving larger zones than terpinen-4-ol against some of the test organisms. The authors concluded that *p*-cymene was correlated with the high level of antimicrobial activity. Penfold and Grant (1925) reported a Rideal-Walker coefficient of 8 for a fraction containing *p*-cymene, and stated that it acted as a synergist when added to other terpenes.

These results suggest that an increase in *p*-cymene in aged tea tree oil might explain the observed increase in bioactivity. However, addition of *p*-cymene to standard tea tree oil to give final concentrations up to 15% did not reduce MIC values (Mann and Markham 1997). In addition, when tested individually, *p*-cymene showed no significant activity in either disc or MIC assays, a result supported by the work of Carson and

Table 5 Composition changes in aged tea tree oil measured by gas chromotagraphy

Component (%)	*Date tested*	
	23/5/96	5/8/96
Terpinen-4-ol	39.89	41.53
1,8-cineole	3.10	3.52
p-cymene	2.78	15.06
γ-terpinene	17.10	4.23
α-terpinene	8.05	0.67
α-terpineol	3.30	3.78

Riley (1995). Other reports have indicated that α- and γ-terpinene possess comparable or greater activity than *p*-cymene (Deans and Svoboda 1989; Carson and Riley 1995). These results contradict the idea that the increase in *p*-cymene alone is responsible for the increased activity and further work is required to establish the correlation between changes in chemical composition as a result of aging and changes in antimicrobial activity.

TESTING OF EFFICACY IN CLINICAL TRIALS

The toxicity of tea tree oil precludes its use internally, but the broad spectrum antimicrobial activity of tea tree oil suggests a variety of external applications, and records of clinical use date from 1930. The oil is claimed to have antiinflammatory and mild anaesthetic properties, and a recent study has reported that terpinen-4-ol and α-terpinene, two of the components of tea tree oil, exhibited a topical antiinflammatory effect in rats (Pongprayoon *et al.* 1997). This long history of use suggests the efficacy and safety of the oil as an antiseptic and the laboratory data outlined in the previous sections of this review supports its antimicrobial properties. However, as highlighted by Cowen (1974), there are many factors which affect antimicrobial efficacy and, unless these are very well understood for each organism, optimum formulation determined by laboratory tests does not always correlate with optimal effectiveness *in vivo.* In addition, interactions, which may be either synergistic or antagonistic, can occur between components and thus it is not sufficient to assume that a product will be effective simply because it contains an ingredient known to be active. This section reviews the published reports of clinical testing of tea tree oil and of products containing tea tree oil; reports are grouped according to the type of infection under investigation.

Bacterial Infections of the Skin

A number of infections of the hair follicles and sebaceous glands, such as pimples, acne vulgaris and furunculosis (boils and abcesses) are caused by bacteria, including *S. aureus, Streptococcus* spp. and *Propionibacterium acnes.* These organisms are sensitive to tea tree oil in laboratory tests (see previous sections) and there have been a number of studies carried out to determine their efficacy in clinical situations.

Feinblatt (1960) discusses the unique property of oils containing terpene derivatives of benzene of penetrating to the subcutaneous layers by mixing with sebaceous secretions and reports the use of such an oil in the treatment of 25 cases of furunculosis (boils). Undiluted ‘cajeput-type’ oil from *Melaleuca alternifolia* was used to paint the infected area two to three times daily, and patients were examined on alternate days. The study included ten untreated controls: five of the controls had similar symptoms at the end of the eight-day study, while the boils of five worsened and were finally incised. Only one of the twenty-five patients treated with the *Melaleuca* oil required incision, and in 15 cases the boil site had completely cleared in eight days. A reduction in symptoms was recorded for the remaining nine participants. Three patients reported slight temporary stinging following application of the oil, but no toxic effects were reported.

Whilst this study provides evidence of the efficacy of tea tree oil in the treatment of furunculosis, the author does not specify the chemical composition of the oil used and it is possible that the oil was a high cineole (eucalyptol) variety. This highlights the importance of specifying the chemical composition of oils used in all studies if comparisons are to be drawn between them, or if the work is to support applications for registration of oils or products containing them.

Bélaiche (1985b) reports the efficacy of the oil in the treatment of two cases of impetigo (one caused by *Staphylococcus* and the other by *Streptococcus*) and three cases of acne caused by *Staphylococcus* spp. Improvement was reported in all cases, but the lack of untreated controls and the small numbers limit the conclusions which can be drawn. In addition, the test oil does not meet the specification of the current ISO standard because of the high concentration of *p*-cymene (16.4%).

A more rigorous study was carried out by Bassett *et al.* (1990) to examine the efficacy of a tea tree oil preparation in the treatment of acne vulgaris, a multifactorial disorder, common especially among teenagers. The bacterium, *Propionibacterium acnes,* which has been implicated as one of the causative factors, is sensitive to tea tree oil in *in vitro* tests (Table 2). Bassett *et al.* (1990) report MIC values of 0.75% or less for 90% of strains tested, while Carson and Riley (1994) report lower values. As a result, the oil has been incorporated in a number of commercially available facewash and pimple gel preparations, generally at concentrations in the range of 0.5–4.0% (w/w or w/v).

Bassett *et al.* (1990) report the results of a single-blind, randomised clinical trial on 124 patients with mild to moderate acne. The efficacy of a 5% tea-tree oil gel was compared to a 5% benzoyl peroxide lotion in terms of its ability to reduce the number of inflamed and non-inflamed lesions over the three-month treatment period. Oiliness, erythema, scaling, pruritis and dryness were also assessed as a measure of skin tolerance.

The results showed that the number of both inflamed and non-inflamed lesions was significantly reduced in both groups, with benzoyl peroxide having a more rapid onset of action. Benzoyl peroxide was significantly more effective than tea-tree oil in reducing inflamed lesions ($P<0.001$ at three months); however, there was, no significant difference between the two protocols in the reduction of non-inflamed lesions. Of the two preparations, tea tree oil was better tolerated than benzoyl peroxide, with fewer reports of side-effects such as skin scaling, pruritis and dryness.

The above study has been criticised because of the lack of a control or placebo group to which the test groups can be compared, and further trials, including a placebo group are needed to confirm the findings of the study. The study was considered single-blind by the authors, because, although neither group was informed of the preparation being used, it was reported that it was not possible to disguise the characteristic smell of tea tree oil. This will remain an issue in all future clinical trials of products containing tea tree oil as the active ingredient.

Tea Tree Oil in the Treatment of Fungal Infections of the Skin and Nails

Fungal infections of the skin and nails tend to be persistent, irritating infections with a high rate of recurrence. Treatment may be oral or topical and is frequently required for many months, as penetration of the drugs to the site of the infection is poor. Side-effects are relatively common, particularly with drugs administered orally. Hence alternative treatments are desirable.

Infections include tinea pedis (also known as athlete's foot), tinea (or pityriasis) versicolor, paronychia (infection of the tissue surrounding the nails) and onychomycosis (infection of the nail). The causative organisms vary from case to case and site to site, but the most common causative organisms are the dermatophytes *Trichophyton rubrum, Trichophyton mentagrophytes, Epidermophyton floccosum* and *Microsporum canis* and the yeast, *Candida albicans.*

Walker (1972) has reported the effectiveness of tea tree oil in sixty patients over a six-year period. Undiluted oil, a 40% solution of tea tree oil with 13% isopropyl alcohol and an ointment containing 8% oil were used to treat a variety of foot ailments. Treatment was applied for periods ranging from one week to up to two-and-a-half years, and a clearing of both infection and symptoms is reported in 63.7% of cases and improvement in symptoms in a further 33.3% of patients. The symptoms which were reported to have been reduced included bromidrosis (foul-smelling secretions), inflammation associated with corns, calluses, bunions and hammertoes, scaling of nails in onychomycosis and symptoms and degree of recurrence of tinea pedis.

Bélaiche (1985b) also reports the effectiveness of tea tree oil in resolving infection of toe-nails in eight cases, with improvement of symptoms in six of these cases. However, both these reports should be considered anecdotal as they are studies of individual cases and do not include controls or comparisons to other treatments.

Double-blind, randomised clinical trials have been carried out by Buck *et al.* (1994) and Tong *et al.* (1992). The former study compares the effectiveness of twice daily topical application of undiluted tea tree oil to that of 1% clotrimazole in the treatment of onychomycosis. Outcomes were measured at one, three and six months by culture, clinical assessment and the patient's subjective assessment. There were no significant differences between the two treatments by any of the above criteria. A full or partial recovery was achieved in approximately 60% of patients in both groups, with less than 20% in either group becoming culture negative at the end of the test period. This study is limited by the lack of inclusion of an untreated control group, but the authors do compare their results to those of studies of other antifungal agents. Cure rates compare favourably with agents such as ciclopiroxolamine, but the two agents tested were both less effective than others including naftitine hydrochloride gel and amorolfine.

The effectiveness of tea tree oil as an antiinflammatory agent is also supported by the work of Tong *et al.* (1992). These authors compare the effectiveness of a sorbolene cream containing 10% tea tree oil to a commercially available tolnafate 1% cream over a four week test period. The study included a control group who were treated with sorbolene cream only. At the end of the study, all three groups had reduced symptoms, but the tea tree and tolnafate groups were significantly better than the placebo group. However, only 30% of the tea tree group had negative culture compared to 21% of the placebo group and 85% of the tolnafate group. It is possible that changes to the tea tree formulation might result in enhanced mycocidal activity, but there is no published data on the kill rate of tea tree oil on the fungi responsible for these infections. This is an area which requires further study.

Infections of the Genitourinary System

Although Humphrey (1930) suggested the suitability of tea tree oil for treatment of vaginal infections, the first report in the literature of its effectiveness in treating such infections is that of Pena (1962). He reported clinical cure following treatment of 130 cases of vaginal infections, due mostly to *Trichomonas* (116 cases) and *Candida albicans* (4 cases). The treatment regime involved douching with a 0.4% solution of *Melaleuca alternifolia* oil and insertion of tampons saturated with a 20–40% solution of the oil. This treatment was reported to be as effective as use of standard antitrichomonal suppositories, and without side-effects of irritation and burning.

Twenty-eight chronic cases of vaginal infection caused by *Candida albicans* were treated by nightly insertion of a pessary containing 20 mg of tea tree oil for a period of three months (Belaiche 1985b). Infection and symptoms were alleviated in 21 cases, with an improvement in symptoms, but persistence of the yeast, in a further three cases. The author also comments that the preparation was well tolerated by all but one of the participants who withdrew from the study early in the first week. Barnes (1990) also reports the alleviation of symptoms of vaginal irritation and burning in a number of women following treatment with a tea tree cream and/or douche. The women in this report all had chronic conditions which had not responded to conventional treatments. Once again, the evidence which suggests that tea tree oil is useful both for its antimicrobial activity and its soothing and pain-relieving effects, is anecdotal, and appropriately designed and controlled clinical trials are needed to establish a sound basis for the marketing of therapeutically useful vaginal products.

AGRICULTURAL APPLICATIONS OF TEA TREE OIL

The role of essential oils within the plants producing them, is primarily one of defence of the plant from attack by other organisms. It is possible that tea tree oil will find many marketable applications in agriculture through exploitation and expansion of this natural function. Essential oils would have the advantage over synthetic chemicals, such as those currently used to control post-harvest pathogens, of being more acceptable both environmentally and to the consumer. Potential applications include control agents of plant pathogens, insect repellants and antifeedants and insecticides. Terpinen-4-ol, the major ingredient of tea tree oil, has been shown to be very active as a repellant of the yellow-fever mosquito, *Aedes aegyptii* (Hwang *et al.* 1985). Bishop and Thornton (1997) demonstrated the ability of tea tree oil to inhibit hyphal growth of fifteen common fungal post-harvest pathogens. Whilst direct contact was more effective, the oil also demonstrated significant antifungal activity in the vapour phase, a characteristic which suggests the possibility of its use as a fumigant for stored crops.

There are very few reports of field trials testing the efficacy of tea tree oil in the control of fungal or viral pathogens of economically important crops. A 1% aqueous solution of tea tree oil was reported to control powdery mildew of greenhouse-grown cucurbits caused by the fungus *Sphaerotheca fuliginea* (Olsen *et al.* 1988). In another study, it was reported that *Nicotiana glutinosa* plants sprayed with 100, 250 and 500 ppm of tea tree oil in distilled water prior to inoculation with the Tobacco Mosaic Virus showed significantly fewer lesions than control plants for 10 days following inoculation (Bishop 1995). Neither of these studies distinguish between microbial inactivation and inhibition of infection through changes to the host plant, and, hence, conclusions cannot be drawn about the mechanism of action. However, these studies indicate the potential of tea tree oil in the treatment of a variety of plant diseases.

OTHER APPLICATIONS OF TEA TREE OIL

The range of possibilities for products containing tea tree oil as an active ingredient is vast. The uses reported in the literature are as diverse as its use as an additive in an aerosol system used for cleansing of air conditioning systems (Ryan 1990), its potential for addition to laundry detergents as an acaricidal agent to destroy mites in bedding and clothing (McDonald and Tovey 1993) and its use in burn preparations for its properties of soothing the damaged tissue, rapid healing, prevention of infection and pain relief (Price 1989).

The literature indicates an interest in the role of plant volatile oils as antioxidants (Deans and Waterman 1993). Essential oils have been shown to act as hepatoprotective agents in aging mammals and to have a positive effect upon docohexanoic acid levels in aging rodent retinas. In a recent study, essential oils of geranium, monarda, nutmeg, oregano and thyme, which contain a number of monoterpenes also present in tea tree oil, demonstrated extensive antioxidant capacities at final concentrations of 0.75 ppm to 100 ppm (Dorman *et al.* 1995). Once the active ingredients in the oils have been elucidated, the potential of tea tree oil as an antioxidant can be assessed.

CONCLUSION

Tea tree oil has a well-established reputation, supported by laboratory data, as an effective, well-tolerated, broad spectrum antimicrobial which possesses a number of advantages over its synthetic counterparts. Results of a limited number of clinical trials have been promising, but further clinical testing, both of standard oils of known chemical composition and of formulated products, is required to enable tea tree oil to broaden its acceptance in the marketplace. Although its antimicrobial activity is well established, little is understood about the way in which it acts on microbial cells. A more rational approach to breeding programs and to the incorporation of tea tree oil in formulated products will be possible once the mode of action of tea tree oil against a range of microbial cell types is elucidated.

REFERENCES

Aboutabl, E.A., Sokkar, N.M., Megid, R.M.A., De Pooter, H.L. and Masoud, H. (1995) Composition and antimicrobial activity of *Otostegia fruticosa* Forssk. oil. *J. Essent. Oil Res.*, 7, 299–303.

Allegrini, J., de Buochberg, M.S. and Maillols, H. (1973) Emulsions d'huiles essentielles fabrication et applications en microbiologie. *Travaux de la Societe de Pharmacie de Montpellier*, **33**, 73–86.
Altman, P.M. (1989) Australian tea tree oil—A natural antiseptic. *Austral. J. Biotech.*, **3**(4), 247–248.
Altman, P.M. (1991) Australian tea tree oil. *Cosmetics and Toiletries Manufacture*, **12**, 22–24.
Atkinson, N. and Brice, H.E. (1955) Antibacterial substances produced by flowering plants. 2. The antibacterial action of essential oils from some Australian plants. *Aust. J. Exptl. Biol.*, **33**, 547–554.
Bagci, E. and Digrak, M. (1996) Antimicrobial activity of essential oils of some *Abies* (fir) species from Turkey. *Flavour and Fragrance Journal*, **11**, 251–256.
Barnes, R. (1990) The "Vaginol" range of formulations containing tea tree oil. *Proceedings of Conference—The Clinical Significance of Tea Tree Oil and Other Essential Oils*. Sydney, December, 1990, pp. 35–42.
Bassett, I.B., Pannowitz, D.L. and Barnetson, R.St-C. (1990) A comparative study of tea-tree oil versus benzoylperoxide in the treatment of acne. *Med. J. Aust.*, **153**, 455–458.
Bélaiche, P. (1985a) Traitement des infections vaginales a *Candida albicans* par l'huile essentielle de *Melaleuca alternifolia* (Cheel). *Phytotherapy*, **15**, 13–14.
Bélaiche, P. (1985b) Traitement des infections cutanees par l'huile essentielle de *Melaleuca alternifolia* (Cheel). *Phytotherapy*, **15**, 15–17.
Beylier, M.F. (1979) Bacteriostatic activity of some Australian essential oils. *Perfumer and Flavorist*, **4**, 23–25.
Biondi, D., Cianci, P., Geraci, C. and Ruberto, G. (1993) Antimicrobial activity and chemical composition of essential oils from Sicilian aromatic plants. *Flavour and Fragrance Journal*, **8**, 331–337.
Bishop, C.D. (1995) Antiviral activity of the essential oil of *Melaleuca alternifolia* (Maiden & Betche) Cheel (Tea Tree) against tobacco mosaic virus. *J. Essent. Oil Res.*, **7**, 641–644.
Bishop, C.D. and Thornton, I.B. (1997) Evaluation of the antifungal activity of the essential oils of *Monarda citriodora* var. *citriodora* and *Melaleuca alternifolia* on post-harvest pathogens. *J. Essent. Oil Res.*, **9**, 77–82.
Brophy, J.J., Davies, N.W., Southwell, I.A., Stiff, I.A. and Williams, L.R. (1989) Gas chromatographic quality control for oil of *Melaleuca* terpinen-4-ol type (Australian tea tree). *J. Agricultural and Food Chem.*, **37**, 1330–1335.
Brown, J.T., Hegarty, P.K. and Charlwood, B.V. (1987) The toxicity of monoterpenes to plant cell cultures. *Plant Science*, **48**, 195–201.
Buck, D.S., Nidorf, D.M. and Addino, J.G. (1994) Comparison of two topical preparations for the treatment of onychomycosis: *Melaleuca alternifolia* (tea tree) oil and clotrimazole. *The Journal of Family Practice*, **38**(6), 601–605.
Carson, C.F., Cookson, B.D., Farrelly, H.D. and Riley, T.V. (1995a) Susceptibility of methicillin-resistant *Staphylococcus aureus* to the essential oil of *Melaleuca alternifolia*. *J. Antimicrob. Chemo.*, **35**, 421–424.
Carson, C.F., Hammer, K.A. and Riley, T.V. (1995b) Broth micro-dilution method for determining the susceptibility of *Escherichia coli* and *Staphylococcus aureus* to the essential oil of *Melaleuca alternifolia* (tea tree oil). *Microbios.*, **82**, 81–185.
Carson, C.F., Hammer, K.A. and Riley, T.V. (1996) *In-vitro* activity of the essential oil of *Melaleuca alternifolia* against *Streptococcus* spp. *J. Antimicrob. Chemother.*, **37**, 1177–1178.
Carson, C.F. and Riley, T.V. (1994) Susceptibility of *Propionibacterium acnes* to the essential oil of *Melaleuca alternifolia*. *Lett. Appl. Microbiol.*, **19**, 24–25.
Carson, C.F. and Riley, T.V. (1995) Antimicrobial activity of the major components of the essential oil of *Melaleuca alternifolia*. *J. Appl. Bact.*, **78**, 264–269.
Chand, S., Luzunzi, I., Veal, D.A., Williams, L.R. and Karuso, P. (1994) Rapid screening of the antimicrobial activity of extracts and natural products. *J. Antibiotics*, **47**, 1295–1304.
Cowen (1974) Relative merits of 'In use' and laboratory methods for the evaluation of antimicrobial products. *J. Soc. Cosmet. Chem.*, **25**, 307–323.
Cruz, T., Cabo, M.P., Cabo, M.M., Jimenez, J., Cabo, J. and Ruiz, C. (1989) *In vitro* antibacterial effect of the essential oil of *Thymus longiflorus* Boiss. *Microbios*, **60**, 59–61.

Deans, S.G. and Svoboda, K.P. (1988) Antibacterial activity of French tarragon (*Artemisia dranunculus* Linn.) essential oil and its constituents during ontogeny. *J. Hort. Sc.*, **63**(3), 503–508.

Deans, S.G. and Svoboda, K.P. (1989) Antibacterial activity of summer savory (*Satureja hortensis L*) essential oil and its constituents. *J. Hort. Sci.*, **64**(2), 205–210.

Deans, S.G. and Waterman P.G. (1993) Biological activity of plant volatile oils. In R.K.M. Hay and P.G. Waterman (eds.), *Volatile Oil Crops: Their Biology, Biochemistry and Production.* Longman Group UK, London, pp. 113–136.

De Groot, A.C. and Weyland, J.W. (1992) Systemic contact dermatitis from tea tree oil. *Contact Dermatitis*, **27**, 279–280.

Dellar, J.E., Cole, M.D., Gray, A.I., Gibbons, S. and Waterman, P.G. (1994) Antimicrobial sesquiterpenes from *Prostanthera* aff. *melissifolia* and *P. rotundifolia. Phytochemistry*, **36**(4), 957–960.

Denyer, S. and Baird, R. (eds.) (1990) *Guide to Microbiological Control in Pharmaceuticals*, Ellis Horwood, Chichester.

Dorman, H.J.D., Deans, S.G., Noble, R.C. and Surai, P. (1995) Evaluation *in vitro* of plant essential oils as natural antioxidants. *J. Essent. Oil Res.*, **7**, 645–651.

Drury, S. (1991) *Tea Tree Oil: A Medicine Kit in a Bottle.* C.W. Daniel Co. Ltd., Essex.

Feinblatt, H.M. (1960) Cajeput-type oil for the treatment of furunculosis. *J. Nat. Med. Assoc.*, **52**(1), 32–34.

Graham, B.M. (1978) The development of Australian legislation for disinfectants. *Aust. J. Hosp. Pharm.*, **8**(4), 149–155.

Griffin, S.G. (1995) Antimicrobial activity of essential oils from Australian flora. B. App. Sc. (Hons.) Thesis. University of Western Sydney, Hawkesbury, Australia.

Griffin, S.G., Markham, J.L. and Leach, D.N. (1998) A modified agar dilution method for the determination of the minimum inhibitory concentration of tea tree oil. *J. Essent. Oil Res.*, Submitted for publication.

Gundidza, M., Chinyanganya, F., Chagonda, L., De Pooter, H.L. and Mavi, S. (1994) Phytoconstituents and antimicrobial activity of the leaf essential oil of *Clausena anisata* (Willd.) J.D. Hook ex. Benth. *Flavour and Fragrance Journal*, **9**, 299–303.

Hammer, K.A. (1996) Potential uses of *Melaleuca alternifolia* (Tea Tree) Oil in the control of yeasts. *Proceedings of the Australian Tea Tree Export and Marketing Ltd. Conference*, October, Sydney pp. 59–62.

Hammer, K.A., Carson, C.F. and Riley, T.V. (1996) Susceptibility of transient and commensal skin flora to the essential oil of *Melaleuca alternifolia* (tea tree oil). *Am. J. Infect. Cont.*, **24**(3), 186–189.

Humphrey, E.M. (1930) A new Australian germicide. *Med. J. Australia*, March 29, 417–418.

Hwang, Y.-S., Wu, K.-H., Kumamoto, J., Axelrod, H. and Mulla, M.S. (1985) Isolation and identification of mosquito repellants in *Artemisia vulgaris. J. Chem. Ecol.*, **11**, 1297–1306.

International Standards Organization (1996) Oil of Melaleuca, terpinen-4-ol type (Tea Tree Oil). *International Standard ISO 4730: 1996(E)*, International Standards Organization, Geneva.

Janssen, A.M., Scheffer, J.J.C., Baerheim Svendsen, A. and Aynechi, Y. (1985) Composition and antimicrobial activity of the essential oil of *Ducrosia anethifolia.* In Baerheim Svendsen, A. and Scheffer, J.J.C. (eds.), *Essential Oils and Aromatic Plants.* Martinus Nijhoff, Dordrecht.

Janssen, A.M., Scheffer, J.J.C. and Baerheim Svendsen, A. (1986) Antimicrobial activity of essential oils: A 1976–1986 literature review. Aspects of the test methods. *Planta Medica*, **53**, 395–398.

Kar, A. and Jain, S.R. (1971) Antibacterial evaluation of some indigenous medicinal volatile oils. *Qual. Plant. Mater. Veg.* XX, **3**, 231–237.

Kazmi, S.J.A. and Mitchell, A.G. (1978) Preservation of solubilised and emulsified systems. II. Theoretical development of capacity and its role in antimicrobial activity of chlorocresol in cetamacrogol-stabilised Systems. *J. Pharm. Sci.*, **67**, 1266–1271.

Knight, T.E. and Hausen, B.M. (1994) Melaleuca oil (Tea tree oil) dermatitis. *J. Amer. Acad. Dermatol.*, **30**, 423–427.

Knobloch, K., Pauli, A., Iberl, B., Weigand, H. and Weis, N. (1989) Antibacterial and antifungal properties of essential oil components. *J. Essent. Oil Res.*, **1**, 119–128.

Kubo, I., Himejima, M. and Muroi, H. (1991) Antimicrobial activity of flavor components of cardamom *Elattaria cardamomum* (Zingiberaceae) Seed. *J. Agric. Food Chem.*, **39**, 1984–1986.

Lassak, E.V. and McCarthy, T. (1983) *Australian Medicinal Plants.* Methuen, Australia.

Lattaoui, N. and Tantaoui-Elaraki, A. (1994) Comparative kinetics of microbial destruction by the essential oils of *Thymus broussonettii, T. zygis* and *T. satureioides. J. Essent. Oil Res.*, **6**, 165–171.

Lorian, V. (1986) *Antibiotics in Laboratory Medicine* (2nd edn.), Williams and Wilkins, Baltimore.

Low, D., Rawal, B.D. and Griffin, W.J. (1974) Antibacterial action of the essential oils of some Australian *Myrtaceae* with special reference to the activity of chromatographic fractions of the oil *of Eucalyptus citriodora. Planta Medica*, **26**, 184–189.

Mann, C.M. and Markham, J.L. (1997) University of Western Sydney, Hawkesbury. Unpublished results.

Mann, C.M. and Markham, J.L. (1998) A new method of determining the MIC of essential oils. *J. Appl. Microbiology*, **84**, 538–544.

Markham, J.L., Leach, D.N., Cornwell, C.P. and Griffin, S.G. (1995) Antimicrobial activity of monoterpenes in essential oils from Australian flora. *Proceedings of the Australian Society for Microbiology Annual Scientific Meeting*, Canberra, September 24–29.

Markham, J.L., Mann, C.M., Leach, D.N. and Southwell, I.A. (1996) The effect of oil chemistry on antimicrobial activity. *Tea Tree Oil Symposium.* August 1996 Abstracts. Wollongbar Agricultural Institute, Wollongbar, pp. 8–9.

McDonald, L.G. and Tovey, E. (1993) The effectiveness of benzyl benzoate and some essential plant oils as laundry additives for killing house dust mites. *J. Allergy Clin. Immunol.*, **92**, 771–772.

Morris, J.A., Khettry, A. and Seitz, E.W. (1979) Antimicrobial activity of aroma chemicals and essential oils. *J. Am. Chem. Soc.*, **56**, 595–603.

Nenoff, P., Haustein, U.F. and Brandt, W. (1996) Antifungal activity of the essential oil of *Melaleuca alternifolia* (Tea tree oil) against pathogenic fungi *in vitro. Skin Pharmacology*, **9**(6), 388–394.

Nguyen, D.C., Truong, T.X., Motl, O., Stránský, K., Presslová, J., Jedlicková, Z. and Serý, V. (1994) Antibacterial properties of Vietnamese cajuput oil. *J. Essent. Oil. Res.*, **6**, 63–67.

Olsen, M.W., Cassells, J. and Cross, D. (1988) Control of *Sphaerotheca fuliginea* on cucurbits with an oil extracted from Australian Tea Tree. *Phytopathology*, **78**, 1595 (Abst).

Patkar, K.L., Usha, C.M., Shetty, H.S., Paster, N. and Lacey, J. (1993) Effect of spice essential oils on growth and aflatoxin B_1 production by *Aspergillus flavus. Lett. Appl. Microbiol.*, **17**, 49–51.

Pena, E.F. (1962) *Melaleuca alternifolia* oil: Its use for trichomonal vaginitis and other vaginal infections. *Obstetrics and Gynecology,* **19**, 793–795.

Penfold, A.R. and Grant, R. (1925) The germicidal values of some Australian essential oils and their pure constituents. Together with those for some essential oil isolates, and synthetics. Part III. *J. Proceedings of the Royal Soc. of NSW.*, **59**, 346–350.

Pongprayoon, U., Soontornsaratune, P., Jarikasem, S., Sematong, T., Wasuwat, S. and Claeson, P. (1997) Topical antiinflammatory activity of the major lipophilic constituents of the rhizome of *Zingiber cassumunar.* I. The essential oil. *Phytomedicine*, **3**(4), 319-322.

Price, J. (1989) The use of tea tree oil in burn treatment products. *Proceedings of Modern Phytotherapy—The Clinical Significance of Tea Tree Oil and other Essential Oils.* Macquarie University, September 17.

Raman, A., Weir, U. and Bloomfield, S.F. (1995) Antimicrobial effects of tea-tree oil and its major components on *Staphylococcus aureus, S. epidermidis* and *Propionibacterium acnes. Letters in Applied Microbiolog y*, **21**, 242–245.

Remmal, A., Bouchikhi, T., Rhayour, K. and Ettayebi, M. (1993) Improved methods for the determination of antimicrobial activity of essential oils in agar medium. *J. Essent. Oil Res.*, **5**, 179–184.

Ryan, R.F. (1990) Oil of *Melaleuca alternifolia* dissolved in liquid carbon dioxide propellant (Bactigas™) used for the control of bacteria and fungi in air conditioning systems. *Proceedings of Conference—The Clinical Significance of Tea Tree Oil and Other Essential Oils.* Sydney, December, 1990, pp. 65–71.

Schilcher, H. (1985) Effects and side-effects of essential oils. In Baerheim Svendsen, A. and Scheffer, J.J.C. (eds.), *Essential Oils and Aromatic Plants.* Martinus Nijhoff, Dordrecht.

Schmolka, I.R. (1973) The synergistic effects of nonionic surfactants upon cationic germicidal agents. *J. Soc. Cosmet. Chem.*, **24**, 577–592.

Scortichini, M. and Rossi, M.P. (1991) Preliminary *in vitro* evaluation of the antimicrobial activity of terpenes and terpenoids towards *Erwinia amylovora* (Burrill) Winslow *et al. J. Appl. Bacteriol.*, **71**, 109–112.

Shapiro, S., Meier, A. and Guggenheim, B. (1994) The antimicrobial activity of essential oils and essential oil components towards oral bacteria. *Oral Microbiol. Immunol.*, **9**, 202–208.

Southwell, I.A. (1993) NSW Agriculture. Unpublished results.

Southwell, I.A., Freeman, S. and Rubel, D. (1997) Skin irritancy of tea tree oil. *J. Essent. Oil Res.*, **9**, 47–52.

Southwell, I.A., Hayes, A.J., Markham, J. and Leach, D.N. (1993) The search for optimally bioactive Australian tea tree oil. *Acta Horticulturae*, **334**, 256–265.

Southwell, I.A., Markham, J. and Mann, C. (1996) Is cineole detrimental to tea tree oil? *Perfumer and Flavorist*, **21**, 7–10.

Tong, M.M., Altman, P.M. and Barnetson, R.St-C. (1992) Tea tree oil in the treatment of *Tinea pedis. Australasian J. Dermatol.*, **33**, 145–149.

Van Doorne, H. (1990) Interactions between preservatives and pharmaceuticals. In Denyer, S. and Baird, R. (eds.), (1990) *Guide to Microbiological Control in Pharmaceuticals.* Ellis Horwood, Chichester.

Villar, A., Recio, M.C., Rios, J.L. and Zafra-Polo, M.C. (1986) Antimicrobial activity of essential oils from *Sideritis* species. *Pharmazie*, **41**, 298–299.

Yousef, R.T. and Tawil, G.G. (1980) Antimicrobial activity of volatile oils. *Pharmazie*, **35**, 698–701.

Walker, M. (1972) Clinical investigation of *Melaleuca alternifolia* oil for a variety of common foot problems. *Current Podiatry*, **2**, 7–15.

Walsh, L.J. and Longstaff, J. (1987) The antimicrobial effects of an essential oil on selected oral pathogens. *Periodontology*, **8**, 11–15.

Williams, A.C. and Barry, B.W. (1991) Terpenes and the lipid-protein-partitioning theory of skin penetration enhancement. *Pharmaceutical Research*, **8**(1), 17–24.

Williams, L.R., Home, V. and Lusunzi, I. (1993) An evaluation of the contribution of cineole and terpinen-4-ol to the overall antimicrobial activity of tea tree oil. *Cosmetics, Aerosols and Toiletries in Australia*, **7**(3), 25–34.

Williams, L.R., Home, V., Zhang, X. and Stevenson, I. (1988) The composition and bactericidal activity of oil of *Melaleuca alternifolia* (Tea tree oil). *Int. J. Aromatherapy*, **1**, 15–17.

Williams, L.R. and Lusunzi, I. (1994) Essential oil from *Melaleuca dissitiflora*: a potential source of high quality tea tree oil. *Industrial Crops and Products*, **2**, 211–217.

10. TOXICOLOGY OF TEA TREE OIL

MICHAEL RUSSELL

Wollongbar Agricultural Institute, Wollongbar, NSW, Australia

INTRODUCTION

For some essential oils, as little as one teaspoonful can be fatal. Wormseed, sassafras, parsley, eucalyptus and camphor have all caused child fatalities in low doses (Tisserand and Balacs 1995). Consequently, guarding against accidentally overdosing with certain essential oils is important, especially with children. Storing essential oils that are known to be harmful out of children's reach and in child-proof bottles is a sensible way to minimise accidental consumption. Nevertheless, it is of vital importance to understand the toxic properties of each specific oil to be certain of the risks involved.

Toxicity can be manifested by many symptoms, beginning with a rash and redness of the skin, extending to organ damage, especially the liver and, in the extreme, resulting in death. Toxicity is generally dose dependant, with oral intake more significant than dermal absorption. Also, acute toxicity results from higher doses than those responsible for longer term chronic problems.

In this chapter, toxicology will be discussed in relation to acute oral and dermal toxicity, to dermal irritancy and to dermal allergy responses. The amount of an essential oil needed to bring about a response in each of these broad categories varies greatly. This variation is due to individual human tolerance (oral or dermal), dose frequency and the oil's intrinsic toxicity.

TEA TREE OIL—*MELALEUCA ALTERNIFOLIA*

Acute Oral Toxicity

Acute oral toxicity of a substance is measured by the LD_{50} test which measures the effect of differing amounts of a test substance given orally to a group of animals (usually rats or mice). The group is of uniform age, size and genetic origin. Hence LD_{50} measures the dosage which is lethal to 50% of the test animals.

The LD_{50} test is probably the prime result for any substance as it determines whether or not the substance can be ingested safely or not. LD_{50} test results are usually expressed in the form of grams (or ml) per kilogram of body weight, thus the heavier the person, the higher a dose needs to be, before it becomes lethal. It is for this reason that children are particularly susceptible to lethal poisoning. To help appreciate the significance of the LD_{50} test, Table 1 converts LD_{50} g/kg results for common essential oils to a lethal dose for a 70 kg adult and a 10 kg child.

Tea tree oil has been reported to have an LD_{50} of 1.9 g/kg (Ford 1988; Tisserand and Balacs 1995; Table 1), 1.9–2.6 ml/kg (or 1.7–2.3 g/kg for an oil of density 0.9) (Bolt 1989d; Altman 1991) and 3.0 ml/kg (or 2.7 g/kg) (Austteam 1995). Tisserand and Balacs (1995), although classifying tea tree oil with oils on the borderline of safe and unsafe, place it at the safe end of that group and state that "we have no reason to suspect any problems with tea tree".

Table 1 Some well known essential oils and components with their LD_{50} values (g/kg) in Rodents, from Tisserand and Balacs (1995) with corresponding calculated doses (gm) for a 70 kg adult and a 10 kg young child

	OIL	*LD_{50}* g/kg	*LD_{50} dose* (g) (70 kg *adult*)	*LD_{50} dose* (g) (10 kg *child*)
GROUP 1	Hydrocyanic acid	0.0001	0.007	0.002
	Boldo leaf oil	0.13	9.1	1.3
	Wormseed oil	0.25	17.5	2.5
	Mustard oil	0.34	23.8	3.4
LD_{50}	Pennyroyal oil (Eur.)	0.4	28	4
0→1	Camphor (chemical)	0.5–15	35–1050	5–150
	Tansy oil	0.73	51.1	7.3
	Thuja oil	0.83	58.1	8.3
	Almond (bitter) oil	0.96	67.2	9.6
GROUP 2	Oil of Wintergreen	1.20	84	12
	Cornmint	1.25	87.5	12.5
	Hyssop oil	1.4	98	14
	Almond (bitter) FFPA*	1.49	104.3	14.9
LD_{50}	Sassafras (braz.)	1.58	110.6	15.8
1→2	Myrrh oil	1.65	115.5	16.5
	Oregano	1.85	129.5	18.5
	Sassafras	1.9	133	19
	Tea tree	1.9	133	19
GROUP 3	Anise			
	Cajeput			
	Coriander			
LD_{50}	Cinnamon bark		between	between
2→5	Eucalyptus		150 and 350	20 and 50
	Neroli			
	Spike lavender			
GROUP 4	Bergamot			
	Chamomile			
	Citronella		above	above
LD_{50}	Geranium		350	50
5→	Lavender			
	Sandalwood			
	Ylang-ylang			

*FFPA = Free From Prussic Acid.

The LD_{50} test for tea tree oil conducted by Bolt (1989d) using OECD method 401 as per the Australian Tea Tree Oil Toxicology Data Sheet No. 5 (Tea Tree Oil Growers of Australia 1989) was performed as follows:

Tea tree oil was diluted with peanut oil and tested at 20%, 25% and 33% concentrations using groups of 5 male and 5 female Sprague Dawley SPF (specific pathogen free) and non-SPF rats. Animals received doses of 1.7–3.0 ml/kg by gavage and were observed for 14 days. The LD_{50} was found to be 2.6 ml/kg in SPF rats and 1.9 ml/kg in non-SPF rats. Ingestion of pure tea tree oil is not recommended.

For the animals surviving the above treatment, the major non lethal reaction was a complete lack of muscular tone in the forelimbs, bloodied noses and weeping (Bolt 1989d).

These results suggest that tea tree oil is less toxic than commonly used oils like pennyroyal (LD_{50} 0.40), wintergreen (LD_{50} 1.20), cornmint (LD_{50} 1.25), basil (LD_{50} 1.40), and bay leaf (W. Indian) (LD_{50} 1.80), (Tisserand and Balacs 1995).

Poisoning

Essential oils in general are a toxic risk especially to children and old people. Cases of child poisoning due to overdose of essential oils such as camphor, cinnamon, citronella, clove, eucalyptus, pennyroyal, sassafras and oil of wintergreen have been reported (Tisserand and Balacs 1995). In America in 1973, 500 cases of camphor intoxication alone were reported. Most of these were young children given camphorated oil instead of castor oil. In one instance, a 16 month old boy died from consuming one teaspoon of 20% camphorated cottonseed oil (Smith and Margolis 1954) thus making about 1 gram of camphor the toxic dose to children. The chemical camphor is up to 60–70 times more toxic to humans (LD_{50} 0.005–0.5 g/kg) than to rodents (LD_{50} 0.5–15 g/kg) (Tisserand and Balacs 1995, Table 1).

As the oral toxicity LD_{50} test is carried out on rodents and not humans the results can be misleading. Take for instance oil of wintergreen (methyl salicylate). In six cases of poisoning from the oil in adult humans, three people died from ingestion of 15 ml, 30 ml and 80 ml and three survived after 6 ml, 16 ml and 24 ml (Stevenson 1937) giving an average oral lethal dose of 0.2 g/kg to 0.3 g/kg for humans. The LD_{50} however, is 1.2 g/kg for rodents suggesting that oil of wintergreen is some three to five times more toxic in humans. According to Opdyke (1977) an oral dose of 4–8 ml of methyl salicylate is considered lethal for a child.

Similarly, eucalyptus oil has shown itself to be fatally toxic to humans in amounts between 30 ml and 60 ml (Gurr and Scroggie 1965) thus making it 4 times more toxic to humans than to rodents. With these figures in mind, one might be tempted to say that LD_{50} tests are of limited value. The lesson here is that with the LD_{50} test, one has a comparative assessment of an essential oil's oral toxicity potential in humans.

Furthermore, with tea tree oil being in the borderline category when tested with rodents (Tisserand and Balacs 1995), it is probably prudent to assume that tea tree oil is orally toxic in reasonably large doses, so oral ingestion should be avoided. This recommendation is even more applicable to children, who should be kept well clear of any container of essential oil including tea tree oil.

This is well illustrated by the known cases of tea tree oil poisoning in humans (Carson and Riley 1995).

Del Baccaro (1995) cites a case of a 17-month-old male child who developed ataxia and drowsiness after ingestion of less than 10 ml of oil of *Melaleuca alternifolia*. With treatment of activated charcoal and observation, the child was discharged 7 hours later.

A similar case involving a 23-month-old boy was reported by Jacobs and Hornfeldt (1994):

> A 23-month old boy became confused and unable to walk thirty minutes after ingesting less than 10 ml of T36-C7, a commercial product containing 100% *Melaleuca* oil. The child was referred to a nearby hospital. His condition improved and he was asymptomatic within 5 hours of ingestion. He was discharged to home the following day.

Adults are also susceptible with one sixty-year-old male, after ingesting half a teaspoonful of oil, reacting with severe dermatitis in addition to feeling quite unwell (Elliot 1993). This was thought to be "systemically induced eczema or a cutaneous reaction to an ingested contact allergen" (Moss 1994).

On the other hand, one adult became comatose for twelve hours and semi-conscious for a further thirty-six hours after ingesting half a cup of neat tea tree oil (Seawright 1993). Despite abdominal pain and diarrhoea which persisted for six weeks, this patient survived a dosage estimated at 0.5–1.0 ml/kg body weight. Hence tea tree oil can be considered safer than many other oils, including camphor, wintergreen and eucalyptus oils.

Acute Dermal Toxicity

The acute dermal toxicity test ascertains whether a specific LD_{50} dose will cause problems if applied dermally. Some chemicals may be as toxic dermally as orally. But tea tree oil, as with most essential oils, is not as toxic dermally as it is orally. This is most likely due to the slow absorption of the oil into the body through the skin, thus allowing organs, such as the liver, time to detoxify it and the kidney to eliminate the metabolites of the toxic material.

Tea tree oil's acute dermal toxicity was assessed by using the OECD method 402 on rabbits as in the Australian Tea Tree Oil Toxicology Data Sheet No. 5 (Tea Tree Oil Growers of Australia 1989):

> An undiluted test sample of tea tree oil was applied dermally at the dose of 2 g/kg to a group of 5 male and 5 female albino rabbits (NZ) and held in contact for 24 hours over an approximate skin area of 175 cm^2. Observations were made for any signs of toxicity, weight change and abnormal behaviour. Apart from slight diarrhoea in one of the animals on day 3, no other signs of toxicity, significant weight loss or abnormal behaviour was noted over a two week observation period. (Bolt 1989c).

However, Villar *et al.* (1994) of the National Animal Poison Control Centre in USA, reports cases involving toxicosis with tea tree oil used in veterinary products for animal dermatitis treatment. It appears that the toxicosis occurred after topical or drenching application of inappropriate or erroneous amounts of the products. The symptoms of the toxicosis were ataxia, incoordination weakness, muscle tremors, behavioural disorders and depression. Treatment was in the form of supportive care and treatment of clinical signs. It appears that while most animals can metabolise the components of tea tree oil, in this case it was at a rate that was below the rate of absorption into the blood stream. Thus, it is advisable to administer tea tree oil preparations and indeed any essential oil preparation strictly according to manufacturer's instructions. As for manufacturers, their product should be appropriately tested and clearly and carefully labelled so as to instruct the purchaser that application must only be for affected skin areas and not over total skin area. The same also applies to products for use on humans.

Dermal Irritation

Toxicology also includes the potential for any compound to cause irritation when applied to the skin. Traditionally this testing has been done on animals. Occasionally, however, some compounds are mild enough to be tested on human volunteers. Before this can occur, the substance tested must be shown to have a low irritation rating on animals.

Skin irritation for tea tree oil has been assessed using the following tests:

(1) Draize Acute Dermal Irritation in the Rabbit (OECD method 404) as in the Australian Tea Tree Oil Toxicology Data Sheet No. 5 (Tea Tree Oil Growers of Australia 1989):

Neat tea tree oil (0.5ml) was applied to both intact and abraded skin under occlusion to 6 albino rabbits (NZ) and observed for erythema (redness) and oedema (swelling) of the skin over a 72 hour period. The Draize index for tea tree oil is 5. Pure tea tree oil may cause dermatitis in some users.

According to Bolt (1989a) tea tree oil, with a score of 5 out of a possible 8, places it in the moderate to severe irritant category. However, literature and reports do not support the result. According to Tisserand and Balacs (1995) tea tree oil is listed as a very mild irritant. Perhaps this can be better understood with the suggestion that the Draize test cannot distinguish well between mild and moderate irritants (Altman 1991; Phillips *et al.* 1972). These authors further suggest that some substances considered unsafe by rabbit tests have been proved non-irritating to human skin. With this in mind, further tests both on rabbits and human volunteers were undertaken.

A more recent repeat of the Acute Dermal Irritation test (OECD 404) on rabbits concluded that "tea tree oil was found to be a mild to moderate irritant at 75%, a minimal irritant at 50% and non-irritant at 25% and 12.5% following a four hour semi-occlusive patch application on intact skin, over a fourteen day observation period" (Rural Industries Research and Development Corporation 1996). The maximum dermal irritation index recorded was 2.3 for the 75% oil after forty eight hours.

(2) 30-Day Dermal Irritation in the Rabbit as in the Australian Tea Tree Oil Toxicology Data Sheet No. 5 (Tea Tree Oil Growers of Australia 1989):

Tea tree oil was diluted to 25% in paraffin oil B.P. and applied to the shaved skin of 6 female (NZ) white rabbits on days 1–5, 8–12, 15–19 and 22–30. The irritation index was found to be 2.2 on observation day 2 and 0 on subsequent observation days. Histological examinations were conducted on skin samples. Although not visibly irritant, superficial pathological changes consistent with localised irritation were observed on microscopic examination.

This test result is consistent with the very mild irritant category into which Tisserand and Balacs (1995) had placed tea tree oil. However, in combination with the Draize method result, more human testing is needed to clarify the issue.

(3) Skin Irritancy Potential in Humans—Clinical Trials:
One clinical trial has been completed (Altman 1991) in which 28 human subjects were each subjected to daily occlusive application of a 0.05 ml sorbolene cream containing 1%, 2.5%, 5% and 10% tea tree oil, plus sorbolene alone onto a predetermined pattern on subject's backs over a 21 day period with 15 daily applications (work days only). Results showed only 5 of 28 subjects exhibited slight irritation. Of these 5 subjects, only one reported slight irritation with 1%, one with 2.5%, 2 with 5% and 2 with 10%. With the exception of subject No. 23, the subjects who reported any skin irritation only did so for 1 or 2 days out of the 15 observation days. Subject 23 reported slight irritation on 11 of the 15 days using the 10% tea tree oil. No one reported irritation using sorbolene cream (placebo). According to Altman (1991) the results show that a cream formulation using up to 10% w/w tea tree oil applied for 5 days out of 7 for three consecutive weeks under occlusive dressing, produced mainly infrequent and mild skin irritation in a small proportion of subjects. Further to this he states "Tea tree oil used topically in dermatological formulations up to 10% w/w would appear to pose little risk of skin irritation when applied under normal conditions".

Furthermore a recent clinical trial (Southwell *et al.* 1997) investigated 25 subjects with occlusive patch tests of 25% tea tree oil in soft white paraffin. The patches were applied daily for 21 days (except weekends). The results indicated that no irritation was present in any of the subjects over the 21 days. In the selection trial, 3 subjects were withdrawn because of an allergic response to tea tree oil. This allergic reaction will be discussed more fully later in the chapter. This trial indicated that tea tree oil even with varying concentrations of cineole (1.5%–28.8%) and terpinen-4-ol (22.6–38.8%) is a non-irritating essential oil. Cineole in the past has been suggested as a mucous membrane and skin irritant (Lassak and McCarthy 1983). This trial not only indicates that cineole is not a skin irritant, but also that tea tree oil cannot be regarded as a skin irritant in concentrations up to 25%. This finding was also confirmed by human dermal irritation studies on 306 subjects which showed no irritation to oil at 25% concentration (Main Camp 1998).

Skin Sensitization and Contact Dermatitis

In recent years there have been several reports in dermatological journals suggesting that tea tree oil causes contact dermatitis. Contact dermatitis is usually caused by sensitization to a substance and at a later stage coming in contact with the same substance. Tea tree oil has been regarded as a non-sensitizing substance (Tisserand and Balacs 1995) and this is certainly the case with guinea pigs. Bolt (1989e) carried out a skin sensitization potential trial with guinea pigs (OECD method 406) using tea tree oil as reported in the Australian Tea Tree Oil Toxicology Data Sheet No. 5 (Tea Tree Oil Growers of Australia 1989):

Groups of 20 albino guinea pigs (HA strain) were tested for sensitization potential using the method of Magnusson and Kligman (1969). The induction procedure consisted of 2 intradermal injections and an epidermal induction application under occlusion for 48 hours. Two weeks after the induction applications the animals were challenged with 30% tea tree oil in petroleum jelly. No sensitization reactions were noted. Tea Tree Oil could therefore be recommended in hypoallergenic preparations.

Yet, there appears to be evidence, in humans at least, that tea tree oil can cause varying allergic responses in certain individuals. Some case studies and clinical trial results follow:

(a) A maximization test was conducted on 22 volunteers with a 1% tea tree oil in a petrolatum formula and produced no sensitization reactions (Ford 1988).
(b) A 43-year-old woman and 10-year-old boy were confirmed by patch challenge testing to have contact dermatitis from tea tree oil (Apted 1991). Individual components of the oil were not tested for allergic response.
(c) A 45-year-old man, with long standing atopic dermatitis, applied tea tree oil which exacerbated the condition after a time (de Groot and Weyland 1992). Although the oil components were listed, no concentrations were given. The constituents listed were more consistent with tea tree oil of the cineole genotype. This genotype was once reported to be readily available in Europe and caused allergic responses on other users as well (van der Valk *et al.* 1994). The patient gave a positive patch response when tested with cineole (eucalyptol) in contrast with the findings of Opyke (1975b), Knight and Hausen (1994) and Southwell *et al.* (1997). This chemical variety is not normally used commercially and would not have conformed to ISO 4730: 1996E (International Standards Organisation 1996).
(d) Two occupational contact dermatitis subjects, one a chiropodist, the other a beautician were said to be allergic to tea tree oil, however no challenge patch test were reported. (de Groot and Weyland 1993).
(e) A 33-year-old woman who had used neat tea tree oil previously with no side effects, developed contact dermatitis confirmed by patch challenge testing with tea tree oil. (Selvaag *et al.* 1994).
(f) Knight and Hausen (1994) conducted a trial on seven patients who were treating pre-existing skin conditions with neat tea tree oil and were challenge patch tested with tea tree oil at 1%. All seven reacted positively. Further to this, they tested the major components of tea tree oil at 1%. Six of seven patients reacted to limonene, 5 to α-terpineol and aromadendrene and one to each of terpinen-4-ol, *p*-cymene and α-phellandrene. However, according to Opdyke (1975a), limonene proved negative on 23 subjects tested for sensitization. Furthermore, Knight and Hausen (1994) found tea tree oil to be a weak sensitizer in guinea pigs as only 3 of 10 responded at 30% challenge after 48 hr. It is interesting to note here that the terpinen-4-ol sensitized human had a previous sensitivity to Peruvian Balsam. These authors acknowledged that they were dealing with a population of tea tree oil sensitised patients with damaged skin and also suggested that "other tea tree oil products containing lower concentrations of the oil and used-on healthy skin may cause no sensitivity reactions."
(g) A 53-year-old patient suffered allergic airborne contact dermatitis from several essential oils after extensive use of the oils in wet dressings, baths and room aerosols (Schaller and Korting 1995). Although the oils used and tested included tea tree oil, only lavender, rosewood and jasmine gave positive patch test responses.

(h) Recent investigations (Southwell *et al.* 1997) looked at the skin irritancy of tea tree oil using 28 human subjects with occlusive tea tree oil 25% patch testing over 21 days. Oils of differing cineole (15–28.8%) and terpinen-4-ol (22.6–38.8%) showed no irritancy except for 3 subjects who showed allergic responses to all the different types of tea tree oils used. These 3 subjects were further tested with the individual constituents or constituent blends at concentrations equivalent to their concentration in a 25% formulation. None of the 3 subjects reacted to the monoterpenoids except for one panellist who reacted to α-terpinene only. However, the sesquiterpene hydrocarbon enriched blends caused reactions in all three subjects. This is consistent with Knight and Hausen (1994) who found aromadendrene, a sesquiterpene hydrocarbon, an allergen. Still more research is needed here.

Cytotoxicity using Human Cell Lines

All of the previous assessments of tea tree oil have been acquired from animal and human testing. Just recently, however, cytotoxic testing of the oil has been carried out. The use of cytotoxic (toxic to cell tissue) *in vitro* (test tube) testing has made testing of substances that are quite toxic, more animal friendly and thus more acceptable. The other advantage is that the same tissues can be used to test many substances and provide a more accurate comparison of the relative toxicity of different substances. This approach is also being used for predictive testing for potential allergens (Krasteva *et al.* 1996). According to Söderberg *et al.* (1996), tea tree oil when tested at 100 µg/ml had a cytotoxic effect on epithelial and fibroblast cells for only the first hour of a 24 hour test duration.

More recently, Hayes *et al.* (1997) investigated tea tree oil cytotoxicity with respect to the major oxygenated monoterpenoid constituents terpinen-4-ol, 1,8-cineole and α-terpineol. The oil tested contained terpinen-4-ol 40%, 1,8-cineole 4%, α-terpineol 4% and results showed that α-terpineol > tea tree oil > terpinen-4-ol > cineole for overall cytotoxicity. When tea tree oil was compared with controls mercuric chloride (highly toxic) and aspirin (very low toxicity) on the same cell types (liver, skin, lymph and bone marrow), ratings were mercuric chloride > tea tree oil > aspirin. The authors state that these cytotoxicity results support the use of tea tree oil in topical applications but not for ingestion purposes.

CINEOLE AND OTHER *MELALEUCA* SPECIES' OILS

According to Tisserand and Balacs (1995), 1,8-cineole is non-toxic, non-irritant and non-sensitising. This being the case, one would expect that high cineole content oils would be similar. However, as mentioned earlier, eucalyptus oil, although fitting into this category, is still regarded as potentially dangerous to small children.

Commercial cajuput oil, presumably the "cajeput" (*Melaleuca leucadendron*) of Tisserand and Balacs (1995) and now known as *Melaleuca cajuputi* (see Chapter 14), is normally a high cineole (50–60%) oil. Tisserand and Balacs (1995) state that this "cajeput" oil has an LD_{50} between 2–5 g/kg, is non-irritant and has negligible risk of skin sensitivity.

Niaouli oil, from *Melaleuca quinquenervia*, also a cineole type oil, has been assessed by Aboutabl *et al.* (1996) using ethanol extracts as having an LD_{50} of 147.5 mg/kg while Tisserand and Balacs (1995) indicate that niaouli oil is safe to be used in aromatherapy and is considered non-irritant and non-sensitizing. Linalool and nerolidol, the major constituents in some chemotypes of *Melaleuca quinquenervia*, are also considered to be safe for aromatherapy use (Tisserand and Balacs 1995) and hence non-irritant and non-sensitising (see Chapter 15).

The only *Melaleuca* oil that is currently considered as a potential carcinogen is the oil of *Melaleuca bracteata* which contains approximately 80% methyl eugenol. According to Randerath *et al.* (1984) and Tisserand and Balacs (1995) this oil should be avoided as currently the oil is considered carcinogenic to rodents. Until such time as it is assessed as non-carcinogenic to humans, this advice should be heeded.

CONCLUSION

Although not unduly toxic when ingested, tea tree should continue to be used mainly for topical applications. It would appear that tea tree oil (*Melaleuca alternifolia*) is non-irritating even when used up to 10% (Altman 1991) or 25% (Southwell *et al.* 1997) on unabraded healthy skin. It is not recommended that it be used on sensitive, dermatitis affected or abraded skin for extended periods. Some people do become sensitized to tea tree oil on rare occasions especially those that have a history of allergic responses. Generally, however, as there was little irritation or sensitization with tea tree oil in formulations up to 25%, products of that concentration or less are recommended.

REFERENCES

Aboutabl, E.A., Abdelhakim, G. and Moharram, F.A. (1996) A study of some pharmacodynamic actions of certain *Melaleuca* species grown in Egypt. *Phytotherapy Research*, **10**(4), 345–347.

Altman, P. (1991) Assessment of the skin sensitivity and irritation potential of tea tree oil. Pharmaco Pty Ltd., Sydney, Australia.

Apted, J.H. (1991) Contact dermatitis associated with the use of tea tree oil. *Australas. J. Dermatology*, **32**, 177.

Austteam, (*ca.* 1995) Technical & Material Safety Data Sheet—Tea Tree Oil, Austteam, Lismore, Australia.

Bolt, A.G. (1989a) Acute dermal irritation of tea tree oil in the rabbit. 13th January. *Report for the Australian Tea Tree Industries Association.* Pharmaceutical Consulting Services, Lindfield, Sydney.

Bolt, A.G. (1989b) Skin sensitization potential of tea tree oil in the rabbit. 13th January. *Report for the Australian Tea Tree Industries Association.* Pharmaceutical Consulting Services, Lindfield, Sydney.

Bolt, A.G. (1989c) Acute dermal toxicity limit test of tea tree oil in the rabbit. 20th February. *Report for the Australian Tea Tree Industries Association.* Pharmaceutical Consulting Services, Lindfield, Sydney.

Bolt, A.G. (1989d) Acute oral toxicity of tea tree oil in the rat. 24th April. *Report for the Australian Tea Tree Industries Association.* Pharmaceutical Consulting Services, Lindfield, Sydney.

Bolt, A.G. (1989e) Skin sensitisation potential of tea tree oil in the guinea pig. 13th January. *Report for the Australian Tea Tree Industries Association.* Pharmaceutical Consulting Services, Lindfield, Sydney.

Carson, C.F. and Riley, T.V. (1995) Toxicity of the essential oil of *Melaleuca alternifolia* or tea tree oil. *Journal of Toxicology—Clinical Toxicology*, **33**(2), 264–269.

De Groot, A. and Weyland, W. (1992) Systemic contact dermatitis from tea tree oil. *Contact Dermatitis*, **27**, 279–280.

De Groot, A. and Weyland, W. (1993) Contact allergy to tea tree oil. *Contact Dermatitis*, **26**, 309.

Del Baccaro, M.A. (1995) *Melaleuca* oil poisoning in a 17-month-old. *Vet. and Human Toxicology*, **37**(6), 557–558.

Elliot, C. (1993) Tea tree oil poisoning [Letter]. *The Medical Journal of Australia*, **159**, 830–831.

Ford, R. (1988) Fragrance raw materials monographs. *Food Cosmet. Toxicology*, **26**, 407.

Gurr, F.W. and Scroggie, J.G. (1965) Eucalyptus oil poisoning treated by dialysis and mannitol infusion. *Australian Annals of Medicine*, **14**, 238–249.

Hayes, A.J., Leach, D.N. and Markham, J.L. (1997) *In vitro* cytotoxicity of Australian tea tree oil using human cell lines. *J. Essent. Oil Res.*, **9**, 575–582.

International Standards Organization (1996) Oil of Melaleuca, terpinen-4-ol type (Tea Tree Oil). *International Standard ISO 4730: 1996(E)*, International Standards Organization, Geneva.

Jacobs, M.R. and Hornfeldt, M.S. (1994) Melaleuca oil poisoning. *Clinical Toxicology*, **32**(4), 461–464.

Knight, T.E. and Hausen, B.M. (1994) Melaleuca oil (tea tree oil) dermatitis. *J. Am. Acad. Dermatol.*, **30**, 423–7.

Krasteva, M., Peuget-navarro, J., Moulon, C., Courtellemont, P., Redziniak, G. and Schmitt, D. (1996) *In vitro* primary sensitization of hapten-specific T cells by cultured human epidermal Langerhans cells-a screening predictive assay for contact sensitizers. *Clinical and Experimental Allergy*, **26**, 563–570.

Lassak, E.V. and McCarthy, T. (1983) *Australian Medicinal Plants*, Methuen Sydney, Australia, p. 97.

Magnusson, H.C. and Kligman, A.M. (1969) The identification of contact allergens by animal assay. The guinea pig maximization test. *J. Invest. Dermatol.*, **52**, 268–276.

Main Camp Tea Tree Oil Group Newsletter (1998), *Tea Tree Oil News*, **1**, 3.

Moss, A. (1994) Tea tree oil poisoning [Letter]. *The Medical Journal of Australia*, **160**, 236.

Opdyke, D.L.J. (1975a) Fragrance raw materials monographs (limonene). *Food and Cosmetics Toxicology*, **13**, 825–826.

Opdyke, D.L.J. (1975b) Fragrance raw materials monographs (eucalyptol). *Food and Cosmetics Toxicology*, **13**, 105–106.

Opdyke, D.L.J. (1977) Fragrance raw materials monographs (methyl salicylate). *Food and Cosmetics Toxicology*, **15**.

Phillips, L., Steinberg, M., Maibach, H.I. and Akers, W.A. (1972) A comparison of rabbit and human skin responses to certain irritants. *Toxicology and Applied Pharmacology*, **21**, 369–382.

Randerath, K., Haglund, R., Phillips, D. and Vijayaraj Reddy, M. (1984) ^{32}P-Post-labelling analysis of DNA adducts formed in livers of animals treated with safrole, estragole and other naturally-ocurring alkenylbenzenes 1. Adult female CD-1 mice. *Carcinogensis*, **5**(12), 1613–1622.

Rural Industries Research and Development Corporation (1996) Acute dermal irritation/corrosion of 75%, 50%, 25%, and 12.5% tea tree oil solutions in the rabbit. *Pharmatox Project Report*, T1836.A, p. 7.

Schaller, M.S. and Korting, H.C. (1995) Allergic airborne contact dermatitis from essential oils used in aromatherapy. *Clin. & Exp. Dermatology*, **20**, 143–145.

Seawright, A. (1993) Tea tree oil poisoning—comment [Letter]. *The Medical Journal of Australia*, **159**, 831.

Selvaag, E., Eriksen, B. and Thune, P. (1994) Contact allergy due to tea tree oil and cross-sensitization to colophony. *Contact Dermatitis*, **31**, 124–125.

Smith, A. and Margolis, G. (1954) Camphor poisoning. *American Journal of Pathology*, **30**, 857–869.

Söderberg, T., Johansson, A. and Gref, R. (1996) Toxic effects of some conifer resin acids and tea tree oil on human epithelial and fibroblast cells. *Toxicology*, **107**, 99–109.

Southwell, I.A., Freeman, S. and Rubel, D. (1997) Skin irritancy of tea tree oil. *J. Essen. Oil Res.*, **9**, 47–52.

Stevenson, C.S. (1937) Oil of wintergreen (Methyl salicylate) poisoning. *J. Med. Sc.*, **193**, 772–788.

Tea Tree Oil Growers of Australia (*ca.* 1989) Australian Tea Tree Oil Toxicology Data, Data Sheet **5**, Tea Tree Oil Growers of Australia, Grafton, Australia.

Tisserand, R. and Balacs, T. (1995) *Essential Oil Safety—A Guide for Health Care Professionals*, Churchill Livingstone, Edinburgh. pp. 45–55, 80, 82, 150, 187, 204, 219.

Van der Valk, P.G.M., De Groot, A.C., Bruynzeel, D.P., Coenraads P.J. en Weijland, J.W. (1994) Allergisch contacteczeem voor 'tea tree'-olie. *Ned Tijdschr Geneeskd,* **138**(16), 823–825.

Villar, D., Knight, M.J., Hanson, S.R. and Buck, W.B. (1994) Toxicity of Melaleuca Oil and Related Essential Oils Applied Topically on Dogs and Cats. *Vet. Human Toxicology*, **36**(2), 139–142.

11. TEA TREE OIL IN COSMECEUTICALS: FROM HEAD TO TOE

DON PRIEST

Main Camp Marketing Pty Ltd, Ballina NSW, Australia

INTRODUCTION

On every continent, in many countries, new cosmetics and personal care products containing Australian tea tree oil (oil of *Melaleuca alternifolia*) are being launched at an ever increasing rate.

Since the pioneering launches of personal care products in the 1980's by companies such as Thursday Plantations, Melaleuca Inc. and Dessert Essence, the 90's has seen an explosion of products containing Australian tea tree oil.

Leading the way is the Body Shop, the tea tree oil range of which has been an outstanding success worldwide. Following closely is L'Oreal in France with the successful Ushuata shampoo and body wash.

Numerous other companies, including Clarins, Yves Rocher, Ella Bache, Boots, Aveda, Unilever, Carter Wallace, Colgate, Benckiser, Reckitt and Colman, Rhone Poulenc, Australian Bodycare and Blackmores Laboratories now market cosmeceutical products containing tea tree oil.

The reason for this is that tea tree oil is not only natural, but is one of the few active ingredients that really works. It is a proven natural antiseptic with broad spectrum activity, effective against a wide range of gram positive and gram negative bacteria, yeast, moulds and fungi. At concentrations of 5% and more, it is used therapeutically world wide for antiseptic, antifungal and acne treatment. At lower concentrations, tea tree oil adds efficacy to a wide range of cosmetics and personal care products—products for use from head to toe, and from baby products to treatments for aged and damaged skin.

In addition to its germicidal properties, there have been numerous anecdotal reports of the anti-inflammatory activity of tea tree oil—including reduction of itch and swelling from insect bites, and erythema in pimples and sunburn. In a clinical study on tinea, Professor Ross Barnetson, one of Australia's leading dermatologists, reported the amelioration of symptoms with a 10% tea tree oil cream but not the underlying fungal infection, and suggested that these results were due to an anti-inflammatory effect rather that anti-fungal action (Tong *et al.* 1992).

Investigating the potential for tea tree oil in oral hygiene, researchers at the Dental Faculty of the University of Tennessee have found that tea tree oil suppresses superoxide which is normally released by neutrophils in the presence of bacterial lipopolysaccharide fragments from cell walls. This superoxide release, a natural defence mechanism against invading micro-organisms, can induce a strong inflammatory response, resulting in erythema, swelling, pain, blisters, and initiation of free radicals which cause skin damage.

The germicidal and anti-inflammatory properties, along with excellent solvency, dermal penetration and green image, form the basis for use of tea tree oil in cosmetics and personal care products which typically contain 0.5–5% of the oil.

An important factor in the growth in popularity of tea tree oil is that a guaranteed supply of ultra pure, high quality, pharmaceutical grade tea tree oil is now available from plantations. High quality tea tree oil is crystal clear, highly active, almost non-irritant and mild in odour.

In this chapter, Australian tea tree oil is seen as an extremely versatile natural ingredient. As one of the few natural ingredients that work, Australian tea tree oil is an important ingredient for formulators of cosmeceutical products.

Its potential for use in a wide range of cosmeceutical products is discussed along with rationales for use. These include hair care (including dandruff), acne and problem skin, lips and face, oral care, shave, after sun, hand and body treatments, cleansers, leg and foot products, natural deodorants and preservatives.

The activity of Australian tea tree oil is dose dependant. At higher doses (>5%) it is used as a treatment for a variety of bacterial and fungal infections, and at lower doses (0.5–3%) to control microorganisms and maintain healthy skin, hair and nails.

Consequently, this chapter outlines the widespread use of Australian tea tree oil in products from head to toe which demonstrate the versatility of this unique ingredient.

HAIR CARE

Hair care products with tea tree oil are usually positioned for oily hair (utilising the solvent properties of tea tree oil), or for anti-dandruff and contain 1–5%.

Recent tests have been carried out on Main Camp's pharmaceutical grade tea tree oil by the University of Western Sydney demonstrate a minimum inhibitory concentration (MIC) of 0.5% against *Pitryosporum ovales*, the fungal cause of dandruff.

In formulating a product range, it may be necessary to position shampoos and conditioners as products for prevention, supplemented by higher strength treatment products designed to allow longer contact time with the scalp.

Treatment and preventative products are also marketed for use against head lice (*Pediculus capitis*). Many insects and mites are either repelled or killed by tea tree oil and the solvent properties of tea tree oil if properly formulated, should assist greatly in removal of the eggs.

FACE AND MOUTH

Cosmetics for problem skin utilise the germicidal activity of tea tree oil against *Propionibacterium acnes* in pimples and acne and other bacteria which may be involved in secondary infection—for example *Staphylococcus* species. The anti-inflammatory activity also provides a calming effect, reducing erythema and swelling around pimples. At 5%, effective treatment products can be formulated, while prevention and maintenance products for daily use, such as night cream, moisturisers, toners and cleansers should normally contain 0.5–1%.

Lip balms and sticks containing 1–3% tea tree oil are particularly effective in the prevention and treatment of sore and cracked lips resulting from sun and windburn.

Australian tea tree oil is increasingly being found in oral hygiene products—as a natural anti-plaque agent in toothpastes and mouthwashes. Although taste can be difficult to mask at concentrations over 0.5%, higher levels are predicted for use in products for sensitive teeth due to efficacy against the microflora of the mouth (Walsh and Longstaff 1987; Carson and Riley 1993), anti-inflammatory action and local anaesthetic properties on mucous membranes.

Obviously shave products would benefit from the germicidal and soothing properties of tea tree oil. The addition of 1% tea tree oil to shave creams, sticks, foams and after shave products will sooth irritated skin caused by the action of shaving.

BODY PRODUCTS

After-sun lotions containing 1% Australian tea tree oil are effective in reducing redness and soreness of sunburn. The inclusion of tea tree oil in sunscreen products could enhance its Sun Protection Factor by suppression of erythema, although the benefit of this might be debatable.

The broad spectrum germicidal activity provides excellent cleansing and deodorancy at 1–2% in bar soaps, liquid soaps, bath oils and shower gels. In addition, the antifungal activity assists in the prevention and control of various fungal-based conditions prevalent in hot, humid climates of Asia, Africa and Central and South America.

Australian tea tree oil is also finding application as a natural deodorant in underarm antiperspirants and deodorants. Usually concentrations of 2–3% are required together with suitable perfume type fixatives to reduce the vapour pressure and extend time of effectiveness. Products include aerosols, sticks, roll-ons, deo-sprays, and talcs.

HANDS

Protective barrier creams (1%) and treatment creams for hands and nails (3–5%) are both applications for tea tree oil. In treatment creams, tea tree oil's antifungal and anti-inflammatory properties can be of great benefit to dry and cracked hands and nails.

FEET AND LEGS

The irritation caused by hair removal products for legs and bikini line—shave, wax and depilatory creams—is substantially reduced by the addition or tea tree oil (1–2%).

Powders, creams and sprays containing tea tree oil (1–3%), can provide excellent deodorancy and assist in prevention of 'athletes foot'.

NATURAL PRESERVATIVE

Finally, formulating tea tree oil into a product at 0.5% or greater may eliminate the need to add a preservative. Studies previously reported by the author demonstrated that Main Camp pharmaceutical grade tea tree oil could be successfully used in a range of formulations as a natural preservative at 0.5% and meet the requirements of both the USP and BP Preservative Challenge Tests.

SUMMARY

In summary, tea tree oil is an effective active ingredient due to its broad spectrum anti-microbial properties, anti-inflammatory action and low toxicity. This versatile ingredient has a place in cosmetic and personal care products from head to toe.

It is an active ingredient that happens to be natural—particularly suited to products for problem skin. It is used in treatment products at higher concentration of 3–5% and as a preventative, deodorant and natural preservative at lower concentrations (0.5–34%).

Modern plantation oil is pure, colourless and low in odour. It can be easily perfumed to be virtually undetectable to the consumer or it can be left unperfumed to capitalise on its natural clean, green, environmental image.

REFERENCES

Carson, C.F. and Riley, T.V. (1993) Anti-microbial activity of the essential oil of *Melaleuca alternifolia. Lett. Appl. Microbiol.*, **16**, 49–55.

Tong, M.M., Altman, P.M. and Barnetson, R.St.-C. (1992) Tea tree oil in the treatment of *Tinea pedis. Australasian J. Dermatol.*, **33**, 145–149.

Walsh, L.J. and Longstaff, J. (1987) The anti-microbial effects of an essential oil on selected oral pathogens. *Periodontology*, **8**, 11–15.

12. FORMULATING FOR EFFECT

JAMES S. ROWE

Technical Consultancy Services Pty Ltd, Rockdale NSW, Australia

INTRODUCTION

The use of tea tree oil in Australia spans many centuries. There is evidence to show that the leaves have been used by aborigines for thousands of years for a variety of ailments. When Australia was discovered by the British, log book entries show the leaves were used as an infusion or tea in an attempt to control the scurvy from which the first fleet suffered and hence the name tea tree oil. Unfortunately *Melaleuca* leaves contain no vitamin C but the name remained. In the 1920's scientists became aware of its antiseptic properties and it was issued to Australian Army personnel during the second world war. With the discovery of antibiotics its use declined until recently. It has now been rediscovered as an effective natural antiseptic with a wide variety of uses in the pharmaceutical and personal care industry.

FEATURES OF TEA TREE OIL

Tea tree oil is a natural product, has a broad spectrum of activity and is environmentally safe with a long history of use. It has excellent antiseptic and wound cleansing properties and a low incidence of skin irritation. These features make it a versatile ingredient in pharmaceuticals, cosmetics, toiletries, pet care and household sanitation.

The oil consists of a complex mixture with over one hundred fractions identified, consisting of a mixture of monoterpenes, sesquiterpenes and terpene alcohols. The germicidal effect is mainly due to terpinen-4-ol although other compounds may act synergistically. Research indicates that the anti-microbial activity increases markedly as the terpinen-4-ol concentration increases up to 35%, then marginally to 40% concentration. No further increase in activity is observed at concentrations in excess of 40% terpinen-4-ol.

The Minimum Inhibitory Concentration (MIC) against commonly encountered gram positive and negative bacteria is typically in the range of 0.5–1.0%. Tea tree oil exerts its action by causing structural damage to the cell wall of the organism followed by denaturation of the cell contents. Unlike antibiotics, there is no evidence of genetically acquired immunity and the oil is effective in the presence of blood, pus, necrotic tissue and mucous discharge.

FORMULATION CONSIDERATIONS

Selection of the Oil

Perhaps the most important consideration in the development of products containing tea tree oil is to select an oil of suitable quality. It has been noted that the activity of the oil depends in part on the terpinen-4-ol content. However high cineole levels have been mistakenly associated with irritation of the mucous membranes. The chemical variants of *Melaleuca alternifolia* have been described as low, intermediate and high cineole forms. For optimal activity, only the low cineole level form should be utilised. On standing and after exposure to light and air, the terpenes convert to *p*-cymene by an oxidative process. Anti-oxidants should thus be considered in formulated products. The pure oil is a clear, mobile liquid. If discolouration occurs then inferior distillation or contamination from the holding vessel should be suspected. The presence of weed or other impurities in the harvest will affect the colour. Odiferous compounds may also result from poor distillation techniques.

The factors most affecting the quality of the oil include post harvest factors such as the conditions of the extraction process, the storage conditions of the oil and most importantly, the formulation and storage of the formulated products. It is this latter consideration on which I now wish to focus.

Solubilisation of Tea Tree Oil

Tea tree oil can be solubilised in water using a variety of surfactants. In our laboratories we found it necessary to use four parts of polysorbate 20 to one part of tea tree oil to produce a clear solution on subsequent dilution with water. This is in contrast to about one to one and a half parts of the surfactant polyoxyl 35 castor oil. The method of preparation is of vital importance. The method can be summarised as follows:

1. The tea tree oil is first mixed with the surfactant and allowed to stand for several minutes.
2. Add small aliquots of water gradually with constant stirring. This results in a thick gel due to hydration.
3. Further gradual additions of water result in a lowering of the viscosity and a clear liquid.

Should the initial addition of water be done too quickly or the tea tree oil not adequately mixed, then at best an opalescent solution results.

The addition of polyhydroxy alcohols such as glycerol or propylene glycol can effect solubilisation without having to go through the initial gelling phase. This can be achieved by triturating the glycerol with the mixture of the surfactant and tea tree oil, prior to the addition of water. The addition of alcohol to the formulation greatly enhances solubilisation producing a clear solution. Typically a range of 5 to 10% alcohol is employed.

Effect of Solubiliser on Antimicrobial Activity

The effects of various surfactants and solubilising agents on the anti-microbial activity of a range of disinfectant and antiseptic products have been well documented in the literature. Our investigations aimed to determine the effect of varying the concentration of surfactants on the anti-microbial activity of tea tree oil. The concentration of tea tree oil remained constant at 0.5% and the surfactants were varied to achieve a ration of Oil : Surfactant of between 1 : 1 to 1 : 5 on a weight to weight basis.

1. Polyoxyl 35 castor oil had no effect on tea tree oil activity when used as the solublising agent in the range of concentrations studied (up to 2.5% surfactant with 0.5% oil). All formulations passed the British Pharmacopoeia (BP) Preservative Efficacy Test with the exception of *Aspergillus niger*. All passed the United States Pharmacopoeia (USP) Preservative Efficacy Test.
2. Increasing the polysorbate 20 : tea tree oil ratio from 1 : 1 to 5 : 1, decreased the activity of tea tree. Higher levels of this surfactant result in failure of the BP/USP test for some bacteria, yeasts and moulds.

These findings are in simple aqueous solution. In formulated products it can be anticipated that the situation will be even more complex.

Effect of Oil Concentration on Antimicrobial Activity

In this series the surfactant was polyoxyl 35 castor oil at a constant concentration of 1.0% to solubilise the tea tree oil. The oil concentration was varied between 0.1% and 1.0%. Again simple aqueous solutions were prepared and subjected to a combined BP/ USP test.

A summary of the results is as follows:

Tea tree oil concentration	Result (combined BP/USP test)
0.1%	Fails some bacteria Fails *Candida* and *A. niger*
0.3%	Passes bacteria Passes *Candida* Fails *A. niger*
0.5%	As for 0.3%
1.0%	As for 0.3%

These results indicate that in colloidal systems and formulated products the nature of the surfactant employed is important with respect to the anti-microbial activity of the oil. It appears that polyoxyl 35 castor oil is superior to polysorbate 20 in that less surfactant is required to achieve satisfactory solubilisation of the oil and it has less effect on the anti-microbial activity of the solubilised oil. Simple aqueous solutions of tea tree oil solubilised with polyoxyl 35 castor oil will pass the USP Preservative Efficacy Test at a level of 0.3%. This level of oil passes the BP test with the exception of *A. niger* as there was no two-log reduction in the count in 14 days. Higher concentrations of tea tree oil up to 1.0% do not appear to improve the activity against *A. niger*.

Effect of Other Additives on Antimicrobial Activity

Other commonly used excipients were examined for their effect on the anti-microbial activity of tea tree oil. Simple 0.5% aqueous solutions, solubilised with polyoxyl 35 castor oil, were again used. The materials examined included EDTA, oil of thyme, propylene glycol and butylene glycol. All materials were examined in a range of concentrations, and it was found that none of the materials examined had any observable effect on the anti-microbial activity of the oil.

Partitioning Effect in Formulated Products

It is well accepted that the anti-microbial activity resides in the aqueous phase of a preserved two phase system and is therefore dependant on the equilibrium concentration of the preservative in this phase. It is expected that tea tree oil will partition between the oil and water phases present in a two component system in accordance with the partition coefficient and the relative ratio of oil and water in the system. To estimate the likely concentration in the aqueous phase of a simple two component system, the partitioning behaviour of tea tree oil between water and Crodamol GTCC (Caprylic/Capric triglycerides, a typical oily component of cream formulations) was investigated. The effect of surfactant concentration was also noted.

Initial aqueous solutions were prepared using polyoxyl 35 castor oil. The ratio of surfactant to tea tree oil varied between 1:1 and 1:3 on a weight basis and no surfactant as a control. The solubilised mixtures were mixed with an equal volume of Crodamol GTCC and the final concentration of tea tree oil in the mixture was 0.5%. The mixtures were mixed on a vortex mixture and allowed to equilibrate over 3 days with occasional shaking. The distribution of tea tree oil in the aqueous phase was determined by GC assay of terpinen-4-ol in the aqueous phase after centrifugation. The results were as follows:

Tea tree oil concentration	With surfactant	Without surfactant
Initial	0.5%	0.5%
Final	0.4%	0.01%

With no surfactant present nearly all the tea tree oil had migrated into the oil phase. In the presence of the surfactant the concentration of the tea tree oil in the aqueous phase remains at about 0.4%. The partitioning appeared to be independent of the concentration of the surfactant in the range studied. This has important implications for formulated products. It is thus essential that appropriate microbial evaluation be undertaken for all tea tree oil products. It is recommended that specific organisms be used depending on the use of the product e.g. *Propionibacterium acnes* and *Candida albicans* for acne and antifungal products respectively.

PACKAGING

Perhaps as important as microbial evaluation is the need for stability testing on the final formulation in the proposed pack for marketing. Glass, whenever possible is the most appropriate for tea tree oil products. In low-density polyethylene (LDPE) a loss of terpinen-4-ol occurs at room temperature after 3 months. At 45°C after 3 months, the terpinen-4-ol is barely detected. Even with low concentrations of tea tree oil the oil migrates through LDPE walls and the solvent attacks the external features on the container such as the label print. High-density polyethylene has been reported to be satisfactory for products containing around 20–25% tea tree oil. The maximum concentration acceptable will be a function of the nature of the product and the wall thickness of the plastic. Deformation has also been reported due to reactions between tea tree oil and plasticising resins. If plastic caps are employed it is important to use impenetrable liners.

STABILITY

The Pure Oil

Tea tree oil is reasonably stable at room temperature when stored in brown glass or stainless steel. It is however sensitive to heat, light and air. On storage there is a drop in terpinen-4-ol, α-terpinene and γ-terpinene and an increase in *p*-cymene. High levels of this compound can indicate poor storage, old oil or bad extraction techniques. The use of antioxidants in long term storage should be considered.

Formulated Products

The importance of adequate stability testing on the final product in the proposed marketing container cannot be overstated. Batches should be stored at 4°C, 30°C, 40°C and possibly cycling temperature conditions and monitored at regular intervals for appropriate physical characteristics. Chemically, it is suggested that terpinen-4-ol, cineole and oxidation products are monitored and preservative efficacy testing is performed at the beginning and end of the stability study.

PRODUCT FAILURE

Reasons for products failing stability testing are many and varied. Most product failures in the past have been associated with the use of packaging incompatible with the product components and poor stability testing prior to product launch. This has resulted in several unsatisfactory products from a physical viewpoint. Problems which have occurred include panelling of bottles and migration of tea tree oil through pack walls resulting in deformation of external decorations such as label print. Regulatory authorities have given little attention to chemical or microbial activity of formulated products as most of the product claims which have appeared are limited to simple antibacterial claims rather than treatment for specific conditions such as candidiasis or acne. However, as clinical testing on formulated products increases there will be increasing attention to the role of formulation in the effectiveness of tea tree oil for specific conditions. The main points to consider in product failure from a performance point of view are: (1) inadequate packaging, (2) inactivation of tea tree oil by solubilisers, surfactants and other excipients, (3) the grade of tea tree oil used and (4) the method of manufacture including the order of addition of the components.

CONCLUSION

The inherent anti-microbial properties of tea tree oil are advantageously offered in formulated products. Because of the properties of the constituents of the oil, great care must be taken with solubilisation, formulation, stability and packaging to ensure maximum utilisation of the bioactivity of the oil.

13. TEA TREE OIL MARKETING TRENDS

RICHARD L. DAVIS

G.R. Davis Pty. Ltd. Warriewood, Australia

INTRODUCTION

Tea tree oil is not a new crop in that it was first discovered by Arthur Penfold in the 1920s (Penfold 1925) and commercial production started soon after.

The oil was sold as pure oil and also in products such as in disinfectants and tooth paste. The oil and products were sold mainly, but not exclusively, on the Australian market.

In the 1950's there was a swing away from natural products towards synthetic preparations, and the demand for tea tree oil fell. The oil was still produced and used as before to a limited extent, but was also used to enhance nutmeg oil and as a source of terpinen-4-ol.

The industry maintained a five to fifteen tonnes per annum production rate until the 1980's by when there was a swing back to natural products. It was at this stage that the tea tree oil industry started to develop into a substantial industry. It is because of the very small production in the previous years, the rapid establishment of plantations and the marketing of a wide range of new tea tree products in the early 1980's, that the marketing of tea tree oil can be regarded as marketing a new product.

STAGES OF DEVELOPMENT

There are normally three stages in the production and marketing of new agricultural products. As soon as it becomes apparent that the product can be grown and marketed, there is a rush to produce it (stage 1). This is often followed by over production, where the demand does not increase as rapidly as production (stage 2). The third stage is that the less efficient and the overcapitalised producers fail and production and demand eventually reach an equilibrium. The price of the oil is then no longer set by supply and demand, but by the lower level of cost of production, distribution etc. including profit margin.

At this stage (early 1997), the industry may be near the end of stage 1. Production has increased greatly. Markets have also increased, but, considering the vast area planted, they might not expand at the same rate as production. Therefore the market trend will be towards lower prices, unless some large scale new uses for the oil are found. If the market does develop at a rate sufficient to absorb the greatly increased production, prices will not fall. The average price for tea tree oil on world markets and estimates of annual production over the last 15 years is shown (Figure 1).

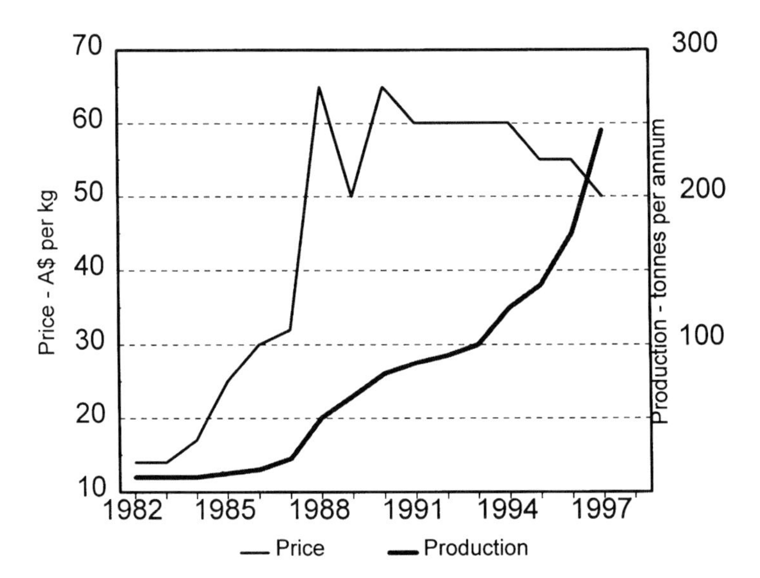

Figure 1 Estimates of tea tree oil average price (in Australian Dollars) and production (in tonnes per annum)

The industry was fortunate in that, in the early years of the new development, one company took a very large proportion of the oil produced. This company has now reduced its intake substantially. The steady growth of the market is encouraging, but the growth in production is such that overproduction is likely. Even a small excess of oil on the market will have a depressing effect on price.

In order to avoid a downward price trend it is necessary to attain a reasonable balance between supply and demand. The desirable and positive way is to increase sales of the oil, with concomitant change in plantation establishment.

MARKETING ADVANTAGES

In recent years tea tree oil sales have increased in sympathy with the increased popularity of natural products. If synthetic products are developed which are very effective in the field now serviced by tea tree oil, the trend might be away from tea tree oil, unless it can be demonstrated that this oil is effective as well as natural. To do this, hard scientific evidence is required. Fortunately, considerable R & D effort is at present being expended to demonstrate the value of the oil.

The history of new products generally is such that unless, after the initial euphoria, they are perceived to be more effective than competing products, they will fade from the market.

Tea tree oil has a number of natural advantages, only some of which are being used by the industry. It has a wide range of uses, mostly antiseptic. It has been demonstrated that the oil is effective in controlling a number of the common human pathogens.

Consequently, in the early stages of its revival it was promoted as a "cure all". While the oil has a good claim to this description, the market is limited for such products. To cover the whole range of its useful capacity, the oil must also be sold for specific purposes where scientific clinical evidence of its efficacy can be given. The market today, except for a small section, is too sophisticated to believe in the "cure all" product.

One of the strengths of tea tree oil is that it is not a narrow single purpose product. Research is showing that different chemotypes of the oil are effective against different pathogens. With considerable variation within the *Melaleuca*-terpinen-4-ol type oil there is scope for producing oils suitable for a number of purposes, thus broadening the potential market for this oil.

SOURCE SPECIES

The industry in general has endeavoured to reduce the genetic potential for tea tree oil by trying to restrict its production to one species, even though at least three species and possibly more, have oil that is almost identical. Even within the one species most commonly grown, limits on the composition of suitable oil, particularly the cineole content, have been advocated, unfortunately successfully, without any scientific evidence to support the limits. This is contrary to the general concept of maintenance of genetic diversity, and eventually must have an impact on good market practice. At this stage of the industry's development it would be inadvisable to restrict genetic potential.

The advantages of working with a genetically variable natural product include the potential for both intra and inter-specific breeding. If the market trend for tea tree oil is to be upwards, a broadening of the range of products is required, and this may be achieved by the development of some specific purpose products.

There are few, if any, major agricultural crops today that have not been substantially modified and improved by selection or breeding. If tea tree oil is to fulfil its potential and so increase its market, selection and breeding will occur. As researchers identify which chemotype of the oil is most effective against a particular pathogen, market development of the chemotype will occur. Chemical modification of the oil is one possible method, but in this age of the veneration of nature, producing a natural oil which does the job is better for marketing purposes. To rule out the opportunity of interspecific breeding by confining tea tree oil to one species would be to limit the industry's potential.

QUALITY OILS

The market for tea tree oil at present has a healthy upward trend, but this will be harder to maintain as the oil displaces other products now being sold. The manufacturers so affected will offer improved products against which tea tree will have to compete. It is therefore essential to keep a positive approach to marketing tea tree oil. Its virtues should be emphasised, yet there is now in the industry an active campaign to rule out top quality oil if it has more than a very small percentage of cineole.

Cineole probably enhances the activity of the oil against certain micro-organisms (Southwell *et al.* 1993), possibly by increasing the oil's capacity to penetrate tissue to reach lesions on which it is acting. Contrary to previous reports, recent scientific evidence suggests that cineole has no significant skin irritant properties. For example Southwell states:

In conjunction with the Skin and Cancer Foundation in Darlinghurst in Sydney, we began investigations into the skin reactions of cineole and tea tree oil. Cineole, under the synonym eucalyptol, had previously been tested on both human and animal subjects and found to cause neither irritation nor sensitisation (Opdyke 1975). Our studies confirmed this finding (Southwell 1996).

Also, further studies show:

There is no evidence to support the current industry misconception that tea tree oils with ultra low levels of 1,8-cineole are superior to oils with higher levels of cineole. Standards specifying a minimum level for terpinen-4-ol and a maximum level for cineole have been interpreted wrongly to mean that terpinen-4-ol and cineole concentrations must be maximised and minimised respectively. Cineole, in concentrations above 15%, is undesirable because of the concomitant decrease in terpinen-4-ol, the active ingredient. However, recent investigations have confirmed that cineole in concentrations up to 15% is not detrimental to the oil, as it is neither a skin irritant nor an antagonist to the activity of the oil (Southwell *et al.* 1996).

As well as the references listed, one has only to consider that there is more than 10 times as much eucalyptus oil as tea tree oil sold in the world, most of which has at least 10 times as much cineole, yet eucalyptus oil is not a skin irritant.

Preliminary work has indicated that some natural tea tree oils of higher cineole content, but still within the ISO Standard, are more effective against certain pathogens. The possibility that antifungal and anti-helminthic activity may be enhanced by cineole has been reported by Southwell *et al.* (1996). If this proves correct, the proponents of very low cineole oils will have painted themselves into a corner.

If the market trend for tea tree oil is to continue upwards, the industry must cease to promote the fear campaign which specifies that one of the main compounds of the oil is dangerous. Such a proposal could be taken up by non tea tree oil competitors to the great disadvantage of the industry.

MARKET OUTLETS

The major outlet for tea tree oil to date is the cosmetic field, as a general disinfectant, and as pure oil for first aid use. Although further development of these markets will occur, expansion is limited because of the difficulty of obtaining regulatory approval to make claims for the efficacy of the oil. Success in this endeavour, which must eventually be achieved, will open new markets for the oil. Registration in various FDA monographs, therapeutic goods registers and pharmacopoeias must be obtained to allow markets to be fully developed. This long and costly process is under way and its outcome will have a substantial effect on market trends.

Nevertheless, to be able to sell oil and make claims in the "pharmaceutical" field will not result in a great increase in markets, unless the oil can be shown to be not only efficacious but at least as good, and preferably better than other products. While the oils' "natural" status is highly regarded in the cosmetic field, the main criterion in the medical field is that the oil must be effective.

STANDARDS

In order to sell pure oil in most countries it is necessary that the oil is in accordance with an authoritative standard. Physical constants for the oil were established in the British Pharmaceutical Codex 1949. Because of the physical standard range, this BPC standard confined tea tree oil to the low cineole form of *Melaleuca* of the terpinen-4-ol type. That is, oil containing not more than about 15% cineole, and the market accepted oil sold on this standard for many years. An Australian Standard was established in 1967 (K175). This standard also established physical constants only, and these were close to the BPC standard. The Australian Standard was revised in 1985 (AS 2782) and the physical constant range altered slightly to accommodate oil from *M. alternifolia* outside the previous range. This enabled the inclusion of natural high quality *M. alternifolia* oil which the previous standard had ruled out. AS 2782 brought in chemical composition for the first time by stipulating a maximum cineole and a minimum terpinen-4-ol content. The standard also increased its definition to include oils from other species of *Melaleuca*, provided the oil conformed to the rest of the standard. This was done because the chemical composition of the oil of some of the species is almost indistinguishable from that of *M. alternifolia*. A further review of the standard in 1995 stipulated limits to a number of compounds in the oil and this same standard has been published by the International Standards Organisation as ISO 4730E (International Standards Organisation 1996). It has also become the Australian Standard. Tea tree oil is marketed in accordance with these standards. While the standards effectively specify the oil, they are also sufficiently broad to allow development of special purpose types of oil.

VALUE ADDED PRODUCTS

While initially the oil was used in the pure form or as a 50% or stronger solution, it was soon recognised that oil in quite low concentration was a very effective antiseptic. This gave rise to a wide range of products containing tea tree oil, and means that a lot of tea tree oil is sold as a minor, though usually the active ingredient of the product. Because of the effectiveness of the oil in low concentration there is good scope for value adding, which is more profitable than selling the oil in the pure form (see Chapters 11 and 12).

Tea tree products available on the market include neat oil, 15% solutions, shampoos, liquid and bar soaps, bath oil, handcreams, antiseptic creams, mouthwashes, toothpastes acne products, tinea creams and powders, vaginitis creams and douches, burn creams and various other health and veterinary care products.

GRADES OF OIL

Any oil which is in accordance with the standards should be sold as good quality Australian tea tree oil. To specify grades within the standard is counter-productive as the instinctive reaction of the buyer is that if there are grades within the standard, some oil within the standard must be inferior.

When specific purpose oils are developed, they will need to be labelled as such, but in the meantime the industry would be ill-advised to downgrade, even by inference, oil which meets the standard. To keep the market on an upward trend, the industry as a whole needs to promote Australian Tea Tree Oil; the essential oil of *Melaleuca* terpinen-4-ol type. To infer that there are lower grades of this oil, an inevitable consequence of stating that there are superior grades, is a bad marketing principle, particularly as there is no credible evidence to show that any oil that conforms with the standard is superior to any other oil that conforms.

To claim that only some ISO standard tea tree oil is pharmaceutical grade is a bad market strategy. Buyers who do not need the oil to be pharmaceutical grade will demand a lower price for non pharmaceutical grade, whereas all oil within the standard is pharmaceutical grade and should be promoted as such.

COMPETITION

The price of tea tree oil on world markets is one of the great challenges confronting the industry. At present, the price is adequate to sustain current methods of production. However, with production increasing rapidly, market expansion needs to keep pace with the rapid increase in planting or prices will fall.

The more successful the industry is in promoting tea tree oil and its products, the more countries will attempt to produce oil to share the market. Thus, even a substantial increase in use of the oil world wide will not necessarily absorb the whole Australian production. In fact, at this stage (early 1997) this is not just a theoretical possibility as production outside Australia is established and growing. Australia, however, has advantages that should ensure pre-eminence on the world market. No other country can claim "country of origin status" and the fact that production methods are so well developed and researched means that Australian-produced oil should remain superior. There is a tendency at present for some sellers to try to increase their market share by denigrating the oil of other producers. It is very unlikely that any substandard oil is offered abroad. If so, market forces will rapidly eliminate this from the market.

If some countries develop large production capacities, the market trend, in terms of price of oil, will almost certainly be downwards. However, a substantial price reduction would result in an upward market trend in volume of oil sold, as one of the restrictions on market expansion at present is the high price of this essential oil.

Some countries have maintained pre-eminence, or even exclusive production of an essential oil, due to peculiar soil or climate conditions or closely guarded plant varieties (e.g. French lavender oil). This is not the case with tea tree oil. Oil is already available from other countries, and seed has been widely distributed. The strategy of establishing a higher standard for a national oil that other countries cannot meet has already been frustrated, at least for the immediate future. There is no sense in setting a higher, i.e. tighter or more restrictive standard, if competing countries can meet it. This is simply counter productive. It again emphasises the futility of offering grades of oil within the standard. If competitors can meet the top grade, then all other grades will be accepted as inferior, and, it is bad marketing if any ISO standard oil is offered as other than finest quality. In any case, it is fundamentally wrong and bad marketing practice to offer grades of a product unless it is quite apparent that there is a difference in quality or efficacy between the grades.

However Australia in particular does have the opportunity to produce unique quality oils.

PRODUCT RESEARCH

The industry is fortunate in that a great amount of research into this essential oil has been, and is still, being carried out. There are already indications that different pathogens are susceptible to tea tree oils of different composition. It is therefore likely that natural oils with the most effective composition for particular purposes can be developed by blending, selection or breeding. The potential for breeding to obtain most effective oils for specific purposes is heavily in Australia's favour. The vast resource of ISO Standard tea tree oil-bearing trees in species other than *M. alternifolia* gives Australia a great advantage in developing specific purpose oils. Research is well under way to capitalize on this advantage.

In 1997 Australia has to recognise that despite discovering and developing tea tree oil, it now, and particularly in the future, has to compete with other countries, particularly the low cost tropical countries. It must be appreciated also that there is no level playing field here. Many of the Asian countries have a cost structure which will allow them to produce tea tree oil at a fraction of our production cost. Furthermore, they might well receive funding from Australia, as well as other developed nations, to help them.

In order to maintain a prominent place in the market, a country must supply the oil at lower cost or higher quality, or provide unique products containing the oil. To compete solely on cost is almost impossible, but it might not be necessary to meet the lowest price if other factors are in that country's favour, although the price cannot afford to be much above the lowest priced good quality oil. Therefore the industry must concentrate on supplying a fairly competitive product, giving excellent service, and ultimately supplying unique specific purpose oils not easily matched by other countries. The market trend not only in degree, but also in direction, will depend on the ability of the industry to achieve these objects. Current R & D in areas of tea tree breeding, agronomy, weed and insect control, quality control, factors effecting oil yield and quality, bioactivity, clinical trials, toxicity, sensitization and product formulation and development will play a big part in meeting these goals. These topics are covered in other chapters in this book.

The industry must continue research into:

1. Development of superior plant stock showing increased growth rate (possibly hybrid vigour) and higher yield of oil. This work is already well advanced.
2. Finding the oil composition most effective against specific pathogens.
3. Developing strains which will produce oil of the required type.
4. Development of more useful products containing tea tree oil.
5. Obtain regulatory approval for tea tree oil and products.

Specific purpose oils might also be produced by modification of oils, either by blending with oils from other species or other genera, or by chemical processes. In the case of chemical modification or blending with other genera the oil might then not qualify to be sold as tea tree oil, but could still be an outlet for tea tree oil in value added products.

If these objectives are achieved, the market trend will be upward.

However, the tea tree oil industry, if it is to consolidate its position on the world market, and thereby attain an upward sales trend, must reverse its present trend of contracting from a broad composition oil to a very narrow composition oil. It must promote all tea tree oil which conforms to the Australian and International Standards as a fine product. It should be mutually supportive and realise that future competition will be from many countries. It must use its great advantage in the diversity of species available to develop superior plant stock exhibiting increased growth rate, increased yield of oil, and specific strains which will be most effective against specific pathogens. Following this path, which is not available to all countries, together with the development of more useful tea tree products which have regulatory approval, will result in a continuing upward trend in the sale of tea tree oil.

REFERENCES

International Standards Organization (1996) Oil of Melaleuca, terpinen-4-ol type (Tea Tree Oil). *International Standard ISO 4730: 1996(E)*, International Standards Organization, Geneva, 8pp.

Opdyke, D.L.J. (1975) Fragrance raw materials monographs (eucalyptol), *Food and Cosmetics Toxicology*, **13**, 105–106.

Penfold, A.R. (1925) The essential oils of *Melaleuca linariifolia* (Smith) and *M. alternifolia* (Cheel). *J. Proc. Roy. Soc. NSW.*, **59**, 306–325.

Southwell, I.A. (1996) Tea tree oil, skin irritancy and bioactivity. *Australasian Aromatherapy Conference, Sydney, March–April 1996*, Sec. 4, 11pp.

Southwell, I.A., Hayes, A.J., Markham, J. and Leach, D.N. (1993) The search for optimally bioactive Australian tea tree oil. *Acta Horticulturae*, **344**, 256–265.

Southwell, I.A., Markham, J. and Mann, C. (1996) Is cineole detrimental to tea tree oil? *Perfumer & Flavourist*, **21**, 7–10.

14. CAJUPUT OIL

JOHN C. DORAN

CSIRO Forestry and Forest Products, Canberra, Australia

INTRODUCTION

Cajuput (also spelt 'cajaput' or 'cajeput') oil, or minyak kayu putih as it is called in Indonesia and Malaysia (minyak is Malay/Indonesian for oil), is obtained from the leaves of *Melaleuca cajuputi* (Plates 4, 18, 19) and some closely related species by steam distillation. Cajuput is most likely a corruption of the Indonesian name for the tree, kayu putih (*kayu*/wood and *putih*/white). The cajuput oil industry has its origins in the natural stands of *Melaleuca cajuputi* subsp. *cajuputi* in the Maluku archipelago of Indonesia, probably in the first part of the eighteenth century. It has been a popular household medication in countries such as India, Indonesia, Malaysia and Vietnam for many generations. The Australian Aborigines used preparations from the leaves of this species in the treatment of aches and pains and inhaled the aromatic vapours from crushed leaves to reduce nasal and bronchial congestion (Aboriginal Communities of the N.T. 1993). It was one of the first products imported to Europe from South East Asia by the Dutch, an activity described in several early eighteenth century publications (Gildemeister and Hoffman 1961 cited in Lowry 1973), because of its reputation as a panacea in the treatment of all kinds of diseases.

PROPERTIES AND USES

Composition

The chemical composition of cajuput oil has been the subject of several studies (Table 1; Ekundayo *et al.* 1987), and wide variation in composition within the principal source species (i.e. within and between populations of *M. cajuputi sens. lat.*) has been found. It appears that different chemical forms of the species predominate in different parts of the natural range (see Table 1) and that their occurrence parallels, in part, the natural distribution of three morphologically variable subspecies, subsp. *cajuputi*, subsp. *cumingiana* (Turczaninow) Barlow and subsp. *platyphylla* Barlow (see below).

Commercial cajuput oil is mainly derived from *M. cajuputi* subsp. *cajuputi*. The oil of this taxon mostly, but not always, contains substantial amounts of 1,8-cineole (3–60%), and the sesquiterpene alcohols globulol (trace-9%), viridiflorol (trace-16%) and spathulenol (trace-30%). Other compounds present usually in significant quantities are limonene (trace-5%), β-caryophyllene (trace-4%), humulene (trace-2%), viridiflorene (0.5–9%), α-terpineol (1–8%), α- and β-selinene (each 0–3%) and caryophyllene oxide (trace-7%). This oil will hereafter be referred to as 'type' oil. The aromatic ether,

Table 1 Variation in the chemical composition of *Melaleuca cajuputi* oil by region and subspecies

Compound	*Australia and Indonesia*[1,2] *ssp.* cajuputi	*Australia*[1,2] *and PNG Chemotype II*[1,2] *ssp.* platyphylla	*PNG Chemotype* 1[1,2] *ssp.* platyphylla	*Indonesia, Thailand, Vietnam*[1,2,3,] *ssp.* cumingiana	*Malaysia*[4] *ssp.* cumingiana	*Vietnam* (1,8-*cineole variant*)[5,6] *ssp.* cumingiana
α-pinene	0.1–1.6	34.3–72.7		1.0–10.3	7.2	3.2–3.8
α-thujene	0–0.2			0–7.8	trace	
β-pinene	0.1–1.6	0.7–1.3		0.1–1.4	1.3	0.8–2.6
α-phellandrene	0–1.9			0.1–2.3	trace	0–0.5
α-terpinene	0–0.3			0–2.8	trace	0.1–0.6
limonene	0.3–5.0	0.1–1.1		0.2–2.8	1.1	4.1–4.8
1,8-cineole	3.2–58.5	0.2–3.2		0.2–6.4		41.1–48.0
γ-terpinene	0–0.8	0.3–1.8		0–19.4	1.0	0–4.6
p-cymene	0.1–0.6	0.2–1.1		0–12.2	0.9	6.8–13.2
terpinolene	0–0.2	0.1–0.8		0–22.3	1.1	0–5.9
linalool	0–0.9	0.1–0.2		0.1–2.3	trace	3.4–3.6
α-elemene					3.2	0–0.1
β-elemene					3.3	0.1–0.3
β-caryophyllene	0.3–3.7	1.9–13.8	0.1–0.7	6.6–43.6	9.2	2.1–2.5
terpinen-4-ol	0–0.5			0–5.6	0.7	1.5–1.6
aromadendrene	0–3.2	1.3–8.8		0–4.8		0–0.1
humulene	0.1–1.8	0.9–7.3	0.1–0.6	3.6–13.5	4.9	0–1.6
viridiflorene	0.5–8.6	0.1–3.3		0.2–2.6		
α-terpineol	0.9–8.3	0.5–3.3		0.4–2.6	0.7	8.7–9.8
β-selinene	0–2.7	0–4.7		0.4–3.8	1.7	0–1.5
α-selinene	0–3.0	0–4.1		0.5–4.8		0–1.5
bicyclogermacrene	0–3.9			0–3.8		
δ-cadinene				0–4.7		
caryophyllene oxide	0.1–7.3	0.1–1.8		0.6–7.3		0–0.3
globulol	0.2–9.2	0.4–6.1		0–1.7		
viridiflorol	0.2–15.6	0.1–2.3		0–1.0		
$C_{15}H_{26}O$	0–9.0	1.2–3.4		0–0.3		
elemol					2.3	
spathulenol	0.4–30.2	0.2–2.5	4.6–9.0	0–8.3	0.6	
isospathulenol					1.7	
γ-eudesmol	0–1.8			0–0.6		0–0.6
α-eudesmol	0–6.4			0–1.6		0–0.7
β-eudesmol	0–6.7				3.2	0–0.7
platyphyllol			21.0–80.0			
cajeputol			0.6–51.0	2.0–17.5[a]	17.0	

[a] reported from Indonesian oil[1,2] only; [1] Brophy and Doran (1998); [2] Brophy and Doran (1997); [3] B. Caruhapattana (1994); [4] Yaacob *et al.* (1989); [5] Motl *et al.* (1990); [6] Todorova and Ognyanov (1988).

cajeputol, appears to be absent from the oil of this subspecies. Oil yield ranges from 0.4% to 1.2% (w/w%, fresh weight).

The essential oil of *M. cajuputi* subsp. *cumingiana* is highly variable in character. The main components reported in oil from Indonesia, Malaysia, Thailand and Vietnam (non- or low-1,8-cineole forms) are γ-terpinene (0–19%), and terpinolene (0–22%). These are accompanied by lesser amounts of the hydrocarbons α-pinene (1–10%), α-thujene (0–8%), α-phellandrene (trace-2%), α-terpinene (0–3%), limonene (trace-3%) and *p*-cymeme (0–12%) as well as the ether 1,8-cineole (0–6%). Oxygenated monoterpenes are represented by terpinen-4-ol (0–6%) and α-terpineol (trace-3%). The major sesquiterpenes detected in the oil are α-, and γ-eudesmol (0–2%), caryophyllene oxide (0–7%), α- and β-selinene (0–4%), viridiflorene (0–3%), humulene (4–14%), aromadendrene (0–5%) and β-caryophyllene (7–44%). There are only small amounts (up to 2%) of the aromatic ether, cajeputol, in oil from Thailand and Vietnam but much larger proportions in oil from Kalimantan, Indonesia and Malaysia (up to 18%). Oil yield ranges from 0.3% to 0.5% (w/w%, fresh weight).

Two separate reports of the composition of cajuput oil from Long An Province of Vietnam (Todorova and Ognyanov 1988; Motl *et al.* 1990) showed typical commercial oils with 1,8-cineole in abundance (41–48%) along with smaller amounts of α-terpineol and other mainly monoterpenoid compounds. Presumably these oil samples were extracted from specimens of subsp. *cumingiana*. However, because of the wide-spread cultivation of *M. cajuputi* over some centuries for its oil, the potential transport of seed of subsp. *cajuputi* across the natural boundaries separating subspecies cannot be ignored.

The essential oils of *M. cajuputi* subsp. *platyphylla* occur in two chemotypes. Typically, the oils contain significant quantities of α-pinene (34–73%), with lesser amounts of 1,8-cineole (0.2–3%), γ-terpinene (trace-2%), *p*-cymene (trace-1%), terpinolene (trace-1%), β-caryophyllene (2–14%), aromadendrene (1–9%), humulene (1–7%), viridiflorene (trace-3%), caryophyllene oxide (trace-2%), globulol (trace-6%), viridiflorol (trace-2%), spathulenol (trace-3%) with an absence of cajeputol. Oil yield of this chemotype is 0.1–0.4% (w/w%, fresh weight). A second chemotype, recently reported from one site in Papua New Guinea (Brophy and Doran 1997), contains the β-triketone, platyphyllol (21–80%), and cajeputol (0.6–51%). The oil yield of this chemotype ranges from 0.2% to 0.6% (w/w%, fresh weight).

The large variation in the chemical composition of commercial cajuput oil is partly a reflection of the substantial natural variation in the oil of *M. cajuputi* but is due also to the use of somewhat similar oil from alternative species (e.g. *M. quinquenervia*, *Asteromyrtus symphyocarpa*) and the common industry practice of blending cajuput oil with oils and individual compounds from other sources (Penfold and Morrison 1950; Lassak 1996).

Colour and Odour

Cajuput oil is mainly a pale yellow mobile liquid sometimes with a greenish-bluish tint. Suggested reasons for this green–blue colouration include the presence of azulene compounds in the higher boiling fractions of the oil, distillation using copper materials, copper chelation with constituents like cajeputol (3,5-dimethyl-4,6-di-O-methylphloroacetophenone) and the addition of suitable dyes to obtain the "correct" colour (Penfold

Table 2 Physicochemical properties of cajuput oil

Properties	*Commercial oil*[1]	*Commercial oil*[2]	*Crude oil*[3]	*Crude oil*[3]
Specific gravity (@15°C)	0.917–0.930	0.915 to 0.926	0.919	0.922
Optical rotation	up to −3°40'	−4°0' to −1°0'	−1°36'	−0°54'
Refractive index (@20°C)	1.466 to 1.472	1.464 to 1.472	1.4703	1.467
Solubility	Soluble in 1 vol. and more of 80% alcohol	Soluble in 1 vol. and more of 80% alcohol	Soluble in 1 vol. and more of 80% alcohol	Soluble in 1 vol. and more of 80% alcohol

[1] Gildemeister and Hoffmann cited in Penfold and Morrison (1950).
[2] Guenther cited in Penfold and Morrison (1950).
[3] Penfold and Morrison (1950).

and Morrison 1950; Lowry 1973). Some cajuput oils with significant amounts of the eudesmols solidify after extraction to a whitish paste. The odour of commercial oil is rather penetrating with a camphoraceous-medicinal aroma, similar to but milder and more fruity than cineole-rich *Eucalyptus* oil (Lawless 1995).

Physicochemical Properties

Reports of the physicochemical properties of cajuput oil are summarised in Table 2.

Uses

Cajuput oil is classified as non-toxic (rodent LD_{50} of 2–5 g/kg, Tisserand and Balacs 1995) and non-sensitizing, although skin irritation may occur at high concentrations (Lawless 1995).

Lassak and McCarthy (1983) described the medicinal uses of cajuput oil:

> The oil is used internally for the treatment of coughs and colds, against stomach cramps, colic and asthma; the dose is one to five drops. It is used externally for the relief of neuralgia and rheumatism, often in the form of ointments and liniments. External application of a few drops on cotton wool for the relief of toothache and earache.

The oil is also reputed to have insect-repellent properties; it is a sedative and relaxant and is useful in treating worms, particularly roundworm, and infections of the genito-urinary system. It is also used as a flavouring in cooking and as a fragrance and freshening agent in soaps, cosmetics, detergents and perfumes (Sellar 1992; Lawless 1995).

The antimicrobial constituents of cajuput oil have been identified as including 1,8-cineole, (−)-linalool, (−)-terpinen-4-ol, (±)-α-terpineol (Nguyen Duy Cuong *et al.* 1994) and 3,5-dimethyl-4,6-di-O-methylphloroacetophenone (or cajeputol) (Lowry 1973).

The use of platyphyllol and similar compounds as sunscreens, bactericides and fungicides has been patented (Joulain and Racine 1994).

SUPPLY SOURCES, QUALITY AND PRICE

Supply Sources

The two principal centres of production of cajuput oil are Indonesia and Vietnam. In Indonesia, there are two sources of 'type' oils. Estimates of production from the natural stands on the Maluku archipelago suggest that a total of some 90 tonnes of oil are produced annually on Buru, Ceram, Ambon and adjacent islands (Gunn *et al.* 1996). Production from an estimated 9,000 ha of Government-owned plantation on Java, established using seed from trees from the Maluku Islands, amounted to approximately 280 tonnes in 1993 (Ministry of Forestry 1995). A minor source of production in Indonesia is southeastern Irian Jaya, where a cineole-rich cajuput oil is produced from *Asteromyrtus symphyocarpa*, a species that was until recently included as part of the *Melaleuca* genus (Craven 1989). No published figures are available on how much of this oil type reaches the market. Production figures for southern Vietnam are not readily available but were estimated by Motl *et al.* (1990) to be in the order of 100 tonnes per annum.

Annual world production of cajuput oil is impossible to quantify accurately, but it would appear to be in the order of 600 tonnes. With a distillery-gate price in Indonesia of about US$7.20 per kg of oil, this indicates an industry value of US$4.3 million before further processing.* However, recent communications with an international buyer of cajuput oil suggest a production of more than twice this figure.

Quality and Prices

There is no accepted International Quality Standard for cajuput oil. As a result, commercial cajuput oils may differ very significantly from one another in a number of characteristics (e.g. chemical composition, colour, smell—see Properties and Uses). The method of grading oil of *Melaleuca cajuputi* subsp. *cajuputi* in the Maluku archipelago of Indonesia, the home of the industry, is described below.

Cajuput oil is purchased and sold in three grades on the Maluku islands (Gunn *et al.* 1996). Grading depends on the location of harvest. Grade 1, the best quality oil, comes from hillside operations and fetches US$8.80–9.60 per kg* at the farm gate. Grade 2, the medium quality class, comes from low-elevation sites and is priced at about US$7.20 per kg*. Gunn *et al.* (1996) were unable to confirm the origins of Grade 3 oil, but presumably this comes from stands of *M. cajuputi* which have a high percentage of trees that produce oil of very low cineole content, such as Gogoria and Wai Geren on Buru Island (Walton 1996). Chemical analysis of oil from stills at various sites showed that the 1,8-cineole percentage of Grade 1 oil was indeed greater than that of Class 2 oil (Table 3).

Plantation-grown oil from Java has an average cineole content of 55% and sold at the factory gate for US$9.40 per kg* in 1997 (Ansorudin 1997). The composition of a sample of oil from a distillery in central Java is given in Table 3.

*These estimates were made in 1996/97 before the devaluation of Asian currencies. In 1998 the distillery-gate price in Indonesia had dropped to about US$1.50 per kg of oil or about 20% of its previous value.

Table 3 Composition of the three grades of cajuput oil found in the market and in the villages on Buru Island (2 grades given here) in comparison with oil from the plantations in Central Java. Quality is partly based on the abundance of 1,8-cineole in the oil which appears to be associated with the locality of harvest in the Maluku Islands

Principal compounds in the oil	*Grade 1 Buru merchant*	*Grade 2 Buru merchant*	*Grade 3 Buru merchant*	*Grade 1 NW Buru villager*	*Grade 2 NE Buru villager*	*Grade 2 Java plantation*
α-pinene	2.1	2.5	3.1	2.1	19.5	3.8
α-thujene	0.3	0.9	1.2	0.2	—	0.8
β-pinene	1.1	0.9	1.1	1.4	8.6	2.5
limonene	5.6	4.9	4.8	5.2	17.4	6.9
1,8-cineole	62.8	41.6	34.0	66.5	21.5	50.7
γ-terpinene	1.2	7.4	5.0	0.9	8.7	3.1
p-cymene	1.3	3.5	5.7	0.5	3.0	1.4
terpinolene	0.6	1.0	0.5	0.3	4.1	1.5
β-caryophyllene	3.7	6.9	7.4	3.3	2.8	4.9
aromadendrene	0.9	1.4	1.7	0.7	1.2	0.9
humulene	1.8	3.9	0.3	1.9	0.1	2.3
viridiflorene	4.5	3.1	2.5	3.8	1.7	3.7
α-terpineol	4.5	3.0	2.4	3.8	1.9	3.8

SOURCE PLANTS

Botanical and Common Names

Family: Myrtaceae

Principal source plant
Melaleuca cajuputi Powell (syn. *M. leucadendron*, *M. minor*)
Common names: kayu putih (Indonesia), cajuput, swamp tea tree, paperbark (Australia), punk tree (USA)

Minor source plants
Asteromyrtus symphyocarpa (F. Muell.) Craven (syn. *Melaleuca symphyocarpa*)
Common name: liniment tree (Australia)
Melaleuca quinquenervia (Cav.) S.T. Blake
Common names: broad-leaved paperbark, belbowrie (Australia), niaouli (New Caledonia)

Description and Distribution of *M. cajuputi*

Melaleuca cajuputi is usually a single stemmed, small tree up to 25 m tall, although it may reach 40 m and 1.2 m in diameter in some situations (Plates 18, 19). It carries dense, dull green foliage with grey to white papery bark. The exact limits of the natural range of *M. cajuputi* are not known as the species has been cultivated in Asia for more than 100 years. The approximate boundaries are given in Figure 1. The latitudinal range is 12°N–18°S and the range in altitude 5–400 m. Three subspecies are recognised, subsp. *cajuputi*,

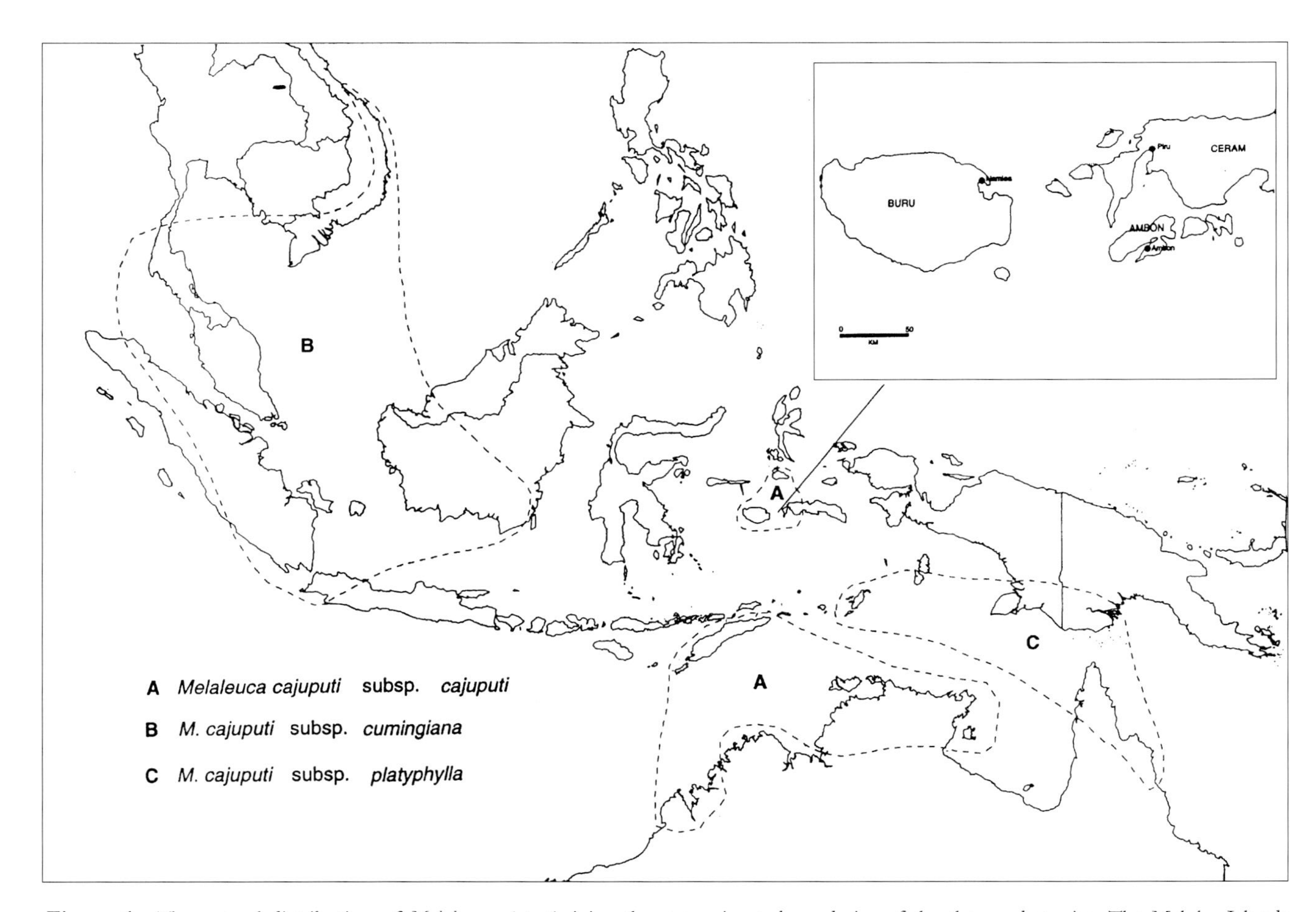

Figure 1 The natural distribution of *Melaleuca cajuputi* giving the approximate boundaries of the three subspecies. The Maluku Islands of Ambon, Buru and Ceram, the home of the cajuput oil industry, are shown in greater detail in the insert to the figure

subsp. *cumingiana* (Turczaninow) Barlow and subsp. *platyphylla* Barlow (Craven and Barlow 1997) and the approximate boundaries of their distribution are given in Figure 1.

The western portion of the distribution, taking in several countries in South East Asia, is geographically disjunct from the Australasian distribution and represents the only natural occurrence of *Melaleuca* beyond the Australasian geophysical boundary (Barlow 1988). Differences in genetic structure between populations east and west of Wallace's line based on allozyme variation led Lum (1993) to the hypothesis that *M. cajuputi* has spread naturally from Australia into South East Asia, aided by the species' propensity for invading disturbed sites.

Ecology of *M. cajuputi*

This is a species primarily of the hot humid climatic zone. Mean annual rainfall ranges from 600–4000 mm with a strong monsoonal pattern and a dry season (consecutive months of <40 mm) of up to 8 months. The species grows in a wide range of situations but most stands are found on low swampy coastal plains often on heavy-textured black soils that are subject to flooding. An exception is populations in Indonesia on the Maluku Islands of Buru, Ceram and adjacent islands. Here extensive and mostly pure populations extend away from the coast to inhabit infertile, gravelly ridges with a subsoil of pale orange to red clay. In wet swamps *M. cajuputi* forms pure forest or mixed open-forest or woodland where it is commonly associated with other tea trees and various eucalypts.

Silvicultural Features and Utilisation

Melaleuca cajuputi is a moderately fast growing tropical tree adapted to both waterlogged and well-drained soils. In addition to cajuput oil, it is a useful source of wood (fuel, posts and poles, household products, construction materials and boat building) and honey and provides shade, shelter and wildlife habitat. *M. cajuputi* is resistant to fire and tolerates exposure to wind. It has the ability to compete with weeds and will regenerate in *Imperata* grassland. It has an extensive root system, and on soils subject to prolonged waterlogging it develops aerial adventitious roots which can form buttresses on the lower trunk. It will coppice and has the ability to reproduce from suckers. Doran and Turnbull (1997) is a useful source of further information.

Genetic Resources

Melaleuca cajuputi remains relatively common throughout its wide geographic range. Genetic variation in many traits is likely, given the range of environments where the species is found and large variation in morphology and oil characteristics. Careful selection of well-adapted and fast-growing seed sources of the high-cineole chemotype will be paramount for plantation development aimed at cajuput oil production. The Australian Tree Seed Centre of CSIRO Forestry and Forest Products, in collaboration with the Indonesian Forest Tree Improvement Research and Development Institute, is assembling range-wide seed collections of *M. cajuputi* suitable for this purpose (Doran and Gunn 1994; Gunn *et al.* 1997).

OIL PRODUCTION

Information on current practices in the cajuput oil industry in Vietnam is not readily available. Discussion here focuses on production methods in Indonesia.

Oil Production from Natural Stands in Indonesia

Traditional land-owners on the Maluku Islands of Buru, Ceram and Ambon are the principal producers in Indonesia of oil from natural stands of *M. cajuputi*.

Most of the oil is produced on Buru Island, a mountainous island of volcanic origin with contrasting vegetation cover. Most ridges and slopes of the northern coastline and those along the major Wai Apu River flowing to the east coast have sparse vegetation comprising of open woodlands and low shrublands of *M. cajuputi* (altitude 30–400 m). There are some 100,000 ha of this vegetation type available for oil production on Buru. The reddish soils of the slopes are frequently gravelly, shallow and slightly to moderately acidic although some populations of *M. cajuputi* on Buru are found on alluvial soils of impeded drainage, which is a more typical environment for this species in Australia and Papua New Guinea. All Buru *M. cajuputi* populations visited during a joint Indonesian/Australian seed collection expedition in 1995 showed evidence of harvesting for essential oil and the ravages of frequent fires (Gunn *et al.* 1996).

Melaleuca cajuputi is also found in western Ceram where it is harvested for oil (Plate 20). The species occurs as an almost pure, continuous stand of some 150,000 ha along the Hoamoal Peninsula. Scattered populations occur elsewhere and also on the three islands between Ceram and Buru, namely Boana, Kelang and Manipa. The distribution of *M. cajuputi* on Ceram is associated with lowland plains and low undulating mountain ridges between 30–150 m above sea level and the soil type resembles that observed on Buru (Gunn *et al.* 1996).

There appear to be only a few scattered stands of *M. cajuputi* on Ambon Island which is wetter and generally more fertile than the other islands. A typical stand occurs on an undulating ridge (altitude about 60 m) near Mt Salahutu on Hitu Peninsula and is relatively small (*ca.* 5 ha). It is comprised of straggly low trees to 4 m, many of which had been coppiced during leaf harvesting for oils when visited in 1995 (Gunn *et al.* 1996).

Leaf Harvesting and Distillation Methods

The harvesting of leaves is a family operation with groups of 2 to 6 people involved in the sequential harvesting of family holdings of some 200 ha of *M. cajuputi*. Coppice growth at 1–2 m tall (6–12 months old) is cut with machette and leaves stripped to fill 20 kg baskets (Plate 20). A skilled cutter can harvest seven baskets per day. The dry season months of May to August are the preferred time of harvest because of reputedly better yield at this time but harvesting may take place throughout the year. After transport to the family still, some producers spread the leaves in the sun for a few days to reduce leaf moisture content.

The still is usually a permanent fixture made from mostly local materials. The distillation vessel (pot) is commonly made from planks butted together with cajuput bark used as a sealant. The lid is made of wood whilst the condenser which comprises a dome and single pipe, sometimes with baffle, is made of mild brass and is purchased by the still owner. Still capacity is usually about 160 kg of dried leaves and cooking time extends for 8 hours. Yield per cook from these stills approximates 3 kg of oil. Unconfirmed estimates (Gunn 1995) suggest there are about 100 family stills operating on Buru, 10–12 on Ceram and one or two operating on each of the islands of Boano, Kelang, Manipa and Ambon.

Oil Production from Plantations in Indonesia

Early attempts to produce cajuput oil on Java and Sumatra, presumably using plants native to that part of the species range, failed because of poor oil quality (Penfold and Morrison 1950). It was not until 1926, when plantations were established at Ponorogo in East Java using seed from Buru Island, that the industry became established on Java (Darmono 1995). Subsequent plantings in East, Central and West Java used seed from Ponorogo. The present extent of *M. cajuputi* plantations on the island is estimated to be 9,000 ha under the control of Perum Perhutani (Forestry Dept) (Ministry of Forestry 1995). Perum Perhutani runs the 12 distilleries on Java. There are 4 major and 8 minor factories producing about 280 tonnes of cajuput oil per year from Government plantations.

Propagation, Planting and Husbandry

Propagation is usually by seed. There is an average of 2,700,000 viable seeds/kg. The seed often germinates readily but the tiny seedlings are easily damaged by overhead watering or rain, or may be killed if the sowing mix dries. Vietnam has adopted the 'bog' technique of watering to avoid these problems in propagating *M. cajuputi*. This involves standing the germination tray permanently in water so that moisture soaks up to the surface which is constantly moist but not flooded. Seed is sown evenly over the surface at the recommended density. An inflated plastic bag is fitted over the container to maintain a moist environment. Once the germinants are sturdy enough to withstand overhead watering (*ca.* 4 weeks), the container is removed from the water and handled normally. The risk of fungal disease is high, so good hygiene is essential. After germination the tiny seedlings can be slow to develop at first, presumably while the roots establish. Once underway, however, they grow quickly and their total nursery period is similar to other fast-growing species like eucalypts. Plantations have also been established with small stump plants, and *M. cajuputi* can be reproduced vegetatively from stem and branch cuttings (von Wulfing *et al.* 1943).

Plantations are established on usually degraded lands using unimproved seedlings at an initial stocking of 5,000 stems per ha. These trees are allowed to grow-on for four years and may be intercropped during the first two years with cassava, corn and peanuts. They are then cut off at 1.1 m above ground level in the first harvest of essential oils. Thereafter the plants are visited annually when coppice shoots greater that 1 cm in diameter are selectively harvested and leaves and twigs stripped into hessian bags for transport to the distillery. In central Java some harvesting for oil production takes place throughout the year. However, peak production is during the period of June to October which coincides with the best months for oil recovery from the leaves.

Yields and Distillation Methods

One hectare of plantation produces about 7.5 tonnes of cajuput leaves annually which in turn produces about 60–65 kg of oil. The industry is a great employer of labour. In one operation alone in Central Java based on 3,200 ha of plantation, 300 local people are engaged in harvesting leaves and a further 70 people are employed at the distillery (Ministry of Forestry 1995).

The cajuput distillation plant of the Gundih forest district, located at Krai in Central Java, is an example of one of the four major plants operated by Perum Perhutani (Forestry Dept). The distillery operates eight, 0.9 tonne capacity pots fed by a steam boiler fuelled by the spent leaves of earlier distillations. A four-hour distillation time is standard. Output for 1993 was 78 tonnes of oil from nearly 9 million tonnes of leaves and twig or a recovery rate of 0.85% (Ministry of Forestry 1995). The composition of a representative sample of cajuput oil from the Krai distillery is given in Table 3.

NEW DEVELOPMENTS

Many of the plantations on Java are considered to be past their prime productivity, because of lack of vigour and consequently low oil yields (Ansorudin 1995). Several factors may be contributing to these low yields, including the genetic quality of the original planting stock, physiological age and the depletion of nutrients through the repeated harvesting of foliage. Perum Perhutani (Forestry Dept) is making plans to gradually replace the existing plantation resource, commencing about 1998. With the interest in establishing new plantations of *M. cajuputi* for oil production has come the interest in increasing the amount and value of oil produced per hectare through selecting, breeding and using genetically better trees. Presently, there are no programs in Indonesia to provide improved planting stock.

The joint Australian (CSIRO)/Indonesian (Agency for Forest Research) *M. cajuputi* seed collecting and oil screening expedition to the Maluku islands in December 1995, supported by COSTAI (Collaboration on Science and Technology Australia/Indonesia), has provided the genetic base for a comprehensive tree improvement program on *M. cajuputi* in Indonesia. Further funding is now being sought to implement a breeding strategy and plan to ensure that planting stock of improved oil yielding capacity is available for the replanting program.

ACKNOWLEDGEMENTS

Much of the data on the cajuput oil industry in Indonesia was collected by two of my colleagues, Brian Gunn and Maurice McDonald. They were involved in seed collecting activities in the Maluku Islands in 1995 in collaboration with counterparts from the Forest Tree Improvement Research and Development Institute, Yogyakarta. Anto Rimbawanto and his father, S. Darmono, assisted with details of development of the cajuput oil industry on Java. I wish to thank Alan Brown, Maurice McDonald, Joe Brophy and Geoff Davis for helpful comments on this manuscript.

REFERENCES

Aboriginal Communities of the Northern Territory (1993) *Traditional Aboriginal Medicines in the Northern Territory of Australia*. Conservation Commission of the Northern Territory of Australia, Darwin.

Ansorudin, M. (1995) Perum Perhutani, K.P.H. Gundih, Indonesia. Personal communication.

Ansorudin, M. (1997) Perum Perhutani, K.P.H. Gundih, Indonesia. Personal communication.

Barlow, B.A. (1988) Patterns of differentiation in tropical species of *Melaleuca* L. (Myrtaceae). *Proc. Ecol. Soc. Aust.*, **15**, 239–247.

Brophy, J.J. and Doran, J.C. (1997) Essential Oils of Tropical *Asteromyrtus*, *Callistemon* and *Melaleuca* Species: in search of interesting oils with commercial potential. *ACIAR Monograph No. 40*, ACIAR, Canberra.

Brophy, J.J. and Doran, J.C. (1998) University of NSW and CSIRO Forestry and Forest Products. Unpublished results.

Caruhapattana, B. (1994) Royal Forest Department, Bangkok, Thailand. Personal communication.

Craven, L.A. (1989) Reinstatement and revision of *Asteromyrtus* (Myrtaceae). *Australian Systematic Botany*, **1**, 375–385.

Craven, L.A. and Barlow, B.A. (1997) New taxa and new combinations in *Melaleuca* (Myrtaceae). *Novon*, **7**, 113–119.

Darmono, S. (1995) Forestry consultant, Bogor, Indonesia. Personal communication.

Doran, J.C. and Gunn, B.V. (1994) Exploring the genetic resources of tropical melaleucas. *FAO Forest Genetic Resources Information*, **22**, 12–24.

Doran, J.C. and Turnbull, J.W. (1997) Australian Trees and Shrubs: Species for Land Rehabilitation and Farm Planting in the Tropics. *ACIAR Monograph No. 24*, ACIAR, Canberra.

Ekundayo, O., Laakso, I. and Hiltunen, R. (1987) Volatile components of *Melaleuca leucadendron* (cajuput) oils. *Acta Pharmaceutica Fennica*, **96**, 79–84.

Gunn, B. (1995) CSIRO Forestry and Forest Products. Personal communication.

Gunn, B., McDonald, M. and Lea, D. (1996) Seed and leaf collections of *Melaleuca cajuputi* Powell in Indonesia and northern Australia, November–December 1995. Unpublished report of Australian Tree Seed Centre, CSIRO Forestry and Forest Products, Canberra.

Gunn, B.V., McDonald, M.W., Lea, D., Leksono, B. and Nahusona, J. (1997) Ecology, seed and leaf collections of Cajuput, (*Melaleuca cajuputi*), from Indonesia and Australia. *IPGRI Plant Genetic Resources Newsletter*, No. 112, 36–43.

Joulain, D. and Racine, P. (1994) Use of derivatives of 6,6-dimethyl 2-acylcyclohex-4-ene-1,3-dione as sunscreens and bactericide and fungicide compositions. *Eur. Pat. Appl. EP* 613,680 (Cl. A61K7/42). (*Chem Abs* (1995) 121: 263279x.).

Lassak, E.V. (1996) Phytochemical Services, Mudgee, Australia. Personal communication.

Lassak, E.V. and McCarthy, T. (1983) *Australian Medicinal Plants*. Methuen, Australia.

Lawless, J. (1995) *The Illustrated Encyclopedia of Essential Oils*. Element Books Ltd, Shaftesbury.

Lowry, J.B. (1973) A new constituent of biogenetic, pharmacological and historical interest from *Melaleuca cajeputi* oil. *Nature*, **241**, 61–62.

Lum, S.K.Y. (1993) Dispersal of Australian Plants Across Wallace's Line: a Case Study of *Melaleuca cajuputi* (Myrtaceae). Unpublished Ph.D dissertation. University of California, Berkeley.

Ministry of Forestry (1995) A guide for field visits, FAO/Govt. of Indonesia Expert Consultation on Non-wood Forest Products, Yogyakarta, 17–27 January 1995. Unpublished document. Ministry of Forestry, Jakarta.

Motl, O., Hodacová, J. and Ubik, K. (1990) Composition of Vietnamese cajuput essential oil. *Flavour and Fragrance Journal*, **5**, 39–42.

Nguyen Duy Cuong, Truong Thi Xuyen, Motl, O., Stránský, K., Presslová, J., Jedlicková, Z. and Serý, V. (1994) Antibacterial properties of Vietnamese cajuput oil. *Journal Essential Oil Research*, **6**, 63–67.

Penfold, A.R. and Morrison, F.R. (1950) Tea tree oils. In Guenther, E. (ed.) *The Essential Oils*, Van Nostrand Co. Inc., New York, Vol. 4, pp. 526–548.

Sellar, W. (1992) *The Directory of Essential Oils*. C.W. Daniel Co. Ltd, Essex.

Tisserand, R. and Balacs, T. (1995) *Essential Oil Safety*. Churchill Livingstone, Edinburgh.

Todorova, M. and Ognyanov, I. (1988) Composition of Vietnamese essential oil from *Melaleuca leucadendron* L., *Perfumer and Flavorist*, **13**, 17–18.

Von Wulfing, W. *et al.* (1943) Note on research on the kayu putih (*Melaleuca leucadendron*) in Ponoroga, Central Java. Unpublished Report in Indonesian Forestry Abstracts, Dutch Literature until about 1960. Abstract 3502. Wageningen: Centre for Agricultural Publishing and Documentation, 1982.

Walton, S. (1996) Preliminary screening of *M. cajuputi* leaf samples for oil quality (1,8-cineole percentage). Unpublished Progress Report to COSTAI. CSIRO Forestry and Forest Products, Canberra.

Yaacob, K.B., Abdullah, C.M. and Joulain, D. (1989) Essential oil of *Melaleuca cajuputi*. Proceedings of the 11th International Congress of Oils, Fragrances and Flavours, New Delhi, India, 12–16 Nov. 1989, Vol. 1, 140, Oxford and IBH, New Delhi.

Plate 18 *Melaleuca cajuputi* subsp. *cajuputi* at Fogg Dam near Darwin, Northern Territory, Australia (D. Jones)

Plate 19 *Melaleuca cajuputi* subsp. *cajuputi* at Flying Fox Creek near Kapalga, Northern Territory, Australia (D. Lea)

Plate 20 Women harvesting leaves of *Melaleuca cajuputi* subsp. *cajuputi* on Ceram Island for the production of cajuput oil (B. Gunn)

Plate 21 *Melaleuca quinquenervia*, tree form, SE Queensland, Australia (I. Holliday)

Plate 22 *M. quinquenervia*, niaouli, tree form, New Caledonia (B. Trilles)

Plate 23 *Melaleuca quinquenervia*, shrub form, Coolum, SE Queensland, Australia (M. Fagg)

15. *MELALEUCA QUINQUENERVIA* (CAVANILLES) S.T. BLAKE, NIAOULI

B. TRILLES, S. BOURAÏMA-MADJEBI AND G. VALET

Laboratoire de Biologie et Physiologie Végétales Appliquées, Université Française du Pacifique, Nouvelle-Calédonie

BOTANICAL AND ECOLOGICAL DESCRIPTION

Melaleuca quinquenervia (Cav.) S.T. Blake (Plates 21–23) is common in New Caledonia, especially on the western coast where it dominates the many flat and hilly savannahs of this region. It has been given several specific names, in particular *M. leucadendron* L., which is now recognised as a complex of ten species and *M. viridiflora* Sol. Ex Gaertner, with which it has often been mistaken (Dawson 1992; This volume, Chapter 1).

Niaouli, which is the common name for *M. quinquenervia* in New Caledonia, generally appears as a 8–10 m tall shrub in its common environment of dry hills exposed to fires and prevailing winds. However in the marshy plains it becomes a handsome tree reaching 20 and sometimes 30 metres with a well-developed trunk (Plate 22). On the other hand, at higher altitude it appears as a tiny shrub less than one metre in height (Plate 23).

This gregarious species displays many different forms and grows mainly in sedimentary or metamorphic soils. It does not thrive on very basic soils with endemic vegetation although these are common in New Caledonia. However it sometimes grows in waterlogged soils near rivers and other water courses.

Melaleuca quinquenervia's main characteristics, besides its general appearance, are its bark and its foliage. As with many myrtaceous species, its very young leaves are silver-grey and satin in appearance. Soon they become leathery with parallel and lateral venation reaching a size of 1–2 cm width and 8 cm length, with colour turning to a dark green. Usually there are 5 parallel veins (3–7). The bark is whitish and thick, scaly and made of several thin layers which can be easily removed from the tree in large fragments. These were traditionally used to cover a dwelling's roof and walls and also to make torches. Although this bark can easily burn, it efficiently protects the trunk wood from fires as it holds rain water that remains deep in the inner layers. Therefore the niaouli tree is perfectly adapted to resist the destructive action of fires occurring in the savannah whereas other plant species are destroyed. This is why the niaouli tree can be considered as fire resistant.

Niaouli blooms are grouped in twos or threes in 9 or 10 cm spikes and gathered in twos or threes along the inflorescence axis (Plate 5). These are also grouped in twos or threes but the axis that bears the blooms can keep growing when the flowering time is ended. Fruits then, quite characteristically, gather either along or at the end of a ramification. The blooms are either creamy-white, white or occasionally pink or red.

The perianth is tiny but there are many stamens (30–40) which are grouped in 5–10 phalanxes and slightly joined together at their lower part. The ovary contains up to 150 tiny seeds. *M. quinquenervia* blooms in December and in June although scarce blossoms may appear all year long depending on the climatic conditions. Their pollen is steadily gathered by bees and their nectar gives a clear flavoured honey that is much appreciated in New Caledonia.

The niaouli tree has a dense high quality wood of a light rosy, sometimes reddish colour similar to some rosaceous plants.

USE

Melaleuca quinquenervia in New Caledonia, is mainly known for its essence that, for decades, has been collected by distillation of its leaves. We shall discuss this subject in more detail when describing the latest work carried out in our laboratory of the Université Française du Pacifique. One must bear in mind that this tree also has a variety of other uses. We have already mentioned the various uses for the bark that are so characteristic of this tree. This bark has been of big help to the island people in old times. It has been plentifully used in traditional dwellings where the walls were made of niaouli bark strips. Double walls were internally padded for efficient heat insulation. When straw was lacking, rooves were covered with large pieces of niaouli bark.

The wood of this tree has been used extensively for carpentry and joinery work. There still exist buildings from last century where frames were made of niaouli wood resistant to common wood-eating insects. This wood has a long-lasting property when it is worked while it is still green and once the bark is removed. It becomes extremely hard after drying and ageing. It is still used for gate poles and fencing cattle paddocks. In the last century it was used for manufacturing carts, cartwheel hubs, anvil blocks, benches and tool hammers.

However it is the tree's foliage that has been used most extensively, by distillation and infusion, for therapeutic inhalation and for making "niaouli tea" respectively. The latter use has now ceased.

THE CHEMISTRY OF *MELALEUCA QUINQUENERVIA* ESSENTIAL OILS

Essential oils are a diverse group of natural products that are important sources of aromatic and flavouring chemicals in food, industrial and pharmaceutical products.

Essential oils are largely composed of terpenes and aromatic polypropanoid compounds derived from the acetate–mevalonic acid and the shikimic acid pathways, respectively. The essential oil composition of plants vary due to genetic and environmental factors that influence genetic expression (Bernath 1986).

Identification of Components

The constituents in each hydrodistilled essential oil were first tentatively identified by peak enrichment and their gas chromatographic (GC) retention indices on two fused

Table 1 Percentage of each chemical class and group in an essential oil of *M. quinquenervia* from New Caledonia

Class	%	*Group*	%
Hydrocarbons	7.8	Monoterpenes	6.9
		Sesquiterpenes	0.9
Oxygenated products	90.6	Terpenols	12.1
		Sesquiterpenols	29.1
		Ethers	49.0
Other	1.6		

silica capillary columns (25 m × 0.25 mm i.d coated with OV-101, film thickness 0.25 μm; 25 m × 0.22 mm i.d. coated with Carbowax 20 M, film thickness 0.25 μm), using a chromatograph (Shimadzu GC-14A) equipped with a Shimadzu C-R4A Chromatopac integrator. Detector and injector temperatures were set at 250°C and 210°C respectively; the oven temperature was programmed from 50°C–200°C at 5°C per min with nitrogen as a carrier gas. The percentage composition was obtained from electronic integration measurements using flame ionisation detection without taking into account relative response factors. Individual oil constituents were identified from their GC retention times relative to standard compounds.

The first analysis showed a complex chemical composition of the essential oils with 1,8-cineole constituting a major proportion of the oil. Only 19 chemical compounds were identified, with a number of unidentified compounds. The compounds which had been identified were classified into two main groups: "Hydrocarbons" and "Oxygenated Products". Acetate and the unknown compounds were classified in an "Other" group. Hydrocarbons contained two main groups: monoterpenes and sesquiterpenes; while oxygenated products contained terpenols, sesquiterpenols and ethers (Table 1).

Monoterpenes were the first compounds to be eluted from the gas chromatographic column. They were more volatile and less polar than the other compounds. This group was easily identified because of relatively simple molecular structures the ready availability of standards.

Subsequently sesquiterpenes were eluted. These natural constituents have more complex structures not as easy to determine as the monoterpene structures. In addition, sesquiterpene resolution was sometimes inadequate.

Following the sesquiterpenes, oxygenated products were eluted. For the New Caledonian *M. quinquenervia* essential oils, this group constituted one or two major chemical components. However, in this group the tree by tree variation was more significant than in the hydrocarbon groups.

Detailed Chemical Composition

Typical results are summarised in Table 2.

In addition to the identified constituents, many unknown components were detected. Numerous peaks appeared in each chromatogram and a large variation in the number

Table 2 Chemical composition (%) of *M. quinquenervia* (Niaouli) oil from New Caledonia

		Chemotype				
		I	II	III	IV	V
Hydrocarbons						
Monoterpenes	limonene	8.2	3.5	0.7	2.2	1.3
	myrcene	0.6	0.7	0	0.1	0
	α-phellandrene	0.1	1.0	0	<0.1	0
	α-thujene	<0.1	1.9	4.5	<0.1	0
	β-pinene	1.2	4.8	0.2	0.6	0.3
	α-terpinene	0.2	3.8	0.1	0.1	0.1
	γ-terpinene	0.4	19.0	3.6	1.5	7.1
	p-cymene	0.5	1.6	13.6	5.2	27.1
	terpinolene	0.2	4.9	0.8	1.0	1.6
Sesquiterpenes	viridiflorene	0.3	0.5	0.3	0.7	0.4
Oxygenated products						
Terpenols	linalol	0.2	6.3	0.5	0.2	0.7
	α-terpineol	11.1	1.5	3.9	20.3	2.7
	terpinen-4-ol	0.8	1.8	3.6	2.1	4.4
Sesquiterpenols	globulol	6.8	18.3	25.5	7.1	17.7
	ledol	0.3	0.3	0.2	0.2	0.2
	trans-nerolidol	0.5	<0.1	6.3	2.5	4.3
	viridiflorol	3.2	3.4	4.1	1.8	2.4
	τ-cadinol	0.4	0.2	3.5	1.6	0.7
	α-cadinol	<0.1	0.1	0.9	0.1	0.2
Ethers	1,8-cineole	59.3	10.5	4.1	35.4	1.3
	epoxycaryophyllene	0.2	0.2	0.4	1.1	0.2
	β-caryophyllene	0.4	0.5	0.5	0.1	0
Acetates	linalyl acetate	0.2	0	0.3	<0.1	2.7
	α-terpinyl acetate	0.9	0	3.7	0.3	0

of peaks was noted sample by sample. The non-polar phase gave from 18 to 102 GC peaks whereas the polar CWX 20M phase gave from 45 to 107 peaks.

These variations illustrate the quality and flavour differences in niaouli essential oil.

The clear, colourless to pale yellow oil consists of a complex mixture of compounds. *M. quinquenervia* exists in five chemotypes.

1,8-Cineole, Chemotype I

The oxygenated compound, 1,8-cineole, usually constitutes 60–75 per cent of this oil giving a refreshing and unique flavour. This main constituent is also the important active principle. The Caledonian Standard recommends at least 60 per cent 1,8-cineole in the oil. As it is important that commercial oils meet this standard, most niaouli oils in use contain 60 per cent 1,8-cineole.

The oil's composition in natural stands of *M. quinquenervia* can vary considerably from region to region, locality to locality, place to place even tree to tree throughout the territory of New Caledonia.

It appears that 1,8-cineole concentration in New Caledonian *M. quinquenervia* essential oils is similar to the mainland Australia cineole chemotype (Brophy *et al.* 1989; Brophy and Doran 1996) and higher than 1,8-cineole concentration in Madagascan and Benin oils which contain 42–60 per cent (Ramanoelina 1992) and 14 per cent (de Souza *et al.* 1994) respectively.

Four Other Chemotypes

The steam distilled oils of most *M. quinquenervia* populations growing in New Caledonia were found to be largely cineole-rich in character but with varying concentrations. Recent work at the Laboratoire de Physiologie Végétales Appliquées, one of the research units of the French University of the Pacific, showed four new chemotypes. The other varieties which contain constituents in greater concentration than 1,8-cineole are:

(a) Chemotype II, with up to 20 per cent γ-terpinene;
(b) Chemotype III, with up to 30 per cent globulol;
(c) Chemotype IV, with up to 30 per cent α-terpineol;
(d) Chemotype V, with 10–20 per cent globulol and 10–30 per cent *p*-cymene.

Niaouli trees were randomly sampled with the viridiflorol chemotype described by Ramanoelina *et al.* (1994) not observed in the western area of New Caledonia. Viridiflorol and nerolidol-rich chemotypes have been reported from Australia (Brophy *et al.* 1989; Brophy and Doran 1996). Chemotype I, rich in 1,8-cineole seems very common as it has been described by Guenther (1950) and other authors.

HYDRODISTILLATION AND HARVESTING

In New Caledonia the resource is not cultivated in plantations. Plants are harvested from natural areas. Some production problems occur however, when the leaves are not harvested from selected trees. The first problem is with oil yield. The current yield is about 0.7 per cent and it is difficult to obtain a yield in excess of this percentage. The second problem is with the concentration of 1,8-cineole. It is necessary to carry out an analysis of the quality of the extracted essential oil from each harvest as the cineole concentration may have fallen. The techniques currently used for extracting essential oils include hydrodistillation, steam distillation, solvent extraction and liquid CO_2 extraction. Hydrodistillation is an excellent method of essential oil extraction as it results in a good yield and recovery of all the essential oil constituents (Charles and Simon 1990). In New Caledonia hydrodistillation is commonly used for extracting niaouli essential oil. Harvesting is carried out manually. Samples of fresh plant material, a mixture of leaves and stems, are used to fill the distillation pot (100 kg fresh weight and 150 litres of water), as shown in Figure 1. The hydrodistillation process lasts 7 hours. Then the essential oil is packaged in 50 ml glass bottles.

The method used in the laboratory was eight hour hydrodistillation of freshly-gathered leaves in 500 ml distilled water with cohobation using all-glass apparatus. The essential oils were dried over anhydrous sodium sulfate, stored at 4°C until analysis.

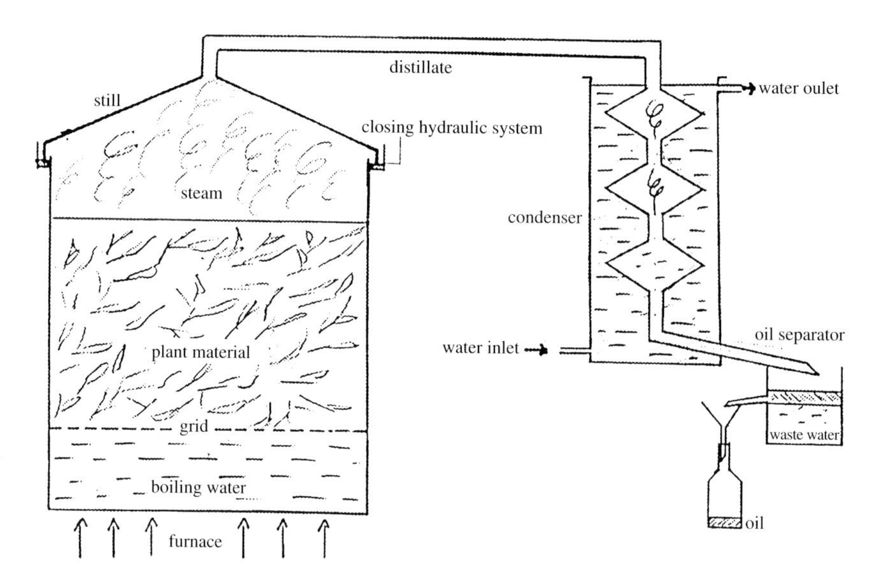

Figure 1 The commercial distillation of *M. quinquenervia* to produce niaouli oil

Currently using laboratory-scale distillation, *M. quinquenervia* leaf yields are classified as follows:

—low oil production: 0–0.5 per cent yield;
—average oil production: 0.5–1.5 per cent yield;
—good oil production: 1.5–2.5 per cent yield;
—excellent oil production: more than 2.5 per cent yield.

The essential oil yield seems to be genetically determined. A current research study of factors affecting oil yield in various soils is showing that the soil type has no effect on the yield. Even soils with a high capacity to retain water have little impact on the yield. In a flat area with impeded drainage, the yield was lower than in a hilly area owing to the water accumulation which has a direct impact on the water content of the leaves.

A study on the relationship between the climate and the essential oil yield showed that there was no significant relationship. However, the yield seemed to decrease over two periods, in August (winter) and in December (high rainfall season). During the other months of the year the fluctuation observed was close to the average yield. However it was noted that the main increase in oil yield seems to occur during summer.

Although this suggests that leaf should be harvested in this period, the tendency in New Caledonia is to harvest throughout the year.

Typical annual production has gradually fallen following the report of Guenther (1950) in which production was stated as 10–30 tonne per year. The 1984 production estimate was 4 tonne (Lawrence 1985) with current estimates of around 7–10 tonne per year for New Caledonia sourced oil. About 19 per cent of this is exported to France and marketed overseas through agents. Today, at current world prices, the annual value of the New Caledonian niaouli oil production is about 7 million CFP (US$67,000).

USES—BIOACTIVITY OF ESSENTIAL OILS OF NIAOULI

Use of Essential Oils

Essential oils are valuable antiseptics with bioactivity against human pathogens. There are two ways of using the niaouli essential oils:

1. Traditional medicine,
2. French Pharmacopoeia regulated preparations.

Use in Traditional Medicine

Essential oils have a high antiseptic capacity. Niaouli oil is mainly used for pulmonary infections especially for colds and bronchitis. The oil can be absorbed as a tea drink (decoction by boiling water with dried leaves), or by inhalation (three drops of commercial oil in steaming water).

Niaouli oil has a high penetration and diffusion capacity which generates efficient immunisation against latent infections.

On the other hand, according to Degrez (1908), aldehyde-free niaouli oil, called "Gomenol", has a cold balsamic flavour. Also, gomenol has low toxicity. On the basis of gram per kilogram (animal) body weight the reported lethal dose for pure liquid niaouli oil is 0.17 compared with 0.34 for collargol (colloidal silver), 0.51 for phenol and 0.79 for formol (formaldehyde) (Degrez 1908). As Gomenol's lethal dose was 3, niaouli oil with the aldehydes removed is relatively innocuous.

French Pharmacopoeia

The French Pharmacopoeia lists six products based on Gomenol (essential oil without aldehydes):

1. Gomenol (AMM 1943);
2. Gomenol soluble (AMM 1950);
3. Gomenol-Syner-Penicilline (AMM 1959);
4. Gomenoleo (AMM 1943);
5. Gomenol rectal (AMM 1996);
6. Gomenol sirop (AMM 1996).

These are listed in the Autorisation de Mise sur le Marché (AMM—equivalent to the USA Food and Drug Administration Monographs).

Bioactivity of Essential Oils

After the essential oil had been obtained from the hydrodistillation of 100 g of freshly-gathered leaves in 500 ml distilled water for 8 hours, each sample was tested on microorganism cultures of *Erwinia*, *Candida albicans* and *Micrococcus*. Each sample of oil came from one tree. The chemical composition was determined before carrying out this antimicrobial measurement.

A 5 mm diameter Whatman circle soaked with essential oil was placed in the centre of a Petri dish containing medium and microorganism. Assay plates were incubated at 27°C (±1°C) for four days. The inhibition zones obtained were measured. Table 3, shows significant but variable bioactivity. Two reference oils, commercial niaouli oil (N) and Gomenol (G), were used to compare the bioactivity of a host of *M. quinquenervia* oil samples with eucalyptus and Australian tea tree oil [Bouraïma-Madjebi *et al.* 1996].

Two effects were illustrated:

1. a bacteriostatic effect shown by an opaque inhibition zone;
2. a bactericidal effect shown by a translucent inhibition zone.

The *Erwinia* bacteriostatic and bactericidal effects were higher than for *Candida albicans.* For the sample tested on *Micrococcus* some activities higher than *Erwinia* bioactivity were

Table 3 Zone of inhibition diameters (mm) for essential oil samples of *M. quinquenervia*: op: opaque; t: translucent; TTE: Australian Tea Tree oil; N: commercial essential oils of *M. quinquenervia*; G: Gomenol (pharmaceutical product); E: eucalyptus oil; Nia, JLD: samples of essential oil of niaouli tested; 8–63: samples of essential oils of *M. quinquenervia*

Sample	Erwinia		Candida albicans		Micrococcus	
	t	*op*	*t*	*op*	*t*	*op*
TTE	2,03	2,1	1,75	2,8	2,04	3,7
N	3,91	1,6	3,5	0,6	2,3	1,1
G	0,11	1	0,32	0,7	0,83	0,23
E	7	11	4	1,3		
Nia	3,3	—	7,3	—	4	—
JLD	3	5	1,3	2,3		
N	5,3	—	3	—		
8	3,5	—	3	—		
15	12	—	3,5	—	2	3,7
16					1,8	2,7
20	2	—	0,7	2		
25	4,3	—	6,7	—		
26					6,3	0,3
30	3	—	5	—		
31	4	—	5,7	—		
32	2,7	—	3,7	—		
33	3	—	1,2	4,7		
35	6	30	7	1		
37	3,7	—	6,3	—	7	3,5
40	17,3	—	4	—	1,5	—
43	5,33	—	2,7	—	8,33	—
54	1	—	11	—	6	—
55	3,7	—	3	—	5	—
60	3,5	—	2,5	—	6	—
62	4	—	1,33	—		
63	6	—	6	3	6	—

observed. Preliminary tests also showed some biostatic and bactericidal activity of the essential oil on *Pseudomonas*, *Klebsiella*, *Escherichia coli* and *Agrobacterium*. Many of the New Caledonian *M. quinquenervia* oils showed bactericidal effects equivalent to or stronger than eucalyptus and tea tree oil.

REFERENCES

Bernath, J. (1986) Production ecology of secondary plant products. In L.E. Craker and J.E. Simon (eds.), *Herbs, Spices and Medicinal Plants: Recent Advances in Botany, Horticulture, and Pharmacology*, Vol. 1, Oryx Press, Phoenix, Ariz, pp. 185–234.

Bouraïma-Madjebi, S., Trilles, B., Valet, G. and Pineau, R. (1996) Valorisation des plantes à essence de Nouvelle-Calédonie. *Rapport de convention, No. 239, Productions Vegetales et Forêts/Direction du Developpement Rural.*, Université Française du Pacifique, Nouvelle-Caledonie, pp. 5–57.

Brophy, J.J., Boland, D.J and Lassak, E.V. (1989) Leaf Essential Oils of *Melaleuca* and *Leptospermum* species from tropical Australia. In D.J. Boland (ed.), *Trees for the Tropics*, ACIAR, Canberra, pp. 193–203.

Brophy, J.J. and Doran, J.C. (1996) Essential Oils of Tropical *Asteromyrtus*, *Callistemon* and *Melaleuca* Species. ACIAR, Canberra, pp. 76–77.

Charles, D.J. and Simon, J.E. (1990) Comparison of extraction methods for the rapid determination of essential oils content and composition of Basil. *J. Amer. Soc. Hort. Sci.*, **115**(3), 458–462.

Dawson, J. (1992) Flore de Nouvelle-Calédonie et Dépendances: Myrtaceae—Leptospermoïdées. *Muséum National d'Histoire Naturelle*, Paris, France, p. 251.

Degrez, F. (1908) Toxicité du Gomenol et pouvoir antiseptique comparé. *Faculté de Médecine de Paris.*

De Souza, S., Ayedoun, M.A., Batonan, A., Ayss, J. and Akplogan, A.B. (1994) Essai d'aromatherapie par l'huile essentielle de niaouli du Benin. *Revue Med. Pharm. Afr.*, **8**(1), 23–34.

Guenther, E. (1950) *The Essential Oils*, Van Nostrand, New York, **4**, pp. 41–44.

Lawrence, B.M. (1985) A review of the world production of essential oils (1984). *Perf and Flav.*, **10**(5), 2–16.

Ramanoelina, P.A.R. (1992) Etude de la variation de la composition chimique de l'huile essentielle de niaouli (*Melaleuca viridiflora* Sol ex Gaeth) de Madagascar. *Thèse de Doctorat des Sciences*, Université d'Aix-Marseille, France, 193pp.

Ramanoelina, P.A.R., Viano, J., Bianchini, J.-P. and Gaydou, E.M. (1994) Occurrence of various chemotypes in niaouli (*Melaleuca quinquenervia*) essential oils from Madagascar using multivariate statistical analysis. *J. Agric. Food Chem.*, **42**, 1177–1182.

16. POTENTIALLY COMMERCIAL MELALEUCAS

JOSEPH J. BROPHY

School of Chemistry, University of New South Wales, Sydney, Australia

INTRODUCTION

The preceding chapters have discussed the commercial species, *M. cajuputi*, *M. quinquenervia* and *M. alternifolia* (including the closely related *M. linariifolia*). Now we present data on other species which have potential for commercial production.

According to the latest research there are upwards of 230 species within the genus *Melaleuca*. Over the last 30 years these have been documented in works by Blake (1968); Carrick and Chorney (1979); Byrnes (1984, 1985, 1986); Barlow (1986, 1988); Barlow and Cowley (1988); Cowley *et al.* (1990); Craven and Barlow (1997), as well as more popular works by Wrigley and Fagg (1993); and Holliday (1989). Of these approximately 230 species (Craven 1998), only 3 (*M. alternifolia*, *M. cajuputi* and *M. quinquenervia*) are at present commonly used for the commercial production of essential oil.

With so many species to choose from it is obvious that during this present century a significant amount of research has been carried out examining the essential oils of the various *Melaleuca* species. This has been carried out both to detail the contents of the various essential oils and also to search for potentially commercial oils.

Research on the essential oils of members of the genus *Melaleuca* at the University of New South Wales has been carried out to both supplement and extend our knowledge of the chemistry of this genus. We have had the advantage of the availability of specimens collected for the current revision of the genus by Craven. To date approximately 180 species have been either examined for the first time or re-examined and the contents of their essential oils documented. During the course of this work some species have shown essential oils that may give promise of commercial use. These species, and the oils obtained from them, are discussed in more detail in this chapter.

The oils were obtained by steam distillation with cohobation in a modified Dean and Stark apparatus. A full description is given in Brophy and Doran (1996). Analysis of the oils was by gas chromatography (GC) and combined gas chromatography–mass spectrometry (GC/MS). Compound identification was, for the most part, by matching of spectra and retention time of various peaks against authentic pure materials or oils of known composition. Usual GC conditions are given in Brophy and Doran (1996).

In choosing species to include in this chapter the three main criteria considered were that the oil, qualitatively, should be interesting and that the oil quantitatively be useful. This latter criterion meant that the oil yield should, all other things being equal, generally be greater than 1% based on fresh foliage. One further restriction which has been placed on the two above criteria is that the species should be a taxonomically defined species. This last restriction has at present excluded close to 100 collections from being considered for this chapter.

As might be expected from so much study of morphological variation, over time there have been different interpretations placed on the data and different names and associations proposed. The species referred to in this chapter follow Craven, though reference in the individual sections is made to previous treatments.

This chapter highlights 15 species of *Melaleuca* and 1 species from the related genus *Asteromyrtus*. The species (*M. acacioides*, *M. alsophila*, *M. bracteata*, *M. citrolens*, *M. dissitiflora*, *M. ericifolia*, *M. leucadendra*, *M. linophylla*, *M. quinquenervia*, *M. squamophloia*, *M. stenostachya*, *M. stipitata*, *M. trichostachya*, *M. uncinata*, *M. viridiflora* and *A. symphyocarpa*) all met the selection criteria and produce either aromatic oils or cajuput type (high 1,8-cineole) oils. *Asteromyrtus symphyocarpa* is included with the *Melaleuca* oils because it was once considered a member of the genus *Melaleuca* before being placed in the genus *Asteromyrtus* (Craven 1988) and it produces an essential oil which is a potential substitute for cajuput oil (Brophy *et al.* 1994).

Of the 15 species of *Melaleuca* included in this chapter, 2 are newly named (*M. stipitata* and *M. squamophloia*) and have had only one report of their chemistry (Brophy and Doran 1996). A further species (*M. linophylla*) has its oil reported on for the first time.

During the course of our recent work on *M. leucadendra* a much larger number of specimens have been examined, encompassing the whole of its range across northern Australia. This has shown that there are two sets of chemical forms of this species. One chemical form is terpenoid and this form occurs in the western half of its range, extending from Western Australia to approximately mid Northern Territory. In the eastern half of its range, from mid Northern Territory to the east coast of Australia, the oil is essentially entirely aromatic. At the sampling location where both forms were found (Kapalga) there appeared to be no interbreeding, with the oils obtained being either terpenoid or aromatic but never a mixture of the two.

There must be considerable doubt about which species is being considered when *M. viridiflora* is mentioned in publications prior to 1968. There was confusion between what is now called *M. viridiflora* and what is now called *M. quinquenervia*, two species that overlap considerably in range and are of similar appearance. Without access to botanical voucher specimens for these oils it is not possible to say just which species is producing the oil, though it is thought that the species producing oils rich in nerolidol and/or linalool, especially coming from New South Wales, is *M. quinquenervia.*

Basically, *M. alsophila*, *M. dissitiflora*, *M. stipitata* and one chemotype of *M. uncinata* produce oils rich in terpinen-4-ol and low in 1,8-cineole. Some of these are lemon scented. *M. bracteata*, *M. squamophloia*, one chemotype of *M. leucadendra*, and one chemotype of *M. viridiflora* produce oils which are aromatic (in either the chemical sense or perfumery sense) while *M. trichostachya*, *M. acacioides*, and some chemotypes of *M. citrolens* produce oils which might have perfumery potential. *M. linophylla*, *M. stenostachya*, and *A. symphyocarpa* all produce oils with similar compositions to cajuput oils.

M. laterifolia subsp. laterifolia, *M. pustulata* and *M. radula* all produce oils in yields of 1–2% in which 1,8-cineole accounts for 70–87% of the oil. It should also be mentioned that both *M. alternifolia* and *M. linariifolia* have chemotypes that are characterised by oils rich in 1,8-cineole. These are usually shunned in favour of the terpinen-4-ol rich oils. *M. nanophylla* gives an oil in 1.7% yield in which the major compounds are the two *β*-triketones flavesone (24–42%) and leptospermone (10–25%) accompanied by *α*-pinene (24–40%) (Brophy 1998). There is also a record of a chemical variety of *M. dealbata* producing an oil in 1.5% yield in which the major compound was leptospermone (70%) (Lassak and Southwell 1977). Oils rich in polyketones such as leptospermone are being used in body care products because of their powerful anti-microbial properties (Joulain 1995).

SPECIES DIGESTS

Melaleuca acacioides

M. acacioides occurs as a shrub or small tree, 4–10 m tall, and may develop a multi-stemmed habit when open-grown. It is found in coastal and sub-coastal (usually saline and seasonally flooded) habitats in far northern Queensland, the north of the Northern Territory and islands of the Torres Strait in Australia and extends to southern Papua. It has potential for production of posts and small poles, fuelwood and windbreaks on difficult sites near the coast.

M. acacioides produced an essential oil, in 0.2–0.4% yield (based on fresh leaves), which was almost entirely composed of sesquiterpenes. The oil has a distinctive pleasant aroma which is associated with the sesquiterpene alcohol fraction. The main components were *β*-selinene (21–30%) and *α*-selinene (53–54%). The next most abundant compounds were selen-11-en-4-ol (6–8%), *δ*-cadinene (0.9–6%), *β*-caryophyllene (1–2%), globulol (0.7–1%) as well as some unidentified oxygenated sesquiterpenes in the range 0.1–3%. Monoterpenes were very poorly represented. Table 1 gives a typical oil analysis.

Table 1 Compounds identified in the essential oil of *Melaleuca acacioides*

Compound	%	*Compound*	%
ethylbenzene	tr	unknown, mw 200	0.1
limonene	0.1	$C_{15}H_{26}O$	0.1
α-copaene	0.1	$C_{15}H_{26}O$	0.2
α-gurjunene	tr	globulol	0.7
β-caryophyllene	1.2	viridiflorol	0.2
β-gurjunene	0.1	$C_{15}H_{26}O$	0.5
$C_{15}H_{24}$	2.0	$C_{15}H_{26}O$	1.8
β-selinene	21.4	$C_{15}H_{26}O$	0.3
α-selinene	53.9	$C_{15}H_{26}O$	0.2
selina-3,7-diene	0.6	$C_{15}H_{26}O$	0.2
δ-cadinene	2.3	$C_{15}H_{26}O$	0.2
β-bisabolene	tr	selina-11-en-4-ol	6.2
$C_{15}H_{24}O$	0.1	$C_{15}H_{24}O$	1.0
$C_{15}H_{24}O$	0.7	farnesol	3.9
$C_{15}H_{24}O$	0.1	$C_{15}H_{24}O$	0.3
methyl eugenol	0.1	$C_{15}H_{24}O$	0.1

This oil has potential in perfumery and also as a source of selinenes which are important components of celery seed oil.

References: (Chemistry) Brophy *et al.* 1987, 1989, Brophy and Doran 1996; (Botany) Barlow 1986, 1988, Byrnes 1985.

Melaleuca alsophila

M. alsophila usually occurs as a small tree to 10 m in height, sometimes multi-stemmed, with white papery bark. It is found in north western Australia and is common on river banks, on the margins of mud flats and in seasonally inundated saline depressions. It was previously a subspecies of *M. acacioides* and could be considered for planting on similar sites.

There appear to be two chemotypes of *M. alsophila*, with varying oil yield. One chemotype is rich in α-pinene and/or 1,8-cineole and gives a low oil yield, while the other chemotype contains significant amounts of neral, geranial and terpinen-4-ol and gives a higher oil yield.

Chemotype I contains α-pinene (8–65%), 1,8-cineole (15–66%) as its major compounds, while there are significant amounts of limonene (1–3%), *E*-β-ocimene (0.8–12%), pinocarvone (0.3–5%), and *trans*-pinocarveol (1–17%). Sesquiterpenes, while numerous, were of little consequence in this oil. The oil yield of this chemotype was 0.04–0.1% on a fresh weight basis.

The second chemotype contained major amounts of neral (2–10%), geranial (2–19%), terpinen-4-ol (13–32%), α-terpineol (1–7%), *p*-cymene (2–40%, the majority > 20%) and geraniol (1–3%). The oil yield of this second chemotype was 0.1–0.6% on a fresh weight basis. Table 2 gives typical analyses for the two chemotypes.

Trees from a further location appeared to be intermediate between these two chemotypes. In this case the major compounds were 1,8-cineole (28–39%), terpinen-4-ol (13–16%), α-terpineol (4–7%), *E*-methyl cinnamate (1–12%) and globulol (1–3%). The oil yield in this case was 1–1.6% on fresh leaves.

This species has the potential to produce oils as a source of α-pinene, cineole, citral or terpinen-4-ol, depending on the genetic material developed for production.

References: (Chemistry) Brophy *et al.* 1987, 1989, Brophy and Doran 1996; (Botany) Byrnes 1985, Barlow 1986, 1988.

Melaleuca bracteata

M. bracteata is typically a large shrub or small bushy tree (5–10 m) but may reach 20 m in height. It has small prickly leaves and dark grey hard bark. *M. bracteata* is one of the most widely distributed species of the genus in Australia occurring in five States. It is frequently found growing along watercourses on rather heavy-textured deep clays. *M. bracteata* makes an excellent shelter tree with potential for small posts and poles.

M. bracteata has been shown to exist in four chemical forms, in yields of 0.1–1.2% based on fresh leaves. These are forms in which the aromatic ethers (I) elemicin, (II) *E*-isoelemicin, (III) *E*-methyl isoeugenol or (IV) methyl eugenol predominate in the oil. In all cases, no matter which component predominates (> 40%), the other three were present in significantly lesser amounts.

Table 2 Compounds identified in the essential oils of *Melaleuca alsophila*

Compound	*Chemotype I* %	*Chemotype II* %	*Compound*	*Chemotype I* %	*Chemotype II* %
isovaleric aldehyde	0.2		acetophenone		0.3
tricyclene	0.6		myrtenal	0.3	
α-pinene	65.0	1.5	*trans*-pinocarveol	1.8	
α-fenchene	0.2		neral	tr	4.5
camphene	0.5		α-terpineol	2.7	1.7
β-pinene	0.1	0.2	borneol	0.7	
sabinene	tr	0.3	geranial	tr	9.2
myrcene	tr	1.0	citronellol		8.5
α-terpinene	0.1	1.0	nerol		0.4
limonene	2.6	1.3	*p*-cymene-8-ol		0.1
1,8-cineole	15.1	tr	phenylpropanol..tent..	0.7	
β-phellandrene		0.4	myrtenol	0.1	
Z-β-ocimene	0.2		*trans*-mentha-1(7),8-dien-2-ol	0.1	
γ-terpinene	0.3	4.8	*trans*-mentha-1,8-dien-6-ol	0.4	
E-β-ocimene	tr		*p*-cymene-8-ol	0.1	
p-cymene	0.8	13.2	4-phenylbutanone	0.1	
terpinolene	0.2	0.4	*cis*-mentha-1,8-dien-6-ol	0.1	
rose oxide		0.4	*cis*-mentha-1(7),8-dien-2-ol	0.1	
cis-linalool oxide		tr	methyleugenol	0.5	
trans-linalool oxide		0.1	*E*-nerolidol		0.1
α,*p*-dimethylstyrene	tr	0.1	$C_{15}H_{26}O$	0.2	
α-campholenic aldehyde	0.2		$C_{15}H_{26}O$	0.2	
citronellal		3.0	$C_{15}H_{26}O$	0.1	
linalool		0.1	$C_{15}H_{26}O$	0.2	
pinocamphone	0.2		globulol	0.8	
pinocarvone	0.3		viridiflorol	0.2	
isopulegol		4.0	$C_{15}H_{26}O$	0.2	
trans-menth-2-en-1-ol		8.2	$C_{15}H_{26}O$	0.3	
β-caryophyllene	0.2		spathulenol	0.2	
terpinen-4-ol	0.1	31.4	$C_{15}H_{24}O$	0.3	
cis-menth-2-en-1-ol		0.6	E,E-farnesol	tr	

In all chemotypes there were lesser amounts (up to a maximum of approximately 30%) of mono- and sesquiterpenes present. Of these compounds, β-caryophyllene, α-farnesene, α-phellandrene and α-pinene appear to be the major contributors. Also present in the oil were small, but significant, quantities of *E*-methyl cinnamate (0.1–9%) which no doubt gives the oil its sweet and fruity odour. The oil yield was generally low (0.1% on a dry weight basis) but one source from a species/provenance trial at Gympie (from seed collected north of Alice Springs) gave an oil yield of 0.7–1.2% on a fresh leaf basis. Table 3 gives typical analyses for the four chemotypes.

The oils from these chemotypes are potential sources of the aromatic ethers elemicin, *E*-isoelemicin, *E*-methyl isoeugenol and methyl eugenol.

Table 3 Compounds identified in the essential oils of *Melaleuca bracteata*

Compound	*Chemotype I* %	*Chemotype II* %	*Chemotype III* %	*Chemotype IV* %
α-pinene		4.9	0.6	0.6
β-pinene	0.1	0.2	tr	tr
sabinene		tr	tr	tr
myrcene	0.1	0.4		tr
α-phellandrene	3.5	12.7	0.1	0.1
α-terpinene	0.1	0.7	tr	tr
limonene	0.3	0.7	1.4	tr
1,8-cineole		0.4		
γ-terpinene		0.1		tr
E-β-ocimene	0.4	1.4		
p-cymene			0.1	0.3
terpinolene	2.7	2.1	tr	tr
mentha-1,3,8-triene	0.8	3.2		
benzaldehyde			tr	tr
α-copaene	0.2	tr		
linalool			0.3	tr
acetophenone			tr	tr
β-caryophyllene	21.4	6.8		
alloaromadendrene	0.1	tr		
methyl chavicol	tr	tr	tr	tr
acetotoluene			0.2	0.6
humulene	1.9	0.5		
α-terpineol			0.2	0.3
germacrene-D	0.5	0.7		
bicyclogermacrene	1.0	0.1		
α-farnesene		2.8		
δ-cadinene	0.4	0.4		
γ-cadinene		tr		
cadina-1,4-diene	tr	tr		
anethole	tr	tr		
unknown, mw148			tr	0.2
p-cymene-8-ol	tr			
Z-methyl cinnamate			tr	tr
methyl eugenol	tr	0.8	17.5	45.7
methyl methoxybenzoate			0.2	0.1
caryophyllene oxide	0.2	0.1		
E-methyl cinnamate	0.1	1.4	2.8	8.6
palustrol	tr	0.1		tr
E-methyl isoeugenol	0.1	0.9	75.9	43.0
unknown, mw240	0.2			
elemicin	57.4	8.8	0.1	0.2
$C_{15}H_{26}O$	0.5	1.5		
Z-isoelemicin	0.1	0.1		
E-isoelemicin	5.6	45.4	0.2	tr

References: (Chemistry) Baker and Smith 1910, Brophy *et al.* 1989, Cosgrove and Thain 1948, Guenther 1950, Aboutabl *et al.* 1991, Penfold and Willis 1954, Brophy and Doran 1996; (Botany) Blake 1968, Byrnes 1986, Carrick and Chorney 1979.

Melaleuca citrolens

M. citrolens is a small tree to 7 m in height or rarely a large shrub with furrowed, firm or slightly papery bark. It occurs in northern Australia usually in open forests, on sandy, stony or loamy soils. It has potential to produce posts and rails and the lemon-scented chemotype may have scope for development as a commercial essential oil.

M. citrolens has been shown to exist in three definite, possibly four, chemical forms. These chemotypes are characterised by the presence of (I) 1,8-cineole, (II) piperitenone, (III) neral and geranial and (IV) citronellal in significant amounts in the oils. The oils are produced in 1.6–3.1% yield (based on fresh leaves).

Chemotype (I) contained 1,8-cineole (34–50%) and terpinolene (10–20%) as major components in the oil. Other compounds present in significant quantities were α-pinene (2–5%), α-phellandrene (2–5%), limonene (2–4%), terpinen-4-ol (0.3–6%), β-caryophyllene (trace-5%), α-terpineol (1–7%) and globulol (1–5%). The oil yield of this chemotype was 1.5–2.4% on an air-dry weight basis.

Chemotype (II) gave an oil in which the major components were 1,8-cineole (8–32%), terpinolene (13–27%) and piperitenone (9–14%). These were accompanied by lesser amounts of α-thujene (4–7%), α-phellandrene (5–9%), γ-terpinene (2–11%), terpinen-4-ol (2–4%) and α-terpineol (2–5%). The oil yield of this chemotype was 2.9–6.1% on an air-dry weight basis.

Chemotype (III) contained neral (7–16%), geranial (9–26%), 1,8-cineole (12–28%) and terpinolene (0.1–7%) as major compounds. The oil yield in this case was 1.3–3.9% on an air-dry weight basis.

Chemotype (IV) contained 1,8-cineole (1–12%), citronellal (20–30%), isopulegol (4–13%), geranial (trace-22%), neral (9–14%) and geraniol (0.7–2%) as major compounds. The oil yield of this chemotype was 1.9–3.1% on an air-dry weight basis. Table 4 gives typical analyses for oils from the four chemotypes.

The latter two oils of this species have potential for development in the perfumery industry as low grade citral and citronellal oils respectively.

References: (Chemistry) Brophy and Clarkson 1989, Brophy *et al.* 1989, Southwell and Wilson 1993, Brophy and Doran 1996; (Botany) Barlow 1986.

Melaleuca dissitiflora

M. dissitiflora is a tall bushy shrub growing up to 5 m in height with erect or spreading branches and papery grey bark. It is found in sandy creek beds in the Flinders Ranges of South Australia as well as elsewhere in the north of that State and in the Northern Territory and western Queensland. It is a useful ornamental and low shelter belt species for dry areas.

M. dissitiflora exists in two chemical forms, obtained in yields of 0.7–2.1% based on fresh leaves. Chemotype (I) contains, as a major compound, 1,8-cineole (63–66%), with lesser amounts of limonene (5–7%), α–pinene (1–2%), terpinolene (3–4%), terpinen-4-ol (1–7%)

Table 4 Compounds identified in the essential oils of *Melaleuca citrolens*

Compound	*Chemotype I* %	*Chemotype II* %	*Chemotype III* %	*Chemotype IV* %
α-pinene	2.7	1.9	0.4	0.5
α-thujene		5.7	0.3	0.4
β-pinene	0.1	0.2	0.1	0.1
sabinene	0.1	0.6	tr	tr
myrcene			1.5	0.5
α-phellandrene	3.5	9.7	0.1	0.3
α-terpinene	0.4	1.2	tr	0.1
limonene	2.9	27	1.9	1.4
1,8-cineole	39.9	26.9	14.2	12.2
γ-terpinene	3.1	7.3	0.9	0.6
E-β-ocimene			tr	tr
p-cymene	1.4	4.6	1.2	1.4
terpinolene	12.1	15.2	2.7	2.8
mentha-1,3,8-triene	tr	0.3		
mentha-1,4-8-triene		0.4		
α,*p*-dimethylstyrene	0.1	0.8	0.1	0.2
trans-sabinene hydrate	tr	tr		
citronellal			0.3	30.1
bicycloelemene	0.8	tr	tr	tr
cis-sabinene hydrate	0.1	0.5		
linalool	tr	tr	0.0	0.2
isopulegol			0.3	5.4
trans-menth-2-en-1-ol		0.1		
isoisopulegol			tr	13.2
terpinen-4-ol	6.1	2.5	tr	0.1
β-caryophyllene	1.4	tr	3.8	3.8
cis-menth-2-en-1-ol		tr		
aromadendrene	1.0		0.1	0.6
alloaromadendrene	1.5		0.7	0.2
citronellyl acetate			2.3	5.2
$C_{10}H_{16}O$	0.4	0.6		
humulene			tr	0.7
neral			16.5	0.9
α-terpineol	5.9	4.9	3.1	2.6
viridiflorene	2.6		0.2	tr
geranial			26.7	tr
bicyclogermacrene	tr	0.2	0.2	tr
citronellol			0.3	3.4
nerol			0.1	tr
p-cymen-8-ol	0.3	2.1	1.9	0.7
geraniol			tr	0.7
piperitenone		12.6		
$C_{10}H_{16}O$		0.1		
palustrol	0.2		tr	tr

Table 4 (*Continued*)

Compound	*Chemotype I* %	*Chemotype II* %	*Chemotype III* %	*Chemotype IV* %
$C_{15}H_{26}O$	0.3		tr	0.9
$C_{15}H_{26}O$	1.0	tr	0.5	0.1
globulol	4.6	0.1	2.0	0.3
viridiflorol	1.0		0.2	0.1
spathulenol	1.4	tr	1.7	1.4
thymol		tr		
$C_{15}H_{24}O$	4.0	0.1	tr	tr
$C_{15}H_{24}O$	tr		1.0	0.1

Table 5 Compounds identified in the essential oils of *Melaleuca dissitiflora*

Compound	*Chemotype I* %	*Chemotype II* %	*Compound*	*Chemotype I* %	*Chemotype II* %
α-pinene	1.7	5.0	$C_{15}H_{24}$		0.3
camphene		tr	$C_{15}H_{24}$	0.1	1.0
β-pinene	0.8	8.0	terpinen-4-ol	3.3	38.0
sabinene	0.3	8.0	aromadendrene		0.5
myrcene	1.0	1.0	*trans*-β-terpineol	0.1	
α-phellandrene	tr	0.5	neryl acetate	0.1	
α-terpinene	0.7	7.0	$C_{10}H_{18}O$	1.5	0.2
limonene	6.1	1.0	α-terpineol	6.8	2.0
1,8-cineole	64.2	5.0	α-terpinyl acetate	0.1	tr
γ-terpinene	3.1	15.0	$C_{15}H_{24}$	0.5	
p-cymene	1.5	8.0	piperitone	0.2	0.1
terpinolene	3.2	3.0	geranyl acetate	tr	0.2
α,*p*-dimethylstyrene	0.2	0.1	*trans*-piperitol	0.5	0.6
linalool	0.1	0.3	*p*-cyme-8-ol	0.1	

and α-terpineol (4–9%). The oil yield of this chemotype was 1.9–2.2%, based on dry leaf. Chemotype (II) contains as a major compound terpinen-4-ol (38–52%). This was accompanied by lesser amounts of α-pinene (2–11%), β-pinene (0.5–15%), sabinene (1–15%), 1,8–cineole (1–8%), γ-terpinene (12–18%), terpinolene (2–4%) and α-terpineol (1–3%). The oil yield of this chemotype, based on air-dry leaf, was 1.4–4.2%. Neither chemotype contained sesquiterpenes in any significant amount. Table 5 lists typical analyses of the two chemotypes.

The cineole form has potential as a eucalyptus type oil. Tea tree oil producers have sought to develop the terpinen-4-ol form as an arid zone source for the commercial "Oil of *Melaleuca*, terpinen-4-ol type".

References: (Chemistry) Brophy and Lassak 1983, Brophy *et al.* 1989, Williams and Lusunzi 1994, Brophy and Doran 1996; (Botany) Byrnes 1986.

Melaleuca ericifolia

M. ericifolia is a shrub to small tree which has corky bark and leaves approximately 1 mm wide and 7–15 mm long. It grows in heath and dry sclerophyll forest, along stream banks and low-lying coastal swamps. Its range is from the Hastings River in NSW to central Victoria in eastern mainland Australia and also in Tasmania (Byrnes 1986; Wilson 1991).

The oil from this species has been known for over a century; the first report being by Joseph Bosisto in 1862 (mentioned in Baker and Smith 1922). Investigations at this time revealed that the oil was "a very limpid and almost colourless oil, partaking much of 'cajuput' flavour; that with age it improved greatly giving more the aroma of flowers". Baker and Smith (1922) reinvestigated the oil of this species and determined that the oil contained 1,8-cineole (<10%) and terpineol (~30%). Penfold and Morrison (1935) could find no α-terpineol but showed that (+)-linalool was the characteristic component. A subsequent investigation by Hellyer (1957) confirmed the presence of linalool at approx. 25%, with 1,8-cineole (21%) and 'terpenes' (approx. 15%) also being present. Later McKern and Willis (1957) stated that the linalool content was 30–40%.

A more recent examination of the oil of this species (Brophy 1998) showed that the oil, isolated in 0.8–2.0%, contained linalool (23–40%), 1,8-cineole (5–26%) and terpinolene (5–25%) as major components. These compounds were accompanied by lesser amounts of α-pinene (0.5–10%), γ-terpinene (2–6%) and α-terpineol (3–5%). Sesquiterpenes did not account for more than 10% of the overall oil, the principal contributors being aromadendrene, globulol and viridiflorol (each <2%). A complete list of compounds detected in oil of this species is given in Table 6.

This species is a potential source of linalool rich tea tree oils.

References: (Chemistry) Baker and Smith 1922, Penfold and Morrison 1935, Hellyer 1957, McKern and Willis 1957; (Botany) Byrnes 1986, Wilson 1991.

Melaleuca leucadendra

The species described here as *M. leucadendra* is only part of the taxon previously described as *M. leucadendron.* This latter description can be regarded as an all embracing term to describe the broad leaved *Melaleuca* species occurring in tropical Australia and into the Melanesian region.

M. leucadendra is frequently a large tree 20–40 m tall with a diameter that may reach 1.5 m. Thin, shiny-green lanceolate leaves, attractive weeping habit and white papery bark are distinguishing features of this species. It is found mainly in coastal and subcoastal areas of tropical Queensland, the Northern Territory and Western Australia but extends inland for up to 350 km along major rivers. It is also found in Papua New Guinea, and in Irian Jaya and on various islands in eastern Indonesia.

M. leucadendra has been shown to exist in two sets of chemical forms which are geographically separated. One chemical form occurs from Western Australia across to approximately the middle of Northern Territory. This chemotype consists entirely of mono- and sesquiterpenes. The second set of chemotypes consists of either methyl eugenol or *E*-methyl isoeugenol chemotypes. This second set of chemical forms occurs

Table 6 Compounds identified in the essential oil of *Melaleuca ericifolia*

Compound	%	*Compound*	%
α-pinene	0.4	alloaromadendrene	0.6
α-thujene	0.6	*trans*-pinocarveol	0.1
β-pinene	0.1	δ-terpineol	tr
myrcene	0.8	neral	0.5
α-phellandrene	1.1	α-terpineol	3.3
α-terpinene	0.9	viridiflorene	0.2
limonene	1.4	β-selinene	0.2
β-phellandrene	0.2	α-selinene	0.2
1,8-cineole	5.6	bicyclogermacrene	0.6
Z-β-ocimene	0.3	calemenene	0.1
γ-terpinene	3.5	geraniol/*p*-cymene-8-ol	0.5
E-β-ocimene	0.8	phenylethyl propionate	0.2
p-cymene	1.1	calacorene	0.0
terpinolene	24.5	$C_{15}H_{24}O$	0.2
mentha-1,3,8-triene	0.1	palustrol	0.1
mentha-1,4,8-triene	0.1	$C_{15}H_{26}O$	0.4
α,*p*-dimethylstyrene	0.4	$C_{15}H_{26}O$	0.1
cis-linalool oxide	0.2	$C_{15}H_{26}O$	0.6
trans-linalool oxide	0.0	globulol	2.1
α-cubebene	0.1	viridiflorol	0.6
bicycloelemene	0.1	$C_{15}H_{26}O$	0.3
α-copaene	0.2	$C_{15}H_{26}O$	0.3
linalool	39.1	spathulenol	1.0
terpinen-4-ol	0.8	$C_{15}H_{24}O$	0.2
aromadendrene	2.4	*E,E*-farnesol	0.3
α-bulnesene	0.2		

from mid Northern Territory eastwards to the Queensland coast. The oil yield of all forms was in the range 0.1–4% based on fresh leaves.

Chemotype (I), from Western Australia, had as its major compounds 1,8-cineole (10–45%), *p*-cymene (5–22%), α-pinene (4–19%), limonene (3–6%) and α-terpineol (6–9%). There was also a significant number of both mono- and sesquiterpenes present in small (<1%) quantities. The oil yield from this chemotype was 0.1–0.5% on a fresh weight basis.

Chemotype (II), which extends into Queensland has two chemical forms. One form contained methyl eugenol (95–97%) as its principal component. The second chemical form contained E-methyl isoeugenol (74–88%) as its major component with lesser amounts of methyl eugenol (6–24%) as a subsidiary component. The oil yield from this set of chemotypes was 1–4% on a fresh weight basis.

There appeared to be no interbreeding between chemotypes (I) and (II); at one location in the Northern Territory (Flying Fox Creek, Kapalga, 12°40'S, 132°19'E) both chemotypes occurred together but there was no sign of the aromatic compounds in the terpenic oil. Analyses of the different chemotypes are given in Table 7 and Table 8.

Table 7 Compounds identified in the essential oil of *Melaleuca leucadendra* Chemotype (I), Western Australia

Compound	%	*Compound*	%
α-pinene	5.7	viridiflorene	1.0
α-fenchene	tr	α-terpineol	8.1
camphene	tr	borneol	0.4
β-pinene	tr	β-selinene	0.1
3-carene	0.2	α-selinene	0.1
myrcene	0.4	bicyclogermacrene	tr
α-phellandrene	0.5	citronellol	0.6
α-terpinene	2.6	*trans*-mentha-1,8-dien-6-ol	tr
limonene	4.3	*p*-cymen-8-ol	0.3
1,8-cineole	31.0	$C_{15}H_{24}O$	0.1
γ-terpinene	17.4	palustrol	0.1
p-cymene	4.6	methyleugenol	1.6
terpinolene	9.6	$C_{15}H_{26}O$	0.2
α,*p*-dimethylstyrene	0.1	$C_{15}H_{26}O$	0.1
$C_{15}H_{24}$	tr	*E*-nerolidol	0.1
α-cubebene	0.1	$C_{15}H_{26}O$	0.2
campholenic aldehyde	0.1	$C_{15}H_{26}O$	0.1
linalool	1.1	globulol	0.3
isopulegol	0.1	viridiflorol	0.3
iso-isopulegol	0.5	$C_{15}H_{26}O$	0.1
fenchone	0.1	$C_{15}H_{26}O$	0.3
terpinen-4-yl acetate	0.1	spathulenol	0.1
terpinen-4-ol	3.1	$C_{15}H_{24}O$	0.1
β-caryophyllene	0.3	γ-eudesmol	0.5
aromadendrene	0.1	α-eudesmol	1.1
α-bulnesene	tr	β-eudesmol	1.0
alloaromadendrene	0.1	$C_{15}H_{26}O$	0.1
$C_{15}H_{24}$	tr	*E,E*-farnesol	0.4
trans-pinocarveol	0.1	$C_{15}H_{26}O$	0.1

Oil from the aromatic ether chemotypes of this species are excellent sources of methyl eugenol and *E*-methyl isoeugenol.

References: (Chemistry) Brophy and Lassak 1988, Brophy *et al.* 1989, Brophy and Doran 1996; (Botany) Blake 1968, Byrnes 1986, Barlow 1988.

Melaleuca linophylla

M. linophylla is a bushy shrub to 4 m in height with a papery bark. The narrow, flat leaves to 6 cm long are a distinguishing feature. This species occurs in creek beds and wet areas in the north west of Western Australia. *M. linophylla* is not widely cultivated but has potential as an ornamental because of its showy flowers.

The leaf essential oil obtained from *M. linophylla* in 1–2% yield based on fresh leaves, was heavily monoterpenic in character. By far the major component was

Table 8 Compounds identified in the essential oils of *Melaleuca leucadendra* chemotype (II), eastern Australia

Compound	*Chemotype IIa* %	*Chemotype IIb* %	*Compound*	*Chemotype IIa* %	*Chemotype IIb* %
E-β-ocimene	tr	0.3	humulene	tr	tr
α-cubebene	tr	tr	germacrene-D	1.2	0.3
δ-elemene	0.1	tr	bicyclogermacrene	0.3	0.1
α-copaene	0.1	tr	δ-cadinene	0.4	0.2
α-gurjunene	tr	tr	cadina-1,4-diene	0.1	0.2
linalool	tr	tr	calamene	0.1	0.3
acetophenone	tr	tr	methyl eugenol	94.6	6.7
β-elemene	tr	tr	*Z*-methyl isoeugenol	0.2	0.7
β-caryophyllene	0.4	0.4	*E*-methyl isoeugenol	0.4	88.0
$C_{15}H_{24}$	0.4	0.4	$C_{15}H_{26}O$	0.4	0.5

1,8-cineole (71–88%). It was accompanied by lesser amounts of limonene (5%), α-pinene (0.2–2%) and *p*-cymene (0.1–2%). With the exception of α-terpineol (2–12%), oxygenated monoterpenes were also present but only in very small amounts. The principal members being terpinen-4-ol (0.1–0.5%), α-terpineol (0.2–0.3%) and the *p*-mentha-1(7),8-dien-2-ols and *p*-mentha-1,8-dien-6-ols (each 0.1–0.3%).

Sesquiterpenes were present in small amounts. The main contributors were the related tricyclic compounds, globulol, viridiflorol and spathulenol (each 0.1–0.4%) and aromadendrene (0.1–0.2%). Table 9 lists a typical analysis.

This oil has potential as a medicinal oil with cineole contents equivalent to BP grade eucalyptus oils.

References: (Chemistry) Brophy and Doran 1996; (Botany) Byrnes 1986, Carrick and Chorney 1979.

Melaleuca quinquenervia

M. quinquenervia is an erect, small to medium-sized tree, normally 8–12 m tall but sometimes reaching 25 m. It has stiff, leathery, lanceolate-elliptic leaves and white or greyish papery bark. The species is found along the east coast of Australia north from Sydney, to southern Papua and in New Caledonia. Niaouli oil is produced from *M. quinquenervia* in New Caledonia and Madagascar.

The species is similar to *M. viridiflora*, and while Blake (1968) separated the two species, Byrnes considered it to be co-specific and reduced it to varietal rank (*M. viridiflora* var. *rubriflora*). As a result of this, care should be taken when dealing with reports from the literature, particularly the pre-1960 literature, in assigning just which species is being dealt with. Traceability to a voucher specimen is probably the only sure solution. It is considered that a significant number of reports on the oil contents, particularly those detailing high nerolidol and linalool oils, probably relate to *M. quinquenervia*.

M. quinquenervia is known to exist in two chemotypes, one chemotype (I) contains large amounts of *E*-nerolidol, while the second chemotype (II) contains a large amount

Table 9 Compounds identified in the essential oil of *Melaleuca linophylla*

Compound	%	*Compound*	%
isovaleraldehyde	tr	α-bulnesene	tr
α-pinene	1.0	methyl benzoate	tr
α-fenchene	tr	alloaromadendrene	0.1
camphene	0.1	methylchavicol	tr
β-pinene	0.6	δ-terpineol	0.2
sabinene	tr	α-terpineol	12.2
myrcene	0.9	geranial	tr
α-phellandrene	0.1	carvone	tr
α-terpinene	0.1	δ-cadinene	tr
limonene	4.2	citronellol	0.2
1,8-cineole	75.6	calemenene	0.1
Z-β-ocimene	tr	*trans*-mentha-1(7),8-dien-2-ol	0.1
γ-terpinene	0.6	*trans*-mentha-1,8-dien-6-ol	0.1
E-β-ocimene	0.2	*p*-cymen-8-ol	tr
p-cymene	0.3	*cis*-mentha-1,8-dien-6-ol	tr
terpinolene	0.2	*cis*-mentha-1(7),8-dien-2-ol	0.1
nonan-2-one	tr	palustrol	0.1
α,*p*-dimethylstyrene	tr	$C_{15}H_{26}O$	0.1
α-cubebene	tr	$C_{15}H_{26}O$	0.1
$C_{15}H_{24}$	tr	$C_{15}H_{26}O$	0.1
α-gurjunene	0.1	$C_{15}H_{26}O$	0.1
linalool	0.4	globulol	0.5
isopulegol	tr	viridiflorol	0.2
isoisopulegol	tr	$C_{15}H_{26}O$	0.1
fenchol	tr	$C_{15}H_{26}O$	0.3
terpinen-4-ol	0.7	spathulenol	0.2
β-caryophyllene	tr	eugenol	0.3
aromadendrene	0.2		

of 1,8-cineole. It is this latter chemotype that produces niaouli oil in New Caledonia. Within this second chemotype we have found trees in which the amount of 1,8-cineole is small and there is a large amount of the sesquiterpene alcohol, viridiflorol.

Chemotype (I) contains *E*-nerolidol (>95%) as its major component. This is accompanied by lesser quantities of *E,E*-farnesol, β-farnesene, β-caryophyllene, linalool, benzaldehyde and 1,8-cineole (all in trace-0.2%). The oil yield of this chemotype, based on fresh leaves was up to 2%.

Chemotype (II) contains 1,8-cineole (50–65%) as its principal component. This was accompanied by lesser amounts of the hydrocarbons α-pinene (2–9%), myrcene (1–2%), limonene (6–8%), terpinolene (0.5–1%), β-caryophyllene (1–3%), aromadendrene (1–2%), α-terpineol (5–10%), viridiflorene (1–2%) and globulol (1–4%). There were a large number of both mono- and sesquiterpenes present in amounts of less than 1%. The oil yield of this chemotype was 1.3–2.4% based on fresh leaves.

As a variant on this chemotype, a group of trees produced an oil in which the major component was viridiflorol (80%). This was accompanied by lesser amounts of α-pinene (10%), limonene (2.5%), viridiflorene (1%), β-caryophyllene (0.4%) and benzaldehyde (0.2%). α-, β- and γ-eudesmol were also present (each <0.2%). The oil yield from this variant was 1.3%, based on fresh leaves. Table 10 lists the oil contents of Chemotype (I) and the viridiflorol variety of Chemotype (II). The other chemotype variety is dealt with under niaouli oil.

In addition to the use of this species for the production of niaouli oil, the oil from the other chemical variants is an excellent source of *E*-nerolidol (for perfumery) and viridiflorol.

References: (Chemistry) Brophy *et al.* 1989, Guenther 1950, Moudachirou *et al.* 1996, Ramanoelina *et al.* 1992, 1994, Jones and Haenke 1937, 1938, Brophy and Doran 1996; (Botany) Blake 1968, Byrnes 1986.

Melaleuca squamophloia

M. squamophloia is a species with limited distribution, it grows on the black soil plains usually in wetter areas of the Darling Downs district in the south-east corner of Queensland. Byrnes (1986) at one time thought that it may be a hybrid between *M. styphelioides* and *M. bracteata.*

There appear to be two chemotypes in this species, which produce oils in 0.4–3.7% yield based on fresh leaves. Both chemotypes are dominated by aromatic compounds, either *E*-isoelemicin or elemicin. Chemotype I produced an oil in which the principal component was elemicin (93–97%). This was accompanied by lesser amounts of the aromatic compounds *E*-isoelemicin (0.1–0.2%), *E*-methyl cinnamate (0.1–2.0%) and the terpenes spathulenol (0.5–2.0%), α-pinene (trace-2.0%), *E*-β-ocimene (0.3–0.5%), linalool (0.2–1.0%) and alloaromadendrene (0.3–1.0%). Chemotype II had as its principal

Table 10 Compounds identified in two chemotypes of *Melaleuca quinquenervia*

Compound	*Chemotype I*	*Chemotype II variant* %	*Compound*	*Chemotype I*	*Chemotype II variant* %
α-pinene		3.9	alloaromadendrene		0.3
camphene		tr	humulene		0.2
β-pinene		1.0	α-terpineol		0.1
myrcene		tr	viridiflorene		2.1
α-terpinene		tr	β-farnesene	0.1	tr
limonene		3.6	β-selinene		0.2
1,8-cineole	0.2	2.6	α-selinene		0.3
γ-terpinene		0.3	δ-cadinene		0.4
p-cymene		0.2	palustrol		0.3
terpinolene		0.1	*E*-nerolidol	95.0	2.9
linalool	0.1	0.1	viridiflorol		66.3
benzaldehyde	0.1	0.2	γ-eudesmol		1.3
terpinen-4-ol		tr	α-eudesmol		2.0
β-caryophyllene	0.1	1.2	β-eudesmol		2.4
aromadendrene		0.2	*E*,*E*-farnesol	0.2	tr

component *E*-isoelemicin (65–79%). This was accompanied by lesser amounts of the aromatic compounds elemicin (12–15%), *Z*-isoelemicin (3.1%) and a dimethoxybenzaldehyde (0.1%).

The main terpenes present in the oil of this chemotype were α-pinene (3–6%), spathulenol (0.5–2.0%), alloaromadendrene (0.1–1.0%) and linalool (0.3–1%).

In all the analyses run there was no oil from a single tree which contained comparable amounts of both elemicin and *E*-isoelemicin, which is thought to indicate that there is no hybridisation between the two chemotypes. Neither was there an appreciable difference in oil yield between the chemotypes. Analyses of both chemotypes are given in Table 11.

Oils from the chemotypes of this species are excellent sources for the commercial production of elemicin and *E*-isoelemicin respectively.

References: (Chemistry) Brophy and Doran 1996; (Botany) Byrnes 1986, Craven and Barlow 1997.

Melaleuca stenostachya

M. stenostachya is a shrub or tree, 4–25 m tall, with a small crown and stiff spreading branches. Bark may be hard or papery. The species occurs in far northern Queensland and north eastern Northern Territory and grows on a wide range of soils including sands, alluviums and skeletals that may be subject to inundation for short periods during the wet season. The wood is used in the round for posts and rails and has potential for fuel.

Table 11 Compounds identified in the essential oils of *Melaleuca squamophloia*

Compound	*Chemotype I* %	*Chemotype II* %	*Compound*	*Chemotype I* %	*Chemotype II* %
α-pinene	1.3	tr	β-selinene		tr
β-pinene	tr	tr	α-selinene		tr
myrcene	tr		bicyclogermacrene	0.3	
limonene	tr	tr	δ-cadinene	0.1	0.2
1,8-cineole		tr	caryophyllene oxide	tr	tr
Z-β-ocimene		tr	*Z*-methyl cinnamate	tr	tr
γ-terpinene		tr	methyleugenol		0.1
E-β-ocimene	0.4	0.3	*E*-methyl cinnamate	2.1	tr
α-copaene	tr	0.2	globulol	0.2	0.1
benzaldehyde	tr	tr	viridiflorol	0.1	0.1
linalool	0.3	0.3	*Z*-methyl isoeugenol	tr	tr
β-caryophyllene	0.1	tr	spathulenol	1.3	1.5
aromadendrene	0.1	tr	*E*-methylisoeugenol		0.4
myrtenal		0.1	elemicin	93.6	13.0
alloaro-madendrene	0.2	0.8	*Z*-isoelemicin		3.4
humulene	tr	tr	dimethoxybenzaldehyde		0.1
viridiflorene	tr	tr	*E*-isoelemicin	0.1	78.5
α-terpineol	tr	tr	methyl eudesmate	0.1	0.1

The essential oil obtained from *M. stenostachya* was monoterpenoid in character. The major comxponents were the ether, 1,8-cineole (53–62%) and the hydrocarbons α-pinene (19–29%), β–pinene (1–2%), limonene (4–6%) and γ-terpinene (0.3–0.5%). Other monoterpenes detected were terpinen-4-ol (0.2–0.5%) and α-terpineol (1–3%).

The major sesquiterpenes detected were β-caryophyllene (2–6%), humulene (0.2–1%), caryophyllene oxide (0.4–1%), globulol (0.4–1%), viridiflorol (0.1%) and spathulenol (0.1–1%). The oil yield, based on fresh leaves, was 0.7–0.9%.

A second collection from north of the Laura River crossing in northern Queensland, contained α-pinene (27–28%), β-pinene (41–44%) and 1,8-cineole (11–13%) as its major components. The oil yield from this source, based on fresh leaves, was 1–1.2%. Table 12 lists a typical analysis of this oil.

The higher cineole oils from this source have potential as medicinal oils of the cajuput, niaouli and eucalyptus type.

References: (Chemistry) Brophy *et al.* 1988, 1989, Brophy and Doran 1996; (Botany) Blake 1968, Byrnes 1986.

Melaleuca stipitata

M. stipitata is a small tree to 6 m in height with light branching, spreading crown and papery bark shedding in small strips. It occurs in dense clumps on shallow loamy soils on shale rises. It is endemic to the Bukbuluk area of Kakadu National Park in the Northern Territory. The performance and range of potential uses of the species in cultivation are unknown.

Table 12 Compounds identified in the essential oil of *Melaleuca stenostachya*

Compound	%	*Compound*	%
α-pinene	28.9	aromadendrene	0.1
camphene	tr	*cis*-menth-2-en-1-ol	tr
β-pinene	1.3	alloaromadendrene	0.1
sabinene	tr	δ-terpineol	tr
myrcene	0.4	humulene	0.7
α-phellandrene	tr	α-terpineol	2.0
α-terpinene	0.1	viridiflorene	tr
limonene	4.2	bicyclogermacrene	tr
1,8-cineole	52.7	Z-α-bisabolene	0.1
γ-terpinene	0.6	calamenene	tr
p-cymene	0.1	caryophyllene oxide	0.4
terpinolene	0.2	methyl eugenol	4.7
rose oxide	tr	$C_{15}H_{24}O$	0.1
α,*p*-dimethylstyrene	tr	globulol	0.6
α-cubebene	tr	viridiflorol	0.1
δ-elemene	tr	$C_{15}H_{26}O$	tr
benzaldehyde	0.1	$C_{15}H_{26}O$	0.1
trans-menth-2-en-1-ol	tr	spathulenol	0.3
terpinen-4-ol	0.2	$C_{15}H_{26}O$	0.5
β-caryophyllene	5.6		

The pleasant, lemon scented oil obtained from *M. stipitata*, obtained in 0.7–3.1% yield based on fresh leaves, was monoterpenic in character. The major components were neral (14%), geranial (30%) and terpinen-4-ol (11%). Other monoterpenes present in significant amounts were α-pinene 1%, β-pinene (0.5%), sabinene (6%), myrcene (2%), α-terpinene (4%), limonene (2%), 1,8-cineole (5%), γ-terpinene (6%), *p*-cymene (1%), terpinolene (1.5%), α-terpineol (5%), nerol (1%) and geraniol (1%).

The major sesquiterpenes were globulol (1%), viridiflorol (0.5%) and spathulenol (0.5%). The oil yield, based on fresh leaves, from 17 trees averaged 1.6% with a range of 0.7 to 3.1%. The species is reported to have bactericidal properties (Doran and Markam 1997). Table 13 gives a typical analysis of this oil.

This product has potential as a lemon scented, low terpinen-4-ol anti-microbial medicinal oil. Some formulations mask the myristic odour of tea tree oil with lemon oils. This oil is a natural mix of both and from the higher yielding forms presents good commercial prospects.

References: (Chemistry, as *M.* sp. 'bukbuluk') Brophy and Doran 1996; (Botany) Craven and Barlow 1997.

Table 13 Compounds identified in the essential oil of *Melaleuca stipitata*

Compound	%	*Compound*	%
α-pinene	0.8	terpinen-4-ol	10.4
α-thujene	0.7	aromadendrene	0.3
β-pinene	0.5	*cis*-menth-2-en-1-ol	0.2
sabinene	6.0	neral	13.7
myrcene	1.8	α-terpineol	5.4
α-phellandrene	0.6	geranial	29.8
α-terpinene	3.4	bicyclogermacrene	0.2
limonene	2.1	unknown	0.1
β-phellandrene	1.0	δ-cadinene	0.2
1,8-cineole	5.3	nerol	0.8
γ-terpinene	5.8	geraniol	0.8
E-β-ocimene	0.4	palustrol	0.1
p-cymene	1.2	$C_{15}H_{26}O$	0.1
terpinolene	1.3	methyl eugenol	0.2
6-methylhept-5-en-2-one	0.1	$C_{15}H_{26}O$	0.1
$C_{10}H_{14}O$	tr	$C_{15}H_{26}O$	0.2
mentha-1,4,8-triene	tr	globulol	0.9
α,*p*-dimethylstyrene	0.25	viridiflorol	0.5
cis-linalool oxide	tr	$C_{15}H_{26}O$	0.2
trans-linalool oxide	tr	$C_{15}H_{26}O$	0.2
α-cubebene	tr	spathulenol	0.5
bicycloelemene	0.1	$C_{15}H_{24}O$	0.1
cis-sabinene hydrate	0.2	$C_{15}H_{24}O$	0.3
trans-menth-2-en-ol	0.1	unknown, mw 208	0.2
$C_{10}H_{16}O$	0.4		

Melaleuca trichostachya

M. trichostachya ranges from a medium-sized shrub of a height of 2 m to a small tree of 15 m, with papery bark. It grows in sandy soils along creeks, rivers and gorges and has a wide distribution in Queensland extending to northern New South Wales and with disjunct populations in South Australia and the Northern Territory. *M. trichostachya* is closely related to and resembles *M. linariifolia*. It is an excellent shade and ornamental tree.

M. trichostachya occurs in at least two chemical forms. One form is high in terpinolene, while the other form is rich in 1,8-cineole and contains significant amounts of terpinen-4-ol.

Chemotype (I) contains 1,8-cineole (45–57%) and terpinen-4-ol (11–16%) as major components. These were accompanied by lesser amounts of α-pinene and α-thujene (each 1–2%), myrcene (1–2%), α-terpinene (1–4%), limonene (4–5%), γ-terpinene (8–12%), *p*-cymene (1–4%) and α-terpineol (5–7%). Sesquiterpenes, both in this chemotype and the following chemotype, did not contribute significantly to the overall oil. The oil yield, based on fresh leaves, was 1.6–2.3%. The two chemotypes are listed in Table 14.

Chemotype (II) contains terpinolene (47–65%) as its major component. This is accompanied by lesser amounts of 1,8-cineole (9–24%), α-pinene (1–3%), limonene (2–4%), terpinen-4-ol (1–3%) and α-terpineol (1–4%). The oil yield of this chemotype, based on fresh leaves, was 1–1.5%.

The potential for this species exists as a high cineole, low terpinen-4-ol oil with moderate antimicrobial activities (Chemotype I) or as a source of terpinolene (Chemotype II). Antimicrobial testing (Markham and Southwell 1996) of an oil of this type has shown anti-microbial activity which, although significant, was not as strong as the terpinen-4-ol rich oils.

Table 14 Compounds identified in the essential oils of *Melaleuca trichostachya*

Compound	*Chemotype I* %	*Chemotype II* %	*Compound*	*Chemotype I* %	*Chemotype II* %
α-pinene	1.7	3.1	fenchol	tr	
α-thujene	1.5		terpinen-4-ol	11.7	1.3
α-fenchene	tr	tr	aromadendrene	0.1	0.2
camphene	tr	tr	α-bulnesene	tr	tr
β-pinene	0.5	0.2	δ-terpineol	0.1	
sabinene		tr	*cis*-piperitol	0.1	
myrcene	1.5	1.2	α-terpineol	5.9	2.1
α-phellandrene		3.8	bicyclogermacrene	0.1	
α-terpinene	1.9	1.2	*p*-cymen-8-ol	tr	0.3
limonene	5.1	2.3	calacorene	0.4	0.1
1,8-cineole	57.3	11.3	caryophyllene oxide	tr	
γ-terpinene	9.1	3.6	globulol	0.1	0.2
p-cymene	1.7	0.5	viridiflorol	0.1	tr
terpinolene	0.7	65.0	spathulenol	0.1	0.1
linalool	tr	1.1			

References: (Chemistry) Brophy and Lassak 1983, Southwell *et al.* 1992, Brophy and Doran 1996; (Botany) Byrnes 1985.

Melaleuca uncinata

M. uncinata is a medium sized broom-like shrub which grows to a height of approximately 3 m. It has probably the widest distribution of all the species of *Melaleuca*, being found across all of the southern half Australia, especially in arid and semi-arid areas. Its terete leaves are approximately 3 mm across and 1.5–6 cm in length. Branches of this shrub are used extensively for brush fencing.

There appear to be 4 chemical varieties of *M. uncinata.* Chemotype I (obtained in 1.1–2.7% yield based on fresh leaves) contains significant quantities of α-, β- and γ-eudesmol and 1,8-cineole. Chemotype II (2.1–4.5% yield based on fresh leaves) contains large amounts of 1,8-cineole and little in the way of oxygenated sesquiterpenes, while Chemotype III (0.7–3.7% yield based on fresh leaves) contains significant quantities of terpinen-4-ol. Chemotype IV (0.25%) contains large amounts of α-pinene and little in the way of sesquiterpenes.

In Chemotype I, 1,8-cineole accounted for 30–60% while α-, β- and γ-eudesmol, in total, accounted for 30–60%. The remainder of the oil was composed of mono- and sesquiterpenes with α-pinene, α-terpineol, globulol and spathulenol being the major contributors.

Chemotype II appears to have a limited distribution on the granitic rocks in the Kalgoorlie-Coolgardie area of Western Australia. In this chemotype 1,8-cineole accounted for 80–85% of the oil, with α-pinene, and α-terpineol each accounting for a further 3–7%. In this chemotype sesquiterpenes are present in only trace or very minor quantities.

Chemotype III differs from the previous two chemotypes in the abundance of terpinen-4-ol present in the oil (24–42%). This is accompanied by significant amounts of the monoterpenes sabinene (3–7%), α-terpinene (4–5%), γ-terpinene (6–9%), *p*-cymene (11–26%) and terpinolene (2–3%). 1,8-cineole (0.4–2%) was quite low. Sesquiterpenes, while present, totalled no more than approximately 10% of the oil.

Chemotype IV has been reported from one collection near Lake Grace in Western Australia. In this chemotype, obtained in 0.25% yield, α-pinene (>85%) is the main component and while sesquiterpenes are reported to be present they are obviously not of any prominence. All four chemotypes are listed in Table 15.

These four chemotypes have potential as sources of eudesmol, 1,8-cineole, terpinen-4-ol and α-pinene respectively, with the terpinen-4-ol type having similar chemistry and antimicrobial properties to *M. alternifolia* (Carson and Riley 1995).

References: (Chemistry) Brophy and Lassak 1992, Brophy *et al.* 1990, Murray 1950, Watson 1943, Brophy 1998; (Botany) Byrnes 1985, Carrick and Chorney 1979.

Melaleuca viridiflora

M. viridiflora is typically a small tree, 5–10 m tall but may attain 25 m in height under favourable conditions. It has leathery, dull-green leaves and pale brown papery bark. The species grows on a wide range of soils on swampy ground close to the coast or sometimes on drier inland sites. It is very common throughout much of northern

Table 15 Compounds identified in the essential oils of *Melaleuca uncinata*

Compound	*Chemotype I* %	*Chemotype II* %	*Chemotype III* %	*Chemotype IV* (Murray 1950) %
α-pinene			1.7	>85
α-thujene	2.5	3.9	1.9	
camphene			tr	
β-pinene	0.7	0.2	0.4	
sabinene	0.1		7.7	
$C_{10}H_{14}$			0.5	
myrcene	0.5	0.8	1.0	
α-terpinene	0.1	tr	5.4	
limonene	4.4	1.0	2.0	
β-phellandrene	0.2		1.0	
1,8-cineole	32.0	84.6	0.5	
γ-terpinene	0.2	0.2	8.6	
p-cymene	0.3	0.3	11.7	
terpinolene	tr	tr	2.4	
mentha-1,3,8-triene			0.1	
menth-1,4,8-triene			0.1	
α,*p*-dimethylstyrene			0.2	
cis-sabinene hydrate ..tent..			0.3	
trans-sabinene hydrate ..tent..			0.4	
trans-menth-2-en-1-ol		tr	1.5	
β-caryophyllene	0.7	tr		
terpinen-4-ol	0.2	1.2	41.5	
aromadendrene	tr			
cis-menth-2-en-1-ol	tr	tr	1.4	
alloaromadendrene	0.3	tr		
cis-piperitol			0.3	
δ-terpineol			0.3	
trans-menth-2,8-dien-1-ol	0.4	tr		
cis-menth-2,8-dien-1-ol	0.4	0.1		
humulene	0.2			
α-terpineol	1.1	6.7	2.0	
borneol			0.3	
bicyclogermacrene	0.6	tr	0.2	
trans-piperitol			0.5	
δ-cadinene			0.2	
cuminal			0.1	
myrtenol			0.2	
p-cymene-8-ol			0.5	
palustrol	0.3		0.1	
$C_{15}H_{26}O$	0.2		0.2	

Table 15 *(Continued)*

Compound	*Chemotype I* %	*Chemotype II* %	*Chemotype III* %	*Chemotype IV* (Murray 1950) %
globulol	0.2	tr	0.2	
viridiflorol	4.4	tr	0.1	
$C_{15}H_{26}O$	0.3	tr	0.3	
spathulenol	1.1	tr	0.5	
γ-eudesmol	12.4		0.5	
α-eudesmol	12.0		0.3	
β-eudesmol	18.0		1.1	

Australia. It also occurs in southern Papua New Guinea and in Irian Jaya. *M. viridiflora* is useful for shelter and amenity planting and has potential to produce fuel, posts and poles. The confusion between this species and *M. quinquenervia* has already been mentioned. The analyses reported here are all from vouched specimens.

There appear to be two chemotypes of *M. viridiflora*, Chemotype (I) being a terpenoid chemotype which showed a large amount of chemical variation, and Chemotype (II) in which the principal component was *E*-methyl cinnamate (Hellyer and Lassak 1968) .

Chemotype (I) seemed to be a variable species. While all the collections of this chemotype were terpenoid in character, three different types of oil have been encountered. One oil, arising from trees grown at Gympie from seed collected north west of Chillagoe, north Queensland, contained, as major components, γ-terpinene (39–47%) and terpinolene (26–33%). These were accompanied by lesser amounts of α-pinene (7–9%), α-phellandrene (2–4%), α-terpinene (7–9%), limonene (1–2%), terpinen-4-ol (0.7–2%) and β-caryophyllene (0.4–1%). The oil yield of this variant, based on dry leaves, was 1.3–2.1%.

A second chemical variant contained large amounts of 1,8-cineole (30–60%) as its major component. This was accompanied by significant amounts of the monoterpenes α-pinene (2–7%), β-pinene (2–5%), myrcene (0.4–2%), limonene (5–10%), α-terpineol (5–8%) and the sesquiterpenes β-caryophyllene (0.5–3%), viridiflorene (1–4%), globulol (1–8%), viridiflorol (3–9%) and spathulenol (3–14%). The oil yield of this variant was 0.4–0.7%, based on fresh leaves.

A third chemical variant contained lesser amounts of monoterpenes, particularly 1,8-cineole (6–12%) and much larger amounts of sesquiterpenes, β-caryophyllene (2–10%), spathulenol (4–14%) and globulol (2–13%). A large number of sesquiterpenes in amounts of less than 3% were present in this oil. The yield, based on air-dry leaves, was 0.4–0.9%.

Chemotype (II) consisted basically of two compounds, *E*-methyl cinnamate (82%) and *E*-β-ocimene (12%). The remainder of the oil was accounted for by 2,4,6-trimethoxyisobutyrophenone (5%), *Z*-methyl cinnamate (0.5%) and linalool (0.6%). The oil yield of this chemotype, based on air-dry leaves, was 4%. Tables 16 and 17 give analyses of the two chemotypes.

Table 16 Compounds identified in the essential oil of *Melaleuca viridiflora* Chemotype I

Compound	*Variant* 1 %	*Variant* 2 %	*Variant* 3 %
α-pinene	9.0	1.7	29.0
camphene		tr	tr
β-pinene	0.1	0.8	0.8
sabinene		tr	tr
myrcene	0.5	tr	tr
α-terpinene	7.3	tr	tr
limonene	1.6	4.6	1.3
1,8-cineole	0.3	48.7	11.6
γ-terpinene	39.0	0.1	0.1
p-cymene	2.9	0.1	0.9
terpinolene	32.6	0.1	tr
δ-elemene		0.2	tr
benzaldehyde		0.1	tr
$C_{15}H_{24}$		0.1	0.1
linalool	tr	0.4	2.1
trans-menth-2-en-1-ol	tr	tr	
terpinen-4-ol	0.8	0.2	1.4
β-caryophyllene	0.8	9.8	2.1
aromadendrene		1.2	0.8
bulnesene		0.5	0.1
alloaromadendrene		0.6	0.5
cis-menth-2-en-1-ol	tr	tr	
humulene	0.2	1.4	0.5
δ-terpineol		0.1	0.1
β-farnesene		tr	tr
$C_{15}H_{24}$			0.1
α-terpineol	tr	6.2	2.1
viridiflorene		tr	0.1
$C_{15}H_{24}$		tr	0.1
$C_{15}H_{24}$		2.1	tr
bicyclogermacrene		2.6	0.1
$C_{15}H_{24}$		0.1	0.1
calamanene		0.1	0.4
calacorene		0.1	tr
$C_{15}H_{24}O$		0.1	0.3
caryophyllene oxide		3.2	1.7
$C_{15}H_{26}O$		0.2	0.2
$C_{15}H_{26}O$		0.1	0.2
$C_{15}H_{24}O$	0.6	0.7	0.4
$C_{15}H_{24}O$		0.2	0.4
$C_{15}H_{24}O$		0.1	0.1

Table 16 (*Continued*)

Compound	*Variant* 1 %	*Variant* 2 %	*Variant* 3 %
globulol		2.0	2.5
viridiflorol		0.3	0.3
spathulenol		3.7	15.5
$C_{15}H_{24/26}O$		6.7	21.4

Table 17 Compounds identified in the essential oil of *Melaleuca viridiflora* Chemotype II

Compound	%	*Compound*	%
tricyclene	tr	β-caryophyllene	0.1
myrcene	0.1	humulene	tr
Z-β-ocimene	0.2	α-terpineol	tr
E-β-ocimene	12.0	*Z*-methyl cinnamate	0.5
benzaldehyde	tr	*E*-methyl cinnamate	81.2
linalool	0.6	2,4,6-trimethoxyisobutyrophenone	4.6
terpinen-4-ol	tr	unknown, mw 192	0.6

There is potential for Chemotype II as a source of methyl cinnamate, though the New England eucalypt, *Eucalyptus olida*, is a better prospect (Curtis *et al.* 1990). The other variants provide sources of α-pinene, γ-terpinene, terpinolene and cineole.

References: (Chemistry) Brophy *et al.* 1989, Lassak and Southwell 1977, Curtis *et al.* 1990, Guenther 1950, Brophy and Doran 1996; (Botany) Blake 1968, Byrnes 1986.

Asteromyrtus symphyocarpa

A. symphyocarpa is a multi-stemmed shrub or small tree in Australia, usually in the height range of 3–12 m, but may reach larger dimensions in Papua New Guinea. This species is adapted to acidic, infertile and periodically waterlogged soils in the lowland tropics. It occurs in the Northern Territory and far northern Queensland. It extends to southern Papua New Guinea and Irian Jaya in Indonesia.

The oil obtained from *A. symphyocarpa* (obtained in 1–1.3% yield, based on fresh leaves) had as principal components 1,8–cineole (39–70%) and α-pinene (7–18%). These were accompanied by lesser amounts of limonene (1–2%), γ-terpinene (0.5–4%), *p*-cymene (0.1–2%), terpinen-4-ol (1–6%), β-caryophyllene (1–6%), α-terpineol (3–5%), globulol (0.5–2%), spathulenol (0.5–1%) and α-, β- and γ-eudesmols (total 1–6%).

This species, under the name *Melaleuca symphyocarpa*, has been the subject of a previous report (Brophy *et al.* 1990), in which the trees examined came from five different sites on Cape York Peninsula and in the Northern Territory. The oil obtained from those trees was both qualitatively and quantitatively similar to that reported here. It differed principally in containing less α-pinene (8–16%), terpinen-4-ol (0.4–1%) and α-, β- and γ-eudesmols (not detected) and containing more 1,8-cineole (45–68%) and β-caryophyllene (4–19%). The oil composition appears to remain

Table 18 Compounds identified in the essential oil of *Asteromyrtus symophyocarpa*

Compound	%	*Compound*	%
α-pinene	8.5	borneol	tr
camphene	tr	δ-cadinene	tr
β-pinene	1.1	γ-cadinene	tr
sabinene	tr	$C_{15}H_{24}$	tr
myrcene	0.5	cadina-1,4-diene	tr
α-terpinene	tr	calamenene	tr
limonene	1.3	caryophyllene oxide	tr
1,8-cineole	67.7	*E*-nerolidol	0.4
γ-terpinene	0.5	$C_{15}H_{26}O$	0.2
p-cymene	0.2	$C_{15}H_{26}O$	0.1
terpinolene	0.1	epiglobulol	0.4
α-copaene	tr	globulol	1.6
α-gurjunene	tr	viridiflorol	0.2
trans-p-menth-2-en-1-ol	tr	$C_{15}H_{26}O$	1.1
fenchone	tr	$C_{15}H_{26}O$	0.1
terpinen-4-ol	0.8	$C_{15}H_{24}O$	0.5
β-caryophyllene	7.6	spathulenol	0.3
aromadendrene	0.4	γ-eudesmol	0.3
α-bulnesene	tr	$C_{15}H_{26}O$	0.1
cis-menth-2-en-1-ol	0.1	α-eudesmol	0.5
alloaromadendrene	0.7	β-eudesmol	0.5
humulene	0.7	$C_{15}H_{26}O$	tr
viridiflorene	0.1	*E*, *E*-farnesol	0.1
α-terpineol	3.4	$C_{15}H_{24}O$	0.1

relatively constant throughout the geographic range of this species. A typical oil analysis is given in Table 18.

Cineole-rich oils of this nature have potential use as medicinal oils similar to cajuput, niaouli and eucalyptus oils.

References: (Chemistry) Brophy *et al.* 1988, 1989, 1990, 1994, Brophy and Doran 1996; (Botany) Craven 1988, Byrnes 1985, Cowley *et al.* 1990.

ACKNOWLEDGEMENTS

Production of the work described in this chapter would not have been possible without the help of the people mentioned below who collected and identified the *Melaleuca* species. Leaf samples were contributed by N. Ashwathappa, Doug Boland, John Clarkson, Lyn Craven, John Doran, Paul Forster, Brian Gunn, Erich Lassak, David Lea, Brendan Lepschi, Maurice McDonald, Jock Morse and John Neldner. Generous help in the isolation of the volatile oils, obtaining their gas chromatographic profiles and checking the past literature was given by Bob Goldsack. I am deeply indebted to them. The analysis of the oils of the species contained in this report was funded as part of a forestry project supported by the Australian Centre for International Agricultural Research (ACIAR).

REFERENCES

Aboutabl, E.A., El Tohamy, S.F., De Pooter, H.L. and De Buyck, L.F. (1991) A comparative study of the essential oils from three *Melaleuca* species growing in Egypt. *Flav. Frag. J.*, **6**, 139–141.

Baker, R.T. and Smith, H.G. (1910) On the Australian Melaleucas and their essential oils, Part III. *J. and Proc. Roy. Soc. NSW.*, **44**, 592–615.

Baker, R.T. and Smith, H.G. (1922) On the Australian Melaleucas and their essential oils, Part VI. *J. and Proc. Roy. Soc. NSW.*, **56**, 115–124.

Barlow, B.A. (1986) Contributions to the revision of *Melaleuca* (Myrtaceae): 1–3. *Brunonia*, **9**, 163–177.

Barlow, B.A. (1988) Patterns of differentiation in tropical species of *Melaleuca* L. (Myrtaceae). *Proc. Ecol. Soc. Aust.*, **15**, 239–247.

Barlow, B.A. and Cowley, K. (1988) Contributions to a revision of *Melaleuca* (Myrtaceae): 4–6. *Aust. Syst. Bot.*, **1**, 95–126.

Blake, S.T. (1968) A revision of *Melaleuca leucadendron* and its allies (Myrtaceae). *Contribution from the Queensland Herbarium*, **1**, 1–114.

Brophy, J.J. (1998) University of NSW. Unpublished results.

Brophy, J.J., Boland, D.J. and Lassak, E.V. (1989) Survey of the leaf essential oils of *Melaleuca* and *Leptospermum* species from tropical Australia. In D.J. Boland, (ed.), *Trees For The Tropics—Growing Australian multipurpose trees and shrubs in developing countries.* ACIAR Monograph No. 10, pp. 193–203.

Brophy, J.J. and Clarkson, J.R. (1989) The essential oils of four chemotypes of *Melaleuca citrolens*. *J. and Proc. Roy. Soc. N.S.W.*, **122**, 11–18.

Brophy, J.J., Clarkson, J.R., Craven, L.A. and Forrester, R.I. (1994) The essential oils of the Australian members of the genus *Asteromyrtus* (Myrtaceae). *Biochem. Syst. and Ecol.*, **22**, 409–417.

Brophy, J.J. and Doran, J.C. (1996) *Essential oils of Tropical* Asteromyrtus, Callistemon *and* Melaleuca *species; In search of interesting oils with commercial potential.* ACIAR Monograph No 40. ACIAR, Canberra.

Brophy, J.J. and Lassak, E.V. (1983) The volatile leaf oils of *Melaleuca armillaris, M. dissitiflora and M. trichostachya. J. and Proc. Roy. Soc. N.S.W.*, **116**, 7–10.

Brophy, J.J. and Lassak, E.V. (1988) The leaf oil of *Melaleuca leucadendra.* L. *Flav. Frag. J.*, **3**, 43–46.

Brophy, J.J. and Lassak, E.V. (1992) Steam volatile leaf oils of some *Melaleuca* species from Western Australia. *Flav. Frag. J.*, **7**, 27–31.

Brophy, J.J., Lassak, E.V. and Boland, D.J. (1987) Volatile leaf oils of the two subspecies of *Melaleuca acacioides* F. Muell. *J. and Proc. Roy. Soc. N.S.W.*, **120**, 135–139.

Brophy, J.J., Lassak, E.V. and Boland, D.J. (1988) Volatile leaf oils of six northern Australian broad-leaved Melaleucas. *J. and Proc. Roy. Soc. N.S.W.*, **121**, 29–33.

Brophy, J.J., Lassak, E.V. and Boland, D.J. (1990) Steam volatile oils of *Melaleuca globifera, M. lateriflora, M. symphyocarpa and M. uncinata. Flav. Frag. J.*, **5**, 43–48.

Byrnes, N.B. (1984) A revision of *Melaleuca* L. (Myrtaceae) in northern and eastern Australia, 1. *Austrobaileya*, **2**, 65–76.

Byrnes, N.B. (1985) A revision of *Melaleuca* L. (Myrtaceae) in northern and eastern Australia, 2. *Austrobaileya*, **2**, 131–146.

Byrnes, N.B. (1986) A revision of *Melaleuca* L. (Myrtaceae) in northern and eastern Australia, 3. *Austrobaileya*, **2**, 254–273.

Carrick, J. and Chorney, K. (1979) A review of *Melaleuca* L. (Myrtaceae) in South Australia. *Journal of the Adelaide Botanic Gardens*, **1**, 281–319.

Carson, C.F. and Riley, T.V. (1995) University of Western Australia. Unpublished results.

Cosgrove, D.J. and Thain, E.M. (1948) Oil of *Melaleuca bracteata* from Kenya. *Bull. Imp. Inst.*, **46**, 46–50.

Cowley, K.J., Quinn, F.C., Barlow, B.A. and Craven, L.A. (1990) Contributions to a revision of *Melaleuca* (Myrtaceae). 7–10. *Aust. Syst. Bot.*, **3**, 166–202.

Craven, L.A. (1988) Reinstatement and revision of *Asteromyrtus* (Myrtaceae). *Aust. Syst. Bot.*, **1**, 373–385.

Craven, L.A. (1998) National Herbarium of Australia. Personal communication.

Craven, L.A. and Barlow, B.A. (1997) New taxa and new combinations in *Melaleuca* (Myrtaceae). *Novon*, **7**, 113–119.

Curtis, A., Southwell, I.A. and Stiff, I.A. (1990) *Eucalyptus*, a new source of E-methyl cinnamate. *J. Essent. Oil Res.*, **2**, 105–110.

Doran, J.C. and Markham, J.L. (1997) CSIRO Forestry and Forest Products, and University of Western Sydney, Hawkesbury. Unpublished data.

Guenther, G. (1950) The Essential Oils, **4**, Van Nostrand Co. Inc., New York.

Hellyer, R.O. (1957) The volatile leaf oil of *Melaleuca ericifolia* Sm. *Aust. J. Chem.*, **10**, 509–512.

Hellyer, R.O. and Lassak, E.V. (1968) The steam volatile constituents of *Melaleuca viridiflora. Aust. J. Chem.*, **21**, 2585–2587.

Holliday, I. (1989) A Field Guide to Melaleucas. Hamlyn Australia, Victoria.

Jones, T.G.H. and Haenke, W.L. (1937) Essential oils from the Queensland flora: IX. *Melaleuca viridiflora. Proc. Roy. Soc. Qld.*, **48**, 41–44.

Jones, T.G.H. and Haenke, W.L. (1938) Essential oils from the Queensland flora: X. *Melaleuca viridiflora. Proc. Roy. Soc. Qld.*, **94**, 95–98.

Joulain, D. (1995) Investigating new essential oils: rationale, results and limitations. In K.H.C. Baser (ed.), *Flavours, Fragrances and Essential Oils.*, Anadolu University Press, Eskisehir, Turkey, **2**, 55–66.

Lassak, E.V. and Southwell, I.A. (1977) Essential oil isolates from the Australian flora. *Int. Flav. Food Add.*, 126–132.

Markham, J.L. and Southwell, I.A. (1996) University of Western Sydney, Hawkesbury and NSW Agriculture. Unpublished results.

McKern, H.H.G. and Willis, J.L. (1957) Some new Australian essential oils. *Perf. and Ess. Oil Rec.*, **57**, 16–17.

Moudachirou, M., Gbenou, J.D., Garneau, F.X., Jean, F.I., Gagnon, H., Koumaglo, K.H. and Addae-Mensah, I. (1996) Leaf oil of *Melaleuca quinquenervia* from Benin. *J. Essent. Oil Res.*, **8**, 67–69.

Murray, K.E. (1950) The essential oil of five Western Australian plants. *J. Proc. Roy. Aust. Chem. Inst.*, **17**, 398–402.

Penfold, A.R. and Morrison, F.R. (1935) The occurrence of linalool in the essential oil of *Melaleuca ericifolia. J. and Proc. Roy. Soc, NSW*, **69**, 171–173.

Penfold, A.R. and Willis, J.L. (1954) The essential oil industry in Australia. *Econ. Bot.*, **8**, 316–336.

Ramanoelina, P.A.R., Bianchini, J.P., Andriantsiferana, M., Viano, J. and Gaydou, E.M. (1992) Chemical composition of Niaouli essential oils from Madagascar. *J. Essent. Oil Res.*, **4**, 657–658.

Ramanoelina, P.A.R., Viano, J., Bianchini, J.P. and Gaydou, E.M. (1994) Occurrence of variance chemotypes in Niaouli (*Melaleuca quinquenervia*) essential oils from Madagascar using multivariate statistical analysis. *J. Agric. Food Chem.*, **42**, 1177–1182.

Southwell, I.A., Stiff, I.A. and Brophy, J.J. (1992) Terpinolene varieties of *Melaleuca. J. Essent. Oil Res.*, **4**, 363–367.

Southwell, I.A. and Wilson, R.W. (1993) The potential for tea tree oil production in northern Australia. *Acta Horticulturae*, **331**, 223–227.

Watson, E.M. (1943–44) The chemistry and chemical exploitation of West Australian plants. *J. Roy. Soc. W.A.*, **30**, 83–104.

Williams, L.R. and Lusunzi, I. (1994) Essential oil from *Melaleuca dissitiflora*: a potential source of high quality tea tree oil. *Industrial Crops and Products*, **2**, 211–217.

Wilson, P.G. (1991) *Melaleuca*. In G.J Harden, (ed.), *Flora of New South Wales*, Vol. 2, NSWU Press, Sydney, pp. 173–179.

Wrigley, J.W. and Fagg, M. (1993) *Bottlebrushes, Paperbarks and Tea Trees and all other Plants in the Leptospermum Alliance*, Angus and Robertson, Sydney.

INDEX

Bold page number indicates photograph

Other volumes in preparation in Medicinal and Aromatic Plants – Industrial Profiles

Ginkgo, edited by T. van Beek
Ginseng, edited by W. Court
Hypericum, edited by K. Berger Büter and B. Büter
Illicium and Pimpinella, edited by M. Miró Jodral
Kava, edited by Y.N. Singh
Licorice, edited by L.E. Craker, L. Kapoor and N. Mamedov
Piper Nigrum, edited by P.N. Ravindran
Plantago, edited by C. Andary and S. Nishibe
Salvia, edited by S.E. Kintzios
Stevia, edited by A.D. Kinghorn
Tea, edited by Y.S. Zhen
Tilia, edited by K.P. Svoboda and J. Collins
Thymus, edited by W. Letchamo, E. Stahl-Biskup and F. Saez
Trigonella, edited by G.A. Petropoulos
Urtica, edited by G. Kavalali